Introduction to Surfactant Analysis

Edited by

D. C. CULLUM
Consultant specialising in Surfactant Analysis
Wirral

BLACKIE ACADEMIC & PROFESSIONAL
An Imprint of Chapman & Hall
London · Glasgow · New York · Tokyo · Melbourne · Madras

Published by
Blackie Academic and Professional, an imprint of Chapman & Hall,
Wester Cleddens Road, Bishopbriggs, Glasgow G64 2NZ

Chapman & Hall, 2–6 Boundary Row, London SE1 8HN, UK

Blackie Academic & Professional, Wester Cleddens Road, Bishopbriggs, Glasgow G64 2NZ, UK

Chapman & Hall Inc., One Penn Plaza, 41st Floor, New York NY 10119, USA

Chapman & Hall Japan, Thomson Publishing Japan, Hirakawacho Nemoto Building, 6F, 1-7-11 Hirakawa-cho, Chiyoda-ku, Tokyo 102, Japan

DA Book (Aust.) Pty Ltd, 648 Whitehorse Road, Mitcham 3132, Victoria, Australia

Chapman & Hall India, R. Seshadri, 32 Second Main Road, CIT East, Madras 600 035, India

First edition 1994

Typeset in 10/12pt Times by Keyset Composition, Colchester
Printed in England by Clays Ltd, St Ives plc

ISBN 0 7514 0025 4

A catalogue record for this book is available from the British Library

Library of Congress Cataloging-in-Publication data

Introduction to surfactant analysis / edited by D.C. Cullum.
p. cm.
Includes bibliographical references and index.
ISBN 0-7514-0025-4
1. Surface active agents—analysis. I. Cullum, D. C.
TP994.I595 1994
668.1—dc20 93-31552
CIP

∞ Printed on acid-free text paper, manufactured in accordance with ANSI/NISO Z39.48-1992 (Permanence of Paper)

Preface

As its title indicates, this book is intended mainly for the scientist who does not have a great deal of knowledge or experience of surfactant analysis. He or she may be a relatively experienced analyst in other fields, entering the surfactants industry for the first time, or someone who has worked with surfactants and now has to analyse them, or a recent graduate with little or no knowledge of either surfactants or analysis. It is this last group for whom the book is chiefly intended, though it is to be hoped that the others will find it worth reading and using. It is certainly primarily an introduction to the subject, but at the same time it includes a sufficiently broad range of methods to be used as a laboratory handbook.

It has been difficult to decide what to include and what to leave out, and not every reader will agree with my choice. For example, the determination of cloud point is omitted because, strictly speaking, it is not analysis. The reader seeking a method is well provided for by BS 6829: Section 4.1: 1988 or the identical ISO 1065:1991, which offer a choice of five. Colorimetric methods for trace amounts of surfactants are not included, because environmental analysis is a specialist area which I felt to be beyond the scope of an introductory text. The literature contains a large number of papers on the subject, and there are relevant chapters in a recent book [1]. The chapter on chromatography does deal with some of the more important trace components in products.

The book falls naturally into three parts. The first four chapters are largely about what to do, the next four about how to do it, and the last four, written by specialists in their respective fields, about the invaluable contribution that modern high-technology instruments can make. The centre of gravity of the book, however, is in classical or 'wet' chemistry, because most surfactant analysts work in small laboratories whose capital budgets are restricted. Most analytical laboratories nowadays have gas and/or liquid chromatographs and at least a modest infrared spectrometer—indeed they would find their task extremely difficult without them—but the cost of high-frequency NMR spectrometers and mass spectrometers cannot be justified except in large research organisations. The manufacturers of beakers and burettes will be in business for a good many years yet. In any case, the great value of high-tech instruments is not that they can do better what wet chemistry can already do, but that they can easily do things which are quite impossible by any other means.

Those interested in such matters will notice that I have used the term 'weight' throughout. There is a fashion for trying to abolish the word, saying mass even when you mean weight. This is just as foolish as saying weight when you mean mass. What a balance measures is weight, and that is what I have called it. I have also used the form '0.1 M' to denote concentrations of reagents, simply because it is the form most often used in practical chemistry, and it is more convenient than the form '$c(C_{27}H_{42}O_2NCl) = 0.004$ mol/l'.

I hope the book will spark off an enthusiasm for the analytical chemistry of surfactants in at least some readers.

D.C.C.

1. Chapters by J. P. Hughes and J. Waters, in *Recent Developments in the Analysis of Surfactants* (ed. Porter, M. R.), Published for the Society for Chemical Industry, by Elsevier Applied Science, London, 1991.

Note: Chapters 9, 10 and 12 represent the views of the authors and are not necessarily the views of Unilever Research.

To John Howard, schoolmaster,
who taught me that chemistry is fun

Acknowledgements

I wish to thank the following organisations and individuals.

The British Standards Institution for permission to refer to material from British Standards BS 3762 and 6829, and, in its role as a member body of ISO, permission to refer to material from a number of International Standards. British Standards may be obtained from BSI Standards, Linford Wood, Milton Keynes MK14 6LE, UK.

The International Organization for Standardization for permission to reproduce diagrams of the mechanical two-phase titration apparatus (Figure 3.2) from ISO 2271: 1989, the continuous liquid–liquid extractor (Figure 5.2) from ISO 6845: 1989 and the apparatus for iodometric determination of oxyethylene groups (Figure 6.1) from ISO 2270: 1989. The complete Standards can be obtained from the ISO member bodies or directly from the ISO Central Secretariat, Case Postale 56, 1211 Geneva 20, Switzerland.

Bio-Rad Laboratories Ltd, Hemel Hempstead, for supplying literature about ion-exchange resins.

Jones Chromatography, Hengoed, Mid-Glamorgan, for supplying literature about the products of International Sorbent Technology and Varian Sample Preparation Products.

Merck Ltd for permission to reproduce illustrations from their catalogue of laboratory apparatus, which appear as Figures 2.1, 3.1, and (adapted) 4.1.

The Royal Society of Chemistry for permission to reproduce the diagram of a surfactant-selective electrode (Figure 3.4) from Development of ion-selective electrodes for use in the titration of ionic surfactants in mixed solvent systems, Dowle, C. J., Cooksey, B. G. and Ottaway, J. M., *Analyst*, **112** (1987), 1299–1302.

The Society for Chemical Industry for permission to reproduce (with slight modification) the diagram of a surfactant-responsive electrode (Figure 3.3) from Chapter 2 of *Recent Developments in the Analysis of Surfactants* (ed. Porter, M. R.), published for the SCI by Elsevier Applied Science, London, 1991.

Professor R. J. Abraham of the Department of Chemistry, University of Liverpool for permission to reproduce the block diagram of a high-resolution NMR spectrometer (Figure 11.3) from *An Introduction to NMR Spectroscopy*, Abraham, R. J., Fisher, J. and Loftus, P., John Wiley and Sons Ltd, Chichester.

Mr J. G. Lawrence, Dr P. E. Clarke and Dr K. Lee, of Unilever Research, Port Sunlight Laboratory, and Dr T. I. Yousaf of the Analytical Department, Procter and Gamble, Newcastle upon Tyne, for writing Chapters 9 to 12.

Dr R. J. Mulder, Analytical Research Manager, Akzo Chemie Research Centre Deventer, for supplying many of the methods described and copies of papers referred to in Chapter 7, and for permission to quote from them.

Dr Roger Platt, Head of Unilever Research, Port Sunlight Laboratory, for permission to use the Laboratory's Research Library.

The staff of the Public Library, Heswall, Wirral and of the Science and Technology Library, Liverpool, for their unfailing courtesy and efficiency.

Contents

Contributors

Dr P. E. Clarke	Unilever Research, Port Sunlight Laboratory, Quarry Road East, Bebington, Wirral L63 3JW
Mr D. C. Cullum	Hillside, Bush Way, Heswall, Wirral L60 9JB
Mr J. G. Lawrence	Unilever Research, Port Sunlight Laboratory, Quarry Road East, Bebington, Wirral L63 3JW
Dr K. Lee	Unilever Research, Port Sunlight Laboratory, Quarry Road East, Bebington, Wirral L63 3JW
Dr T. I. Yousaf	Analytical Department, Procter & Gamble Ltd, Whitley Road, Longbenton (PO Box Forest Hall No. 2), Newcastle upon Tyne NE12 9TS

1 Introduction

1.1 Why analyse surfactants?

This book aims to tell the reader something about how to analyse surfactants and products containing them. The reader may well ask why one should want to do so in the first place. The answer, always, is: to find out something true, or as close to the truth as possible, about the material to be analysed. There are several compelling reasons why one should want to do this. They are the same for surfactants as for anything else, and they all boil down to one fundamental: money.

Long ago the present author had the privilege of serving on the BSI Committee that dealt with the analysis of glycerol. The Committee spent a great deal of time, years in fact, going into the minutest details of the various methods, to ensure that they were as accurate, precise and reliable as they could be made. All this cost a great deal of time and money. The reason was that glycerol was, and is, an expensive commodity, and companies manufacturing, selling, buying and using it felt, quite rightly, that they had to know a great deal about the material they were dealing in, in order to avoid losing or wasting money. Surfactants are also expensive commodities, and the arguments that applied to glycerol apply with equal force to them. The following are some of the principal reasons for needing to analyse them.

1. It is very desirable to confirm the identity of raw materials and to check that they comply with specification with respect to, for example, active content, chain-length distribution or presence and nature of impurities, i.e. to make sure you are getting what you are paying for, or supplying the customer with what he is paying for. Failure to do so can and probably will cost money.
2. Quality-assurance procedures on manufactured products nearly always require confirmation that active levels are within specification, because an out-of-specification product may result in needlessly high production costs, customer dissatisfaction or even legal action. Sometimes it may be necessary to confirm that the composition complies with the demands of legislation, and it seems likely that this need will increase with the passage of time. And when things go wrong with a production process, trouble-shooters will not get very far unless they can call upon first-class analytical support.
3. In research and development, it is very often necessary to characterise

the product of a synthesis or to monitor the output of a new manufacturing process; indeed it is true that any chemically based research and development programme can only be as good as its analytical support.

4. Many organisations consider it necessary to analyse competitors' products, to discover whether a new product really does embody new technology, to detect unadvertised formulation changes, to try to understand how a product works, to expose unsupportable advertising claims or even to try to make a copy of the product. In the competitive world of modern technology, knowing what the opposition is up to is essential to survival.
5. Finding out why a particular batch of a product does not work.

The reader will no doubt be able to think of others.

If analysts are to do a particular analysis properly, they need to know why they are being asked to do it. If the reason is not clear to the person directing the analytical operation, it is likely that time and effort will be wasted in doing work to generate more information than is required, or even the wrong sort of information altogether. If this person is not the originator of the sample, which is usually the case, he or she must seek complete clarification about why the analysis is necessary, and exactly what information is required. An analysis to the *n*th decimal place is a waste of time if a quick go/no-go test is sufficient, but the latter is equally a waste of time if a precise quantitative determination is required; and a determination of the length and distribution of alkyl and ethylene oxide chains is a waste of time if the client wants only to be reassured that the sample is an alkyl ether sulphate. The reason for the analysis nearly always defines what has to be done. If the person who submitted the sample cannot define exactly what he or she needs to know, or give a convincing reason why the analysis is required, then very probably the analysis is not going to serve any very useful purpose at all. In order to do his job effectively, therefore, the analyst or the individual responsible for the analysis must seek answers to the following questions.

Who wants the analysis done?
Exactly what information is required?
Why? What use will be made of the information?
In what form and in how much detail should the results be reported?
To whom should they be copied?
How and where did the sample(s) originate, and is there plenty of it?

It was once said in another context that the biggest difficulty was not answering the clients' questions, but getting them to ask the right questions in the first place. Analytical clients should please take note.

Finally, if the aim of the game is to find out something true, it follows that the analyst must be competent. More than one senior scientific manager is on record as saying that anyone can do analysis. It is true that there are many people who can do analysis badly, but those who can do it really well are few and far between.

1.2 Sampling

Let us now refine slightly the definition of the purpose of analysis given in section 1.1: the purpose of analysing a sample is to find out something true, or close to the truth, about the composition of the bulk material from which it was drawn.

This simple and obvious fact is not always appreciated by those who submit samples for analysis. If the analyst asks 'What do you want to know?', the answer will usually be something like 'How much *x* has it got in it?', 'it' being undefined not only in the reply but also in the respondent's mind. In the great majority of cases, 'it' is actually a large or very large quantity from which the sample has been drawn, and if the sample does not have the same composition, within close limits, as the bulk, the analysis will not merely be pointless, it will be misleading; the client would be better off without it. As much thought and care must be given to obtaining a representative sample as to the analysis itself. If the sample is not truly representative it is a complete waste of time, and therefore of money, to analyse it at all.

Sampling is a science in its own right, and there are many chapters and even whole books on the subject. ISO 8212 [1] provides some useful general guidelines (sections 1.2.3 to 1.2.6 are based on some of them) and is a good starting-point, although it does not deal with all possible cases or with surfactants and surfactant products that are not detergents. It deals only with sampling of manufactured products (individual solid products in the form of bars, etc., particulate solids, pastes and liquids) from assemblies of small and large packages, which are defined by weight or volume. Sampling from bulk containers is not covered, nor is the question of withdrawing a representative test portion in the laboratory. In all cases, ISO 8212 requires the sample units (bars, packets, jars, etc.) to be taken with the aid of tables of random numbers, which are included in the Standard together with instructions for their use. The Standard also includes a large number of illustrations of devices for sampling and sample division.

It refers to a number of other International Standards, notably ISO 607 [2], covering methods of sample division, and ISO 8213 [3], covering the sampling of particulate solids of a wide range of particle sizes, and the reader is strongly advised to consult these.

It is not possible to deal comprehensively with sampling here, but the following section provides a basic guide.

1.2.1 Sampling devices and sample dividers

The following are a few of the most commonly used devices for obtaining samples from bulk and for dividing large samples of powders into smaller samples. For a more complete description, with illustrations, refer to ISO 8212 and 607 [1,2].

Sampling spears or core samplers are tubes, usually metal, and are of various patterns. The simplest consist simply of a tube cut lengthwise to remove one half or one third of the circumference, forming an open channel of semicircular or horseshoe-shaped cross-section. The device has a transverse handle at one end and may be a metre or more in length. More sophisticated designs can be closed, either by withdrawing the spear into a sheath or by means of a rotary cover, concentric with the sampling tube, which is twisted into position to protect the sample. Another type consists of a closed tube with large holes at regular lengthwise intervals. Sampling spears are used for sampling powders, by inserting the spear into the bulk material as far as possible and then withdrawing it laden with the powder.

Auger samplers are Archimedean screws and are used for sampling compact materials such as soap. They may be simple open augers or enclosed in a tube. The act of screwing the auger into the bulk material drives fragments of the sample up to the surface.

Conical sample dividers are of many designs. The simplest consists of a metal or plastic cone resting in a circular base whose circumference is partitioned into a large number of collection recesses. Powdered samples are poured on to the cone through a centrally located funnel, so that the powder flows down the sides of the cone and into the collection recesses, with a more or less equal amount entering each recess. The contents of one or a few recesses are taken and the rest discarded. In rotary conical dividers the powder flowing down the cone falls past a single- or double-ended wedge-shaped container which rotates as the sample is poured, so that a proportion is collected and the rest runs to waste. The angle of the wedge determines what fraction is collected; for example, an 18° single-ended wedge would collect 5% of the material. In another version the cone itself rotates and has a slot cut in it. Powder falling through the slot is collected and the rest runs to waste.

Rotary sample dividers or *rifflers* may or may not use a cone or dome to spread the sample. With one type, the powder is poured through a funnel and falls either directly or down a stationary cone into a set of jars attached to a rotor, so that the sample receivers themselves rotate and similar proportions of the sample fall into all the jars. Another major variant employs a dome which rotates at high speed and flings the powder into an array of stationary receivers.

1.2.2 Sample containers and labelling

The sample container must be clean, dry and airtight. Its function is not only to protect the sample from contamination by extraneous matter such as atmospheric moisture and carbon dioxide or airborne particulate materials, but also to prevent changes in composition through loss of volatile components, and the escape of toxic or flammable vapours. It should be filled as full as possible, leaving minimal head-space, and the closure must be put on securely.

It is very good practice, and one which should be mandatory in all laboratories, to label the container correctly before the sample is put into it. The label should carry the full name of the contents (not a set of initials) and sufficient additional information to identify the sample uniquely. This may be a batch number, an experiment number, a date, or anything else unique to that individual sample, i.e. information which distinguishes it from all other samples, past, present or future, of the same material. Any analyst receiving a sample not so identified is fully justified in throwing it away without further inquiry, and if he does so there will be no second offenders. Replicate samples should be called A, B, C, etc., to distinguish them from one another.

Correct labelling is a serious matter. There are cases on record of all kinds of dire consequences of incorrect or inadequate labelling, including one instance of multiple deaths [4].

1.2.3 Individual solid products

Individual solid products are tablets or cakes, e.g. of soap. The sample should consist of up to twenty of the bars or cakes. They must be stored in an airtight container. They are then cut into eighths by three mutually perpendicular cuts passing through the centre of each bar, and diametrically opposite eighths are taken. These are sliced, grated or ground, mixed and used to fill completely a 500 g airtight container.

1.2.4 Particulate solids

Particulate solids are powders, flakes, chips, etc., e.g. detergent powders, soap chips.

Sampling from bulk powders consisting of a single finely divided substance is not likely to present sampling problems. Dry-mixed compositions such as detergent powders containing ‘post-dosed’ ingredients may separate on storage and particularly during transportation. They should be remixed if possible before sampling, and a core sampler (section 1.2.1) should be used. This is a device which is pushed through the bulk material and withdraws a long, thin, rod-shaped sample. Several samples should be drawn, from different regions of the bulk container and particularly from various depths. These can be mixed before the analytical test portions are taken, or they can be analysed individually for some component that can be quickly and accurately determined. If the analyses are closely similar, it may be acceptable to do the full analysis on only one or two of them.

According to ISO 8212 [1], packages of up to 5 kg are sampled by taking sufficient numbers and dividing the whole of each sample by using a stationary or rotary sample divider (ISO 607) [2]. One fraction is taken from each sample to give a combined sample totalling approximately 1 l, which is stored in an airtight container.

Packages of over 5 kg are sampled by using a probe or core sampler holding 0.5 l. The samples are mixed and a 1 l final sample is obtained by means of a sample divider.

1.2.5 Pastes

Sufficient packages are taken, and their contents are mixed, either with a spatula or by mechanical stirring, preferably at 20–30°C. Similar-sized portions are removed from all the packages with a scoop or cup (packages up to 1 kg or 1 l) or with a probe appropriate to the viscosity (packages over 1 kg or 1 l) and combined to give a total of about 1 l, which is then mixed and stored in an airtight container.

1.2.6 Liquids

Solutions can be mixed to homogeneity, and will then remain homogeneous in most circumstances. But cycling between low and high temperatures can cause layering or precipitation, and if there is an appreciable head-space in the container, volatile components can escape from solution, with a reduction in concentration at or near the surface. The author has had personal experience of a case in which the concentration of a sample submitted for analysis was not much more than 80% of that of the bulk from which it was drawn, although that is an extreme case. The material was an aqueous solution of a volatile solute in a drum with excessive head space, and the contents were not mixed before sampling. Bulk solutions should be remixed shortly before sampling, and the airtight sample container should be completely filled.

ISO 8212 [1] directs that small packages (up to 5 kg or 5 l) be shaken and that equal quantities be taken from all to give a total of about 1 l, which is then mixed. The contents of large packages (over 5 kg or 5 l) are mixed by shaking, rolling, mechanical stirring, etc., and are then sampled by sampling tube, combined and mixed.

1.2.7 Sampling in the laboratory

There is usually a second sampling process, namely the withdrawal from the primary sample of a secondary sample or test portion, upon which the actual analysis will be carried out. All samples should be remixed by an appropriate procedure before a test portion is withdrawn. Some samples will require special procedures because of their physical nature; for example, bars or sticks, if they have not been sampled in accordance with ISO 8212 [1], may need to have their outer surface removed and to be sampled with a cork-borer, and in the case of extrudable pastes supplied in collapsible tubes, the first few grams to be extruded should be discarded. Solid samples that may pick up atmospheric moisture should be weighed from (not into) a weighing bottle. Liquid samples that may pick up atmospheric moisture or lose volatile components should be weighed from a weighing pipette. Care should be taken to obtain large enough samples of dry-mixed powders, which should then be remixed before taking a sub-sample. Mixers are available that gently tumble the sample to avoid breaking up structured particles. Subsampling is conveniently done by means of a riffler or similar type of device that depends on pouring the powder on to a rotating cone or dome and collecting the spun-off material in an array of beakers or jars.

Even the way in which the test portion is withdrawn from the sample container, which is a sampling process, can create problems. The following is a true story.

The sample had a plasticine-like consistency and contained a fairly high proportion of water. It consisted of 'particles' varying in size from spherical lumps 2 cm or more in diameter down to a fine dust. Two operators analysed it by the same well-established method, and obtained very different results. After much frustrating re-examination, it was found that one analyst had tipped the test portion from the jar into a beaker and the other had removed it on the end of a narrow spatula. The first had obtained a preponderance of large lumps, the other of fine powder. The lumps had a much higher water content than the powder. Evidently the composition of the test portion depended to a major extent on the manner in which it was removed from the container. How could a test portion be taken that would reasonably represent the bulk from which it had been drawn? The solution to this problem was not obvious. However, it was established that the material could be dried at a low temperature without harming it chemically. Multiple samples were drawn from the bulk and dried to constant weight, the weight and water content of each being recorded. The dried samples were ground and analysed separately, and the sampling error was found to have disappeared.

Unless the analytical procedure directs otherwise, it is good practice to take as large a sample as practicable, dilute it to a known volume and take a suitable measured portion for analysis.

1.3 Features peculiar to surfactant analysis

Surfactant analysis differs from analysis in other fields in a number of respects, the most obvious of which is that the substances to be measured are nearly always ill-defined. To begin with, they are rarely single homologues. They may contain chains of polymerised ethylene or propylene oxide, and such chains never have a single length. For example, an ethoxylated alcohol with nominally three molecules of ethylene oxide contains appreciable amounts of all possible chain lengths from zero to seven or eight ethylene oxide units. The active constituent of a typical surfactant is therefore a mixture, not a clearly defined single substance. It is quite usual for three or four homologues to be present, and there may be as many as twenty or thirty different but chemically related components in an alkylene oxide derivative. Such mixtures do not have well-defined molecular weights. In all volumetric and some gravimetric analyses the molecular weight must be known in order to turn an experimental result into a percentage of a component, and it is often necessary to use an empirical molecular weight supplied by the manufacturer. This may not be very reliable, because mixtures of substances from natural sources do not always have a very constant composition.

Mixtures of a few, or many, components that are chemically similar but show a spread of carbon or alkylene oxide chain lengths do not crystallise, and extracted solids are therefore often sticky or waxy. This is not of much practical consequence, but a difficulty due to the same root cause is that solvent extraction procedures may preferentially extract shorter or longer chains. An extreme instance is fatty alcohol ethoxylates. Petroleum ether extracts the parent alcohol and monoethoxylates from aqueous ethanol, but the solubility in petroleum ether decreases rapidly as the number of ethylene oxide groups increases, and triethoxylates and higher ethoxylates are extracted to only a small extent or not at all.

Other unusual difficulties stem from the very fact that surfactants are surface active. For example, the majority of surfactants are foaming or emulsifying agents. This can create problems with emulsions in liquid–liquid solvent extractions. It is frequently not possible to use a simple water–solvent pair such as water–petroleum ether, and a low-molecular-weight alcohol or an electrolyte is usually included to prevent or reduce emulsification. Also, if a solution for analysis foams during mixing, the concentration of surfactant will be higher in the foam than in the bulk solution, and at low concentrations this can cause serious errors. Mixing must be done by some method that minimises foam formation, e.g. by stirring or by repeated gentle inversion of the flask.

Another consequence of surface activity is the fact that at low concentrations in aqueous solution, adsorption losses on glassware can be very substantial. At concentrations of 10^{-4} mol/l and lower, particularly in the case of cationic surfactants, it can be beneficial to rinse all glassware with a solution of roughly the concentration to be used and to allow it to drain without rinsing, before making accurate dilutions.

Yet another special feature is that in contact with ion-exchange resins, in an aqueous medium, there is a hydrophobic interaction between the hydrocarbon chains of surfactants and the organic matrix of the resin, which can cause substantial semipermanent retention on the column at the concentrations used in ion chromatography. This does not occur, however, in the largely non-aqueous media and at the concentrations used in the simple ion-exchange separation procedures described later.

1.4 Calculations

1.4.1 Volumetric calculations

In a volumetric analysis, a known weight of sample is titrated with a reagent of known concentration until an end-point of some sort is reached, corresponding more or less exactly with 100% reaction between sample and reagent. The titration volume then has to be converted to a percentage of X, where X is the component being determined. All volumetric calculations are based on an equation linking purity of sample, volume and concentration of titrant, molecular weight of the material being determined, the number of molecules of titrant that react with one molecule of this material, and the weight of sample taken. This equation is derived as follows.

The weight of X titrated in any volumetric analysis is given by the volume of titrant used multiplied by the weight of X titrated by 1 ml.

One millilitre of titrant contains $C/1000$ mol of titrant, where C is its concentration in mol/l, and reacts with $C/1000N$ mol of X, where N is the number of mol of titrant consumed by 1 mol of X.

The weight of X titrated by 1 ml of titrant is this number of moles multiplied by M, the molecular weight of X. Then:

$$\%\ \text{X in sample} = V \times \frac{C}{1000} \times \frac{M}{N} \times \frac{100}{W}$$

where W is the weight of sample actually titrated.

All volumetric calculations make use of this equation. If any five of the six parameters are known, the sixth can be found by simple rearrangement. For example, the amount of sample required to give a reasonable-sized titration with a titrant of any given concentration is found by interchanging % X and W.

In analyses where a measured excess of titrant is added and back-titrated, V is the net titration, i.e. the amount of reagent added minus the back-titration, assuming equal concentrations. If the concentrations are not equal, V is given by

$$\text{ml reagent added} - V'q$$

where V' is the volume of the back-titration and q is the volume of the reagent that reacts with 1 ml of back-titrant.

In some analyses V is the difference between the titration on the sample and a blank titration.

If a sample is weighed and then diluted one or more times using pipettes and volumetric flasks, the weight of sample actually titrated is given by

$$W = W' \times \frac{\text{product of volumes of all pipettes}}{\text{product of volumes of all flasks}}$$

where W' is the amount of sample weighed out.

So, for example, if W' g of sample is weighed out and diluted to D ml, and then A ml of that solution diluted to E ml and B ml is taken for titration:

$$W = W' \times AB/DE$$

1.4.2 Calculation of molecular weight (1)

This is an illustration of the second principal use of the equation derived in section 1.4.1. Suppose a known weight of a material X of known percentage purity is titrated with a reagent of known concentration and for which N is known. To find the molecular weight of X, the equation is rearranged thus:

$$M = W \times \frac{\%\,\mathrm{X}}{100} \times \frac{N}{V} \times \frac{1000}{C}$$

For example, suppose a sample of 0.5878 g of an anionic surfactant was dissolved in water and diluted to 250 ml, and that 10 ml of the solution was titrated with exactly 0.004 mol/l benzethonium chloride. Suppose the sample had been prepared in such a way that it could be assumed to be 100% pure for all practical purposes, and that the titration volume was 16.7 ml. Then:

$$M = 0.5878 \times \frac{10}{250} \times \frac{100}{100} \times \frac{1}{16.7} \times \frac{1000}{0.004} = 352.0$$

1.4.3 Calculation of molecular weight (2)

This is concerned not with volumetric analysis but with the sort of data that might emerge from a chromatogram or a mass spectrum. Suppose a sample consisted of five components in the following proportions. The molecular weights are consistent with an ethylene oxide adduct of dodecyl alcohol, but the proportions are not.

Component	A	B	C	D	E
% by weight	10	23	34	23	10
Molecular weight	186	230	274	318	362

An analyst needing to calculate the mean molecular weight of the mixture might do so by the following obvious but incorrect method.

$$\text{Wrong: } M = (\%\,\text{A} \times \text{MW A} + \%\,\text{B} \times \text{MW B} + \ldots + \%\,\text{E} \times \text{MW E})/100$$
$$= [(10 \times 186) + (23 \times 230) + (34 \times 274) + (23 \times 318) + (10 \times 362)]/100 = 274.0$$

This is equal to the central value of the original distribution and might be thought to be self-evidently correct because the distribution is symmetrical. But it cannot be, because 10% of A contains almost twice as many molecules as 10% of E, so that the mean molecular weight must be less than the central value.

$$\text{Right:} \quad M = 100 \Big/ \left(\frac{\%\,\text{A}}{\text{MW A}} + \frac{\%\,\text{B}}{\text{MW B}} + \frac{\%\,\text{C}}{\text{MW C}} + \frac{\%\,\text{D}}{\text{MW D}} + \frac{\%\,\text{E}}{\text{MW E}}\right)$$

$$= 100/[(10/186) + (23/230) + (34/274) + (23/318) + (10/362)] = 264.7$$

The error in the first value was 3.5%.

1.4.4 Concentrations of volumetric solutions

All the volumetric methods described in later chapters include an instruction such as 'titrate with 0.1 M hydrochloric acid'. The subsequent calculation then includes a term *C*, the exact concentration of titrant in mol/l. Why not use the known value, 0.1? Because a reagent which is nominally 0.1 M or 1.0 M may actually be slightly weaker or stronger than the nominal concentration. In some laboratories it is the practice to prepare solutions that are exactly 0.1 M, 1.0 M, etc., but this seems unnecessarily laborious. In order to prepare such a solution it is necessary first to prepare a solution of approximately the desired concentration, standardise it very accurately, then dilute or strengthen it, and finally restandardise it to make sure the adjusted solution is correct. The adjustment is unnecessary when insertion into the equation of the concentration found by the first standardisation is all that is needed.

Solutions made by the careful use of the vials of concentrated reagents offered by most laboratory suppliers may generally be assumed to be of the nominal concentration, but for very accurate work it is advisable to standardise them against a primary standard.

1.5 Accuracy and precision

The accuracy of an analytical method may be defined as the extent to which its results coincide with the truth. If the mean of a large number of measurements closely coincides with the truth, the method may be said to be accurate. Accuracy in this sense is really freedom from bias. Bias is primarily a characteristic of the procedure, although it may be influenced by the operator.

The precision of a method is the level of agreement that can be obtained between replicate results. It is a measure of the ability of the method to discriminate between similar samples. For example, if in the hands of a particular operator a method could give replicates with a spread of say 0.3%, it would be able to discriminate between samples differing by about that amount, the exact level of discrimination depending on the number of replicate measurements. There is an inherent limiting value for the discriminating power of every method, upon which no analyst, however skilled, can improve, but the skill of the operator is the main factor. There can be no doubt that a careful, skilled operator can consistently achieve better precision than a careless, unskilled one.

Thus, for example, a method which in the hands of a skilled operator gave replicate results scattered over a range of say 2% relative, i.e. 2% of the measured value or one part in fifty, but with a mean lying very close to the true value, would be accurate but not precise. A method whose results scattered over a range of 0.1% but with a mean 2% too low would be precise but not accurate. A method capable of giving results with a low degree of scatter and with a mean within say 0.2% of the truth would be both accurate and precise.

ISO 5725 [5] defines precision as 'the closeness of agreement between mutually independent test results obtained under stipulated conditions', and identifies two aspects of it, which it terms 'repeatability' and 'reproducibility'. Repeatability is superficially similar to what is here called precision, being a measure of the agreement between replicate results obtained by a single operator on the same sample, using the same apparatus in the same laboratory over a short period of time. But this concept seems to be based on the assumption that all operators are equal, and they are manifestly not. Reproducibility is a measure of the level of agreement between the results obtained by different operators on the same material but using different apparatus in different laboratories and possibly at different times. This concept does recognise that all operators are not equal.

In the author's opinion, the most important single factor influencing the precision of a method is the amount of care used in carrying it out.

None of the classical volumetric methods mentioned in later chapters is biased to any significant degree. As a rule of thumb, a good operator using reasonable care should be able to obtain duplicate results agreeing to within 0.2–0.3% with most of these methods, although a few are less good. All operators should be able to achieve this.

The use of large samples and large titrations not only minimises sampling errors but also reduces the relative errors in weighing and titration, and is to be preferred to the use of small samples and small titrations, circumstances permitting. A weighing error of 1 mg on a 10 g sample is 0.01% relative; on a 0.1 g sample it is 1% relative. An error of one drop, or about 0.05 ml, on a 25 ml titration is 0.2% relative; on a 5 ml titration it is 1% relative.

1.6 Analysis of competitors' products

1.6.1 Non-analytical sources of information

Certainly in the case of consumer products, but perhaps not in the case of industrial products, it is usually possible to discover useful information about the likely composition of competitors' products from readily available non-analytical sources. Having first ascertained what information is required and why, it is worth considering the following before starting any experiments.

The claims made on the pack itself or in promotional literature may give clues about the basis for 'new improved' claims. If the same brand is on sale in the USA, the pack will carry a list of the ingredients. This is not always as helpful as one might wish, because some of the ingredients may be identified only by their manufacturers' trade marks, but it is nevertheless often worth the time and trouble to obtain a pack of the USA product.

Press and television advertising claims may also give useful clues. If the claims major on the bleach being in the 'purple dots', the product probably does not contain anything very novel in the surfactant area.

Patents too can be useful sources. If the manufacturer has been patenting novel surfactants or mixed surfactant systems, the inventions disclosed are likely to turn up in a product sooner or later. Similarly, the manufacturer's recent publications in the scientific literature will provide pointers to the areas of technology in which their research activities are concentrated, and may suggest where to start the analysis.

1.6.2 Date codes

Packages often carry a date code. If this can be interpreted, it will disclose the age of the sample, and this can be very important. An unfamiliar pack may be very old as well as very new, and there can be little point in analysing a five-year-old sample. A second piece of information that may be disclosed is the place of manufacture, if the manufacturer has more than one factory. This could be of interest to anyone trying to piece together a clear picture of the manufacturer's operations.

Cracking date codes is not usually very difficult, but the code-breaker needs plenty of data, in the shape of a large number of samples known to span a period of several years. These should be arranged in order as accurately as possible, for example by comparing them against a library of collected samples displaying changes in pack design with time. If starting from cold, this library must be compiled by collecting packages on a monthly basis over a period of years and recording date of purchase of each. The order of purchase dates is not necessarily the order of manufacturing dates, but will approximate to it if the samples are all bought from the same high-turnover outlet. The library itself will of course contribute some date codes.

Information that may be contained in the date code includes the following. Not all of these items are necessarily present. Characters may be letters as well as numbers.

Year: usually two characters, not necessarily adjacent, possibly reversed. Not necessarily present at all; manufacturer may use a work cycle of arbitrary length, say 250 or 1000 days.

Month: two characters, not necessarily adjacent, possibly reversed, identified by the fact that they show only twelve or thirteen different values.

Week: if a month code is present, the week code will be a single character with not more than five values. Otherwise two characters, again not necessarily adjacent, possibly reversed, one showing only five different values.

Day: may be day of week, of month, of year, of arbitrary work cycle.

Others: single characters may identify a production line or a manufacturing plant.

There are no rules for code-breaking. What is needed is a large collection of samples, patience and a lively imagination.

1.6.3 Samples

A few points about samples of competitors' products are worth mentioning.

First, when separate samples of a competitor's product are submitted to different laboratories or operators for, say, analysis and performance evaluation respectively, it is imperative to make absolutely sure that they are samples of the same formulation. Failure to do so could result in catastrophically misleading information being delivered to the originator of the work. It is best to do the analysis and the evaluation on the same individual pack if possible, then there can be no doubt that that is the performance of that formulation.

Secondly, always obtain plenty of the product, or make sure the client does so. Materials are cheap, labour is expensive. Labour will be wasted if an analysis has to be abandoned because there is no sample left.

Finally, if the purpose of the analysis is to discover something about a new consumer product, make sure as far as possible that the sample analysed really is the new product. If the packaging design is new, that is normally sufficient evidence. If it is not, check the date code to confirm that the sample is of very recent manufacture. If it does not have a date code, or the date code cannot be interpreted, obtain the samples from a large and busy supermarket where the turnover of stock is likely to be rapid, rather than from a small corner shop, which may sell only a few packs a year. Getting hold of new industrial products is more difficult.

1.7 National and international Standards

British Standard BS 6829 [6] covers the analysis of surfactant raw materials, and British Standard BS 3762 [7] covers the analysis of formulated detergents. Some but not all parts of these Standards are identical with International Standards. Similarly, the International Organization for Standardization (ISO) publishes a large number of standard analytical methods for detergents, many but not all of which are identical with parts of the British Standards. Appendix 1 lists corresponding titles from the two bodies. In addition there are several British Standards dealing with soaps. The American Society for Testing and Materials (ASTM) also publishes annually a collection of standard methods [8], but they do not correspond to any significant extent with the ISO or BS methods. A list of them is given in Appendix 2.

It will be obvious that the scope of these sets of standard methods is somewhat limited, dealing as they do with only the most common constituents of detergent products. Some of the methods are, however, applicable to a wider range of materials than those specified. The existing

British and International Standards and ASTM methods will by no means satisfy all the needs of the surfactants analyst, but it is recommended that their methods should be used wherever they are appropriate to the sample under examination and the information required.

References

1. ISO 8212: 1986 (E). Soaps and detergents—Techniques of sampling during manufacture. International Organization for Standardization, Geneva.
2. ISO 607: 1980 (E). Surface active agents and detergents—Methods of sample division. International Organization for Standardization, Geneva.
3. ISO 8213: 1986 (E). Chemical products for industrial use—Sampling techniques—Solid chemical products in the form of particles varying from powders to coarse lumps. International Organization for Standardization, Geneva.
4. Lidbeck, W. L., Hill, I. B. and Beeman, J. *J. Amer. Med. Assn.*, **121** (1943), 826–827.
5. ISO 5725: 1986 (E). Precision of test methods, Part 1. Guide for the determination of repeatability and reproducibility for a standard test method by interlaboratory tests. International Organization for Standardization, Geneva.
6. British Standard BS 6829: Analysis of surface active agents (raw materials). British Standards Institution, Milton Keynes.
7. British Standard BS 3762: Analysis of formulated detergents. British Standards Institution, Milton Keynes.
8. Annual Book of ASTM Standards, 1991, Section 15, Volume 15.04. Soap, Polishes, Leather, Resilient Floor Coverings. American Society for Testing and Materials, Philadelphia, USA.

2 Surfactant types; classification, identification, separation

2.1 Definitions

Surfactant. For the purpose of this book, a surfactant is a substance which in solution, particularly aqueous solution, tends to congregate at the bounding surfaces, i.e. the air–solution interface, the walls of the containing vessel and the liquid–liquid interface if a second liquid phase is present. As a result, the various interfacial tensions are reduced.

The reason for this behaviour is that the surfactant molecule contains two structurally distinct parts, one of which is hydrophilic while the other is hydrophobic. Oil-soluble surfactants have an oleophilic and an oleophobic part. In the great majority of surfactants, the hydrophobic part is a hydrocarbon chain, which usually has an average length of 12 to 18 carbon atoms and may include an aromatic ring. A single molecule in aqueous solution seeks the surface, because its hydrocarbon 'tail' is repelled by the water, and it tends to remain there, with the hydrophobic part above the surface and the hydrophilic part below, i.e. in the water phase. Further molecules seek the surface or the walls of the vessel until both are full, at which point further additions result in the formation of micelles, which are clusters of molecules arranged with the hydrophobic parts towards the centre and the hydrophilic parts on the outside.

Surfactants are classified according to the nature of their hydrophilic parts.

Anionic surfactant: a surfactant in which the hydrophilic part carries a negative charge. Examples: soaps, $RCOO^-$; alkyl sulphates, $ROSO_3^-$. For typical cations see section 2.2.1.

Nonionic surfactant: a surfactant in which the hydrophilic part is uncharged. Examples: acyl diethanolamides, $RCON(C_2H_4OH)_2$; ethoxylated fatty alcohols, $R(OC_2H_4)_nOH$.

Cationic surfactant: a surfactant in which the hydrophilic part carries a positive charge. Examples: alkyltrimethylammonium salts, $RN^+(CH_3)_3$; alkyldimethylbenzylammonium salts, $RN^+(CH_3)_2CH_2C_6H_5$. For typical anions see section 2.2.3.

Amphoteric surfactant: a surfactant in which the hydrophilic part contains both positive and negative charges. Examples: alkylaminopropionates, $RNH_2^+(CH_2)_2COO^-$; alkylbetaines, $RN^+(CH_3)_2CH_2COO^-$. It is possible for amphoteric surfactants to have more than one charge of either sign, or to lose one charge by addition or removal of a proton.

Weakly acidic and weakly basic surfactants. Anionic surfactants containing carboxy groups (e.g. soaps) are salts of weak acids and are fully ionised only in alkaline media (pH about 10 or higher). In acid media (pH about 4 or lower) they exist in the carboxylic acid form, which is virtually nonionised. They are thus anionic only in alkaline solution, and nonionic in acid solution. Similarly, cationic surfactants containing primary, secondary or tertiary nitrogen (fatty amines, amine oxides and amine ethoxylates) are weak bases. They are protonated and therefore ionised only in acid media, and deprotonated, i.e. uncharged, in alkaline media. They are thus cationic only in acid solution, and nonionic in alkaline solution. At intermediate pHs, both types are partially ionic and partially nonionic. This behaviour has important consequences in analysis. For no obvious reason, weakly acidic surfactants are usually regarded as anionic, but weakly basic surfactants are usually regarded as nonionic. In this book they are treated as cationic.

In nonaqueous media weak acids are fully ionised above the point of neutrality, and weak bases are fully ionised below it. The point of neutrality is conventionally taken to be at pH 7, but may be higher or lower in practice.

2.2 Common types of surfactants of all four classes

The number of different surfactant molecules that have actually been made at some time or another must run into hundreds or possibly thousands, and the number of theoretically possible surfactant molecules may run into millions. Fortunately only a few kinds find any practical use. In a book of this size it is possible to deal only with substances the analyst is reasonably likely to come across, and the following tables list examples of the most frequently encountered types of surfactants. The behaviour of many excluded compounds will in practice follow that of listed materials with similar structural features.

R may be assumed to contain 12 to 18 carbon atoms except where otherwise stated. Hydrocarbon chains derived from vegetable oils or animal fats contain even numbers of carbon atoms, including the carboxy carbon in the case of fatty acids, but those obtained from petrochemical sources contain both odd and even numbers.

2.2.1 Common types of anionic surfactants

The following list includes most of the frequently encountered anionic surfactant types. Only the structure of the anion is shown. The most commonly used cations are Na^+, NH_4^+ and triethanolammonium,

$NH^+(C_2H_4OH)_3$, but K^+, Mg^{2+} and other alkanolammonium ions are also found. Petroleum sulphonates may have Ca^{2+} or Ba^{2+}. A few specialised materials have more complex organic cations.

Sulphates

Alkyl sulphates	$ROSO_3^-$
Alkylether sulphates	$R(OC_2H_4)_nOSO_3^-$
Sulphated alkanolamides (also sulphated ethoxylated derivatives)	$RCONHC_2H_4OSO_3^-$
Monoglyceride sulphates	$RCOOCH_2CHOHCH_2OSO_3^-$
Sulphated alkylphenol ethoxylates (typically R = C_9H_{19}–, n = 3 or more)	$RC_6H_4(OC_2H_4)_nOSO_3^-$

Sulphonated hydrocarbons

Alkylbenzene sulphonates	$RC_6H_4SO_3^-$
Alkane sulphonates	RSO_3^-
Alpha-olefin sulphonates (R′ = mainly $-CH_2-$ and $-C_2H_4-$)	$RCH{=}CHR'SO_3^-$

Olefin sulphonates are complex mixtures as a result of migration of the double bond during sulphonation and hydration across the double bond, resulting in the formation of hydroxyalkane sulphonates and sultones (internal esters) and the so-called disulphonic acids (actually sulphatosulphonic acids). These are more fully discussed in chapter 5.

Alkylnaphthalene sulphonates (typically R = C_4H_9–)	$RC_{10}H_6SO_3^-$
Petroleum sulphonates, e.g.	dialkylbenzene sulphonates $R_2C_6H_3SO_3^-$ (R = $C_{10}H_{21}$–$C_{14}H_{29}$)
and	dialkylnaphthalene sulphonates $R_2C_{10}H_5SO_3^-$ (R = C_9H_{19})

Alkyl and alkylether sulphates and all types of sulphonated hydrocarbons always contain a small proportion of unsulphated or unsulphonated starting material.

Sulphonated esters

Acyl isethionates	$RCOOC_2H_4SO_3^-$
Fatty ester α-sulphonates (R′ = CH_3– to C_4H_9–)	$RCH(SO_3^-)COOR'$

Monoalkylsulphosuccinates ($R = C_{12}H_{25}$ or longer)	$ROOCCHSO_3^-$ | CH_2COO^-
Dialkylsulphosuccinates (R and $R' = C_6H_{13}$– to C_8H_{17}–)	$ROOCCHSO_3^-$ | $R'OOCCH_2$

Sulphonated amides

Acyl methyltaurates	$RCON(CH_3)C_2H_4SO_3^-$
Alkyl sulphosuccinamates	$RNHCOCHSO_3^-$ | CH_2COO^-

Carboxylates

Soaps	$RCOO^-$
Alkyl ethoxy carboxylates	$R(OC_2H_4)_nOCH_2COO^-$
Acyl sarcosinates	$RCON(CH_3)CH_2COO^-$

Phosphates

Alkyl phosphates	$RO.PO.(OH)_2 + (RO)_2POOH + (RO)_3PO$
Alkylether phosphates	$RO(C_2H_4O)_nPO(OH)_2 + [RO(C_2H_4O)_n]_2POOH + [RO(C_2H_4O)_n]_3PO$

Alkyl and alkylether phosphates are generally mixtures. The monoester usually contains some diester and may contain some triester. The diester usually contains both mono- and triester. They are often supplied as the free acids and both are then likely to contain free phosphoric acid.

2.2.2 *Common types of nonionic surfactants*

Non-nitrogenous

Fatty alcohols	ROH
Fatty alcohol ethoxylates	$R(OC_2H_4)_nOH$
Alkylphenol ethoxylates	$RC_6H_4(OC_2H_4)_nOH$
Polyethylene glycol esters (ethoxylated fatty acids)	$RCO(OC_2H_4)_nOH$
Sorbitan esters (also ethoxylated derivatives)	$RCOOCH_2CHOHC(CHOH)_2CH_2$ (ring closed by –O– between C and CH_2)

Alkyl polyglycosides ($n = 0$ to 4)

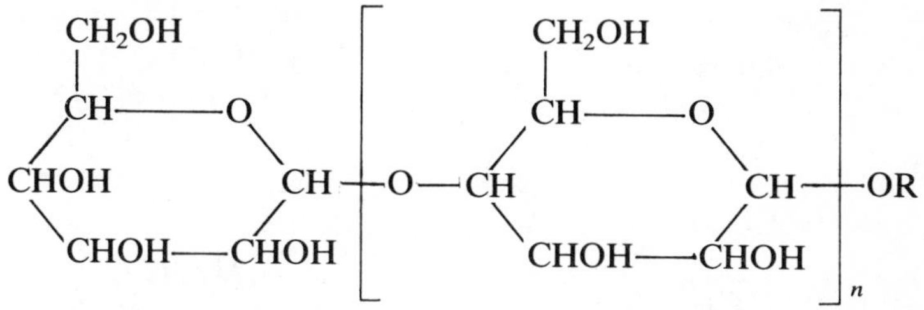

Sucrose mono and polyesters

Also fatty acid mono- and diesters of ethylene and propylene glycols, mono- and diglycerides and ethoxylated derivatives.

Alkanolamides. Usually ethanolamides, sometimes isopropanolamides.

Monoalkanolamides	$RCONHC_nH_{2n}OH$
Dialkanolamides	$RCON(C_nH_{2n}OH)_2$
Ethoxylated monoalkanolamides	$RCONHC_nH_{2n}(OC_2H_4)_xOH$
Ethoxylated dialkanolamides	$RCON\langle^{C_nH_{2n}(OC_2H_4)_xOH}_{C_nH_{2n}(OC_2H_4)_yOH}$

Polymers of alkylene oxides. Block copolymers of ethylene and propylene oxides, $HO(C_2H_4O)_x(C_3H_6O)_y(C_2H_4O)_zH$ and $HO(C_3H_6O)_x(C_2H_4O)_y(C_3H_6O)_zH$, and variants of these in which the outer blocks are random copolymers of ethylene and propylene oxide in varying proportions. One of the terminal hydrogen atoms can be replaced by an alkyl chain. More complex types derived by alkoxylation of glycerol or ethylene diamine also exist.

In block copolymers the polypropylene block(s) constitute the hydrophobic part and the polyethylene block(s) the hydrophilic part. Random copolymers do not have identifiable hydrophobic or hydrophilic parts.

2.2.3 *Common types of cationic surfactants*

Quaternary ammonium salts. Only the structure of the cation is shown. The most commonly used anions are Cl^-, Br^-, CH_3COO^- and $CH_3OSO_3^-$, the last resulting from the use of dimethyl sulphate as a quaternising agent. Weak bases are shown in the deprotonated form; the protonated form has the anion of whatever acid was used for protonation.

Monoalkyltrimethylammonium salts	$RN^+(CH_3)_3$
Dialkyldimethylammonium salts	$R_2N^+(CH_3)_2$
Benzalkonium salts	$RN^+(CH_3)CH_2C_6H_5$
Alkylpyridinium salts	$RN^+C_5H_5$
Ethoxylated quaternary salts, e.g.	$(C_2H_4O)_xH$ | RN^+CH_3 | $(C_2H_4O)_yH$
Imidazolinium salts, e.g.	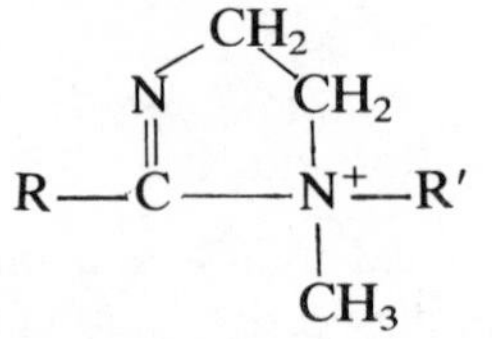

Weak bases. These are often classified as nonionics; they are cationic only in acid solution, and nonionic in neutral or alkaline solution.

Alkylamines	RNH_2
Amine oxides	$RN(CH_3)_2 \rightarrow O$
Dialkylmethylamines	$RR'NCH_3$
Ethoxylated amines	$RN\begin{cases}(C_2H_4O)_xH \\ (C_2H_4O)_yH\end{cases}$ $(x + y = n = 2\text{–}50)$

2.2.4 *Common types of amphoteric surfactants*

Because both the acid and the basic groups can be either weak or strong, four classes of amphoteric are distinguished here, without considering those with more than one functional group of one type or the other. If one were to consider individually all the possible surfactants with two or more acid or basic groups, including those in which one group was weak and the other strong, the range of possible compounds would become enormous and their analytical chemistry very complex. Fortunately, most of these more complex structures exist up to the present only as laboratory curiosities, but diglycinates and dipropionates do exist as articles of commerce.

For ease of reference, the four basic classes are referred to throughout this book as WW, WS, SW and SS amphoterics, where W means weak and S strong, and the first letter refers to the acid group and the second to the basic group. This is the author's own convention and is not generally recognised.

Carboxylates with weakly basic N (WW). Structures shown are of the zwitterionic, i.e. internally neutralised, forms. These gain a proton in acid solution to give the fully protonated (cationic) form, electrically neutralised by the anion of the acid. They lose a proton in alkaline solution to give the fully deprotonated (anionic) form, electrically neutralised by the cation of the base.

Alkylglycinates	$RNH_2^+CH_2COO^-$
Alkylaminopropionates	$RNH_2^+CH_2CH_2COO^-$
Alkyldiglycinates (X^+ = any cation, e.g. Na^+)	$RNH^+(CH_2COO^-)_2 + X^+$
Alkylaminodipropionates (X^+ = any cation, e.g. Na^+)	$RNH^+(CH_2CH_2COO^-)_2 + X^+$
Imidazoline derivatives (many possibilities)	e.g. $RCONHCH_2CH_2NH^+(CH_2COO^-)CH_2CH_2OH$

Imidazoline derivatives are sometimes shown as ring structures. This is now generally thought to be incorrect; ring opening occurs during synthesis and the products are derivatives of a monoamide of ethylene diamine, as shown here.

Carboxylates with strongly basic N (WS). These are also shown as zwitterions. They gain a proton in acid solution to give the fully protonated (cationic) form, but cannot lose a proton.

Alkyldimethyl betaines	$RN^+(CH_3)_2CH_2COO^-$
Alkylamidobetaines	$RCONHC_nH_{2n}N^+(CH_3)_2CH_2COO^-$

Sulphonates (sulphobetaines) with weakly basic N (SW). The structure shown is a zwitterion. It can lose a proton in alkaline solution to give the fully deprotonated (anionic) form, but cannot gain a proton.

Amphopropyl sulphonates	$RCONH(CH_2)_2NH^+(CH_2CH_2OH)CH_2CHOHCH_2SO_3^-$

Sulphonates (sulphobetaines) with strongly basic N (SS). Having neither weak acid nor weak base groups, these can exist only in the zwitterionic form, whatever the pH of the solution.

Hydroxypropyl sulphonates or hydroxysultaines	$RCONH(CH_2)_3N^+(CH_3)_2CH_2CHOHCH_2SO_3^-$

Commercial amphoterics often contain more than one surface active species.

In a few cases the formulae given above are oversimplified, in the sense that commercial materials contain other related compounds which may complicate their analysis. The complexity of α-olefin sulphonates has already been mentioned. With such mixtures, conversion of a titration volume to an active percentage is not straightforward, because it is debatable which constituents should be considered as actives, and the effective molecular weight is uncertain. A more complicated case is that of imidazoline and imidazolinium derivatives. The chemistry involved in the synthesis of these materials is very complex, and it seems likely that all the commercial materials contain a mixture of several different active structures [1,2].

The lists given in sections 2.2.1 to 2.2.4 are far from comprehensive and are intended only to illustrate some of the most frequently encountered species. Specialised products, e.g. cationics and amphoterics containing hydrazinium ($-NHNR_3^+$), phosphinium (R_3P^+-) and sulphonium (SR_2^+-) groups have been excluded.

2.3 Classification of surfactants using indicators

ISO 2271: (1989) [3] is a method for the determination of anionics by titration with a cationic surfactant. It uses a mixed indicator system in a chloroform–water solvent system. One indicator reacts with excess anionic to colour the chloroform layer pink, and the other reacts with excess cationic titrant to colour the chloroform layer blue. It is possible to use this mixed indicator to classify surfactants, by taking advantage of the following differences.

1. Sulphates and sulphonates are anionic at all pHs. Carboxylates are anionic only at high pH.
2. Quaternary ammonium salts are cationic at all pHs. Weak bases are cationic only at low pH.
3. WW amphoterics are anionic at high pH and cationic at low pH.
4. WS amphoterics (carboxybetaines) are zwitterionic at high pH and cationic at low pH.
5. SW sulphobetaines (with weakly basic nitrogen) are zwitterionic at low pH and anionic at high pH.
6. SS sulphobetaines (with quaternary nitrogen) are zwitterionic at all pHs.

The classification is done by adding 1 ml of the mixed indicator (disulphine blue V and dimidium bromide) to two small portions of a solution of the surfactant, one acid and the other alkaline, adding a few

Table 2.1 Classification of surfactants using indicators

Class of surfactant	Colour of chloroform	
	Acid	Alkaline
Sulphobetaine with weakly basic N	Colourless	Pink
Carboxylate	Colourless	Pink
Anionic (sulphate or sulphonate)	Pink	Pink
Amphoteric excluding betaine (WW)	Blue	Pink
Quaternary ammonium salt	Blue	Blue
Betaine, amine, amine oxide	Blue	Colourless
Sulphobetaine with strongly basic N	Colourless	Colourless
Nonionic	Colourless	Colourless

millilitres of chloroform to each and shaking. The colours of the two chloroform layers identify the class (see Table 2.1).

This simple procedure is capable only of identifying the predominant class of surfactant present in a mixture, and its usefulness in practice, at least in this simple form, is confined to classification of unidentified samples known to contain only one class of active, e.g. eluates from ion-exchange systems. Its scope can, however, be extended by using the following procedure.

1. Make a solution of the unknown in 80% methanol, to contain about 0.01 mol/l of total surfactant.
2. Pass 10 ml through a column containing 2 ml of a strongly acidic cation exchange resin (see chapter 4) in 80% methanol and wash with 5 ml of the same solvent.
3. Pass another 10 ml through a column containing 2 ml of a strongly basic anion exchange resin in the hydroxide form in 80% methanol and wash with 5 ml of the same solvent.
4. Evaporate both solutions to a low volume, dilute to about 10 ml with water and carry out the indicator test on small portions of each of them in both acid and alkaline solution. The solution from the cation exchange column contains only anionics, soaps, SW and SS sulphobetaines and nonionics, and the solution from the anion exchange column contains only cationics with both weakly and strongly basic nitrogen, betaines, SS sulphobetaines and nonionics. WW amphoterics are retained by both columns.

If desired fatty acids and weak bases can be extracted before this is done, which makes the test more specific but also more cumbersome.

There are many other tests for the classification of surfactants, and Rosen and Goldsmith [4] give a comprehensive selection. However, it is usually more convenient to proceed to a separation by ion exchange and examine the isolates for the four main classes.

2.4 Classification by acid and alkaline hydrolysis

The different classes of surfactant respond in various ways when exposed to hydrolytic conditions. This provides a basis for a broad classification.

2.4.1 General

Acid hydrolysis is carried out by boiling under reflux in aqueous hydrochloric or sulphuric acid solution. Boiling for 2 h in 1.0 M hydrochloric acid or 0.5 M sulphuric acid will completely hydrolyse alkyl sulphates, alkylether sulphates, sulphated alkylphenol ethoxylates and the sulphate groups of alkanolamide sulphates, but other functional groups may take much longer. It is possible to use more dilute acid, but the hydrolysis time is then longer; in 0.1 M acid, several hours' boiling may be needed. Excessive foaming often occurs in the early stages, and this can be avoided by heating on a steam bath for 20–30 min before starting to boil. If acid hydrolysis is done in alcoholic solution, any fatty acid liberated may esterify with the solvent, and subsequent alkaline hydrolysis is then necessary.

Amides in general can be hydrolysed, but with difficulty. *N*-acyltaurates and methyltaurates, sulphosuccinamates and sarcosinates are hydrolised by boiling under pressure for 6 h at 150–160°C in 6 M hydrochloric acid [5]. Alkanolamides require rather less extreme conditions; boiling for 8 h with 6 M hydrochloric acid is adequate.

Alkaline hydrolysis is usually carried out in ethanolic solution, usually under mild conditions, typically boiling for 1 h or less in 0.5 M alkali, although high-molecular-weight esters may require more concentrated alkali or a longer boiling time (see Saponification Value, chapter 3). Hydrolysis with aqueous acid is of fairly general application, but hydrolysis with alcoholic alkali is better for carboxylate ester groups.

The sulphonate group itself is generally very resistant to hydrolysis, but in the case of fatty ester α-sulphonates the sulphonate group is removed by several hours' boiling with 1 M alkali. Acid hydrolysis, however, has no effect.

ISO 2869 [6] and ISO 2870 [7] deal with the determination of alkali- and acid-hydrolysable detergents respectively. ISO 2869 prescribes boiling 25 ml of a 0.003–0.005 M solution for 30 min with 5 ml of aqueous 10 M sodium hydroxide. ISO 2870 prescribes boiling a similar sample for 3 h with 5 ml of aqueous 5 M (490 g/l) sulphuric acid. In both cases the hydrolysed solution is neutralised and titrated with benzethonium chloride (see section 3.5). In the acid treatment, $[H^+]$ is only 1.67 M, and this may be insufficient for complete hydrolysis of the more resistant titratable species. Both methods should be used where appropriate, but ISO 2870 may not be universally valid, and if it is used for products containing amides, it is advisable to use the device of doing a second experiment with a longer hydrolysis time. Hydrolysis of unknowns is best done at least in duplicate, the second sample being hydrolysed for 1 h longer than the first. If the results show that further hydrolysis has occurred during this time, repeat, again in duplicate, but for longer times, until the second experiment confirms that hydrolysis is complete.

The main classes respond as outlined below. It is necessary to identify and quantify the hydrolysis products, and methods are given in chapter 5.

2.4.2 *Alkyl sulphates, alkylether sulphates and sulphated alkylphenol ethoxylates*

Acid. All sulphates yield an alcohol, a sulphate ion and a hydrogen ion, regardless of the structure of the rest of the molecule.

$$ROSO_3^- + H_2O \rightarrow ROH + SO_4^{2-} + H^+$$

Alkyl sulphates yield the fatty alcohol. Alkylether and ethoxylated alkylphenol sulphates yield the ethoxylated alcohol and ethoxylated alkylphenol respectively.

Alkali. Sulphates are not hydrolysed by alkalis under mild conditions.

2.4.3 *Sulphated mono- and diglycerides*

Acid. The sulphate group is removed with liberation of a hydrogen ion, as with other sulphates, and the ester linkage is also hydrolysed, with formation of fatty acid and glycerol.

$$RCOOCH_2CHOHCH_2OSO_3^- + 2H_2O \rightarrow RCOOH + CH_2OHCHOHCH_2OH + SO_4^{2-} + H^+$$

Alkali. Conversion to soap and glyceryl sulphate takes place.

$$RCOOCH_2CHOHCH_2OSO_3^- + OH^- \rightarrow RCOO^- + CH_2OHCHOHCH_2OSO_3^-$$

2.4.4 *Sulphated monoalkanolamides and ethoxylated derivatives*

Acid. Conversion to fatty acid, alkanolammonium ion and sulphate takes place. Boiling for 8 h with $[H^+]$ 6 mol/l is required to hydrolyse the amide linkage.

$$RCONHC_2H_4OSO_3^- + H_2O \rightarrow RCOOH + NH_3^+C_2H_4OH + SO_4^{2-}$$

Alkali. The amide group in monoalkanolamides is resistant to alkalis, but is hydrolysed by boiling for 6 h in 1 M potassium hydroxide in ethylene glycol. The sulphate group is not affected by this treatment.

2.4.5 *Sulphonated esters*

Acid. Conversion to carboxylic acid and alcohol takes place. The sulphonate group may be on either of these. Thus acyl isethionates yield a fatty acid and isethionate ions, mono- and di-alkyl sulphosuccinates yield fatty alcohols and sulphosuccinic acid, and α-sulphonated fatty esters yield the α-sulphonated fatty acid and a low-molecular-weight alcohol.

$$RCOOC_2H_4SO_3^- + H_2O \rightarrow RCOOH + HOC_2H_4SO_3^-$$

$$\begin{array}{l} ROOCCHSO_3^- \\ \quad\;\;| \\ R'OOCCH_2 \end{array} + 2H_2O \rightarrow ROH + R'OH + \begin{array}{l} HOOCCHSO_3^- \\ \quad\;\;| \\ HOOCCH_2 \end{array}$$

$$\begin{array}{l} RCHCOOR' \\ \;| \\ SO_3^- \end{array} + H_2O \rightarrow \begin{array}{l} RCHCOOH \\ \;| \\ SO_3^- \end{array} + R'OH$$

Alkali. The results are similar, but fatty acids appear as soap. Alkaline hydrolysis is generally to be preferred for esters, but fatty ester α-sulphonates undergo rapid hydrolysis of the ester group accompanied by slower hydrolysis of the sulphonate group, so this is not a practical method for their determination.

2.4.6 *Sulphonated amides*

Acid. Taurates, methyltaurates and alkyl sulphosuccinamates are hydrolysed by boiling for 6 h under pressure at 150–160°C with 6 M hydrocholoric acid.

Taurates and methyl taurates yield fatty acid and the aminosulphonic acid, methyltaurine, with the amino group protonated.

$$RCON(CH_3)C_2H_4SO_3^- + H_2O \rightarrow RCOOH + CH_3NH_2^+C_2H_4SO_3^-$$

Alkyl sulphosuccinamates yield a salt of a fatty amine and sulphosuccinic acid.

$$\begin{array}{l} RNHOCCHSO_3^- \\ \qquad\quad| \\ \quad\;\; CH_2COO^- \end{array} + H_2O + 2H^+ \rightarrow RHN_3^+ + \begin{array}{l} HOOCCHSO_3^- \\ \quad\;\;| \\ CH_2COOH \end{array}$$

Alkali. Amides are resistant to alkaline hydrolysis, but see the comment in section 2.4.4.

2.4.7 Sulphonated hydrocarbons

Sulphonated hydrocarbons are not hydrolysed by either acids or alkalis.

2.4.8 Carboxylates

Sarcosinates are hydrolysed under the same conditions as methyltaurates to give a fatty acid and sarcosine. Other carboxylates are not hydrolysed by either acids or alkalis.

2.4.9 Alkyl phosphates and alkylether phosphates

Phosphates are not hydrolysed by either acids or alkalis.

2.4.10 Nonionic esters

This category includes polyoxyethylene glycol esters, sorbitan esters, fatty acid esters of glycols and glycerol, including ethoxylated derivatives, and sugar esters.

Acid. Conversion to fatty acid and alcohol takes place, i.e. polyoxyethylene glycol, sorbitan, glycol or glycerol, sugar.

Alkali. Conversion to soap and alcohol takes place. Alkaline hydrolysis is of more practical use.

2.4.11 Alkanolamides and ethoxylated derivatives

Acid. Boiling for 8 h at $[H^+]$ 6 mol/l leads to conversion of both mono- and dialkanolamides to fatty acid and alkanolammonium ion.

$$RCONHC_2H_4OH + H_2O + H^+ \rightarrow RCOOH + NH_3^+C_2H_4OH$$

Alkali. Contrary to the general belief, dialkanolamides are readily saponified by boiling with 0.5 M ethanolic alkali, but monoalkanolamides are little affected by this treatment. Both types are hydrolysed by boiling for 6 h with 1 M potassium hydroxide in ethylene glycol, which boils at 197°C. They are converted to soap and the corresponding alkanolamine. Ethoxylated derivatives behave similarly.

2.4.12 *Quaternary ammonium salts*

Acid. The quaternary ammonium group is not hydrolysed by acids, but ester or amide groups in the alkyl chain may be hydrolysed.

Alkali. Conversion to olefin and tertiary amine or pyridine takes place.

$$RCH_2CH_2N^+(CH_3)_3 + OH^- \rightarrow RCH{=}CH_2 + N(CH_3)_3 + H_2O$$

There are, however, apparently no reports in the literature of the use of this hydrolysis for analytical purposes.

2.4.13 *Imidazolinium salts*

Acid. Imidazolinium salts are not hydrolysed by acids under mild conditions.

Alkali. Ring opening occurs, with formation of an ethylenediamine derivative.

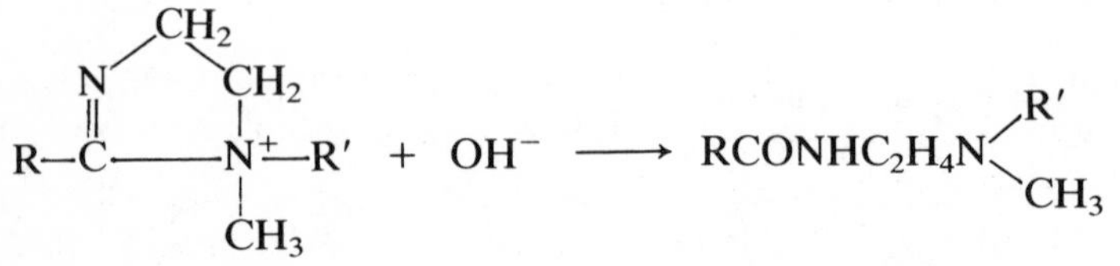

The product is a tertiary base, which is cationic in acid solution and can still be titrated with sodium dodecyl sulphate.

2.4.14 *Amphoterics*

Alkylglycines, alkylaminopropionates and alkyldimethyl betaines are not hydrolysed by either acids or alkalis. Imidazoline derivatives are hydrolysed by acids to fatty acid and an ethylenediamine derivative. Alkylamidobetaines are converted by acids to fatty acid and a short-chain dimethylbetaine.

To classify an unknown surfactant, either acid or alkaline hydrolysis or both are carried out, and the reaction products are examined. Procedures for this examination are discussed later; quantitative analysis of hydrolysis products is usually necessary. Comparison against the previously mentioned lists will in most cases identify the class of surfactant. Mixtures may present difficulties.

A systematic scheme for identifying a very wide range of surfactants has been devised by Rosen and Goldsmith [4].

2.5 Elemental analysis

Elemental analysis is useful only in the analysis of unknowns. Because the range of elements that occur in surfactants is limited the following tests are usually all that are necessary. It may safely be assumed that carbon, hydrogen and oxygen are present. Nitrogen, sulphur and phosphorus are the only other non-metals of interest.

2.5.1 Metals

It is very unlikely that metals other than sodium, potassium and magnesium will be present, although calcium or barium may be found. Determine the ash content. If there is no ash, no metals are present. Boil a sample with a little sodium carbonate. If the solution remains clear, magnesium and other alkaline-earth metals are absent. If it smells of ammonia, an ammonium salt is present.

Metals are most conveniently determined by atomic absorption or X-ray fluorescence spectroscopy. Flame photometry has gone out of fashion, but it is a cheap and effective way of determining sodium and potassium. If necessary sodium, potassium and divalent metals can be determined by standard text-book methods. For example, sodium can be determined by precipitation as sodium uranyl acetate and weighing. Potassium can be determined by precipitation with sodium tetraphenylborate and weighing, or by adding a measured excess of the same reagent and back-titrating with thallium. Magnesium can be determined by precipitation as magnesium ammonium phosphate, filtering, igniting and weighing as magnesium pyrophosphate. There are many other methods in the literature.

2.5.2 Determination of nitrogen

Nitrogen is determined by the classic Kjeldahl method, in which the nitrogen is converted to ammonium ions, distilled from an alkaline solution and determined by acid titration. Various catalysts have been used for the conversion to ammonium ions. The one suggested here works satisfactorily.

The Kjeldahl apparatus (Figure 2.1) consists of a long-necked flask, a double neck, a tap funnel, a splash head, a connecting tube, a vertical condenser, a delivery tube and a receiving flask. Semi-automatic apparatus is available which works in exactly the same way but is more convenient. The digestion (step 4) must be done in a well-ventilated fume cupboard.

1. Weigh a sample containing 0.02–0.03 g nitrogen and place it in the Kjeldahl flask. Liquids and pastes may be weighed into a small plastic

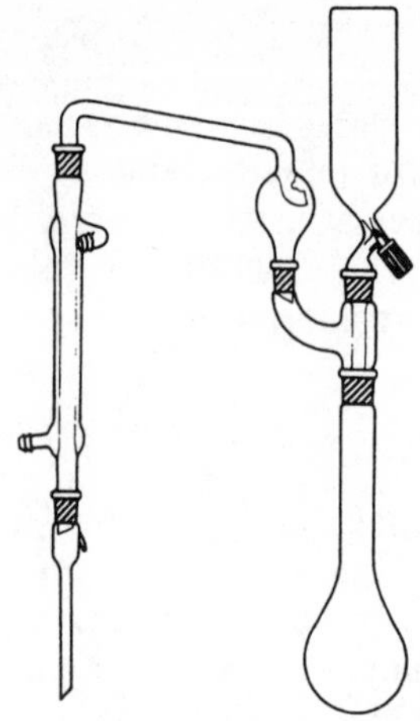

Figure 2.1 Kjeldahl distillation apparatus (reproduced by kind permission of Merck Ltd).

capsule which is then transferred to the flask. In this case a blank test must be performed on a similar capsule.

2. Add five selenium catalyst tablets, 25 ml concentrated sulphuric acid (caution: corrosive) and two or three glass beads.
3. Set up a blank experiment in the same way but omitting the sample. If the sample was weighed in a capsule, include a similar capsule in the blank.
4. Clamp both flasks with their necks inclined at about 45° and heat over low bunsen flames until fumes appear above the acid solution. Continue heating until the liquid becomes clear, then for a further 30 min. Heat the blank for the same time as the sample. Allow to cool.
5. Slowly and carefully add 100 ml water. It is advisable to cool the flask in running water while doing this.
6. Assemble the distillation apparatus as shown, ensuring an adequate water supply to the condenser.
7. Place 30.0 ml 0.1 M hydrochloric acid in a 500 ml conical flask and place it in position so that the flask is tilted and the delivery tube reaches to the lowest point. Add enough water to ensure that the delivery tube dips at least 2 cm below the surface.
8. Check that the tap of the tap funnel is closed and place 125 ml 10 M sodium hydroxide (caution: caustic) in it.
9. Slowly run the alkali into the distillation flask. Close the tap while there is still about 1 ml alkali left.
10. Heat the flask until the contents boil, using a bunsen or a heating mantle. Distil for 20 min or until at least 125 ml distillate has been collected, whichever comes later.

11. Open the tap funnel and turn off the heat. Remove the receiving flask and rinse the delivery tube inside and out with a few millilitres of water, collecting the water in the receiving flask.
12. Titrate to bromocresol green or methyl red with 0.1 M sodium hydroxide.

$$\% \text{ nitrogen} = (V_b - V_s) \times \frac{C}{1000} \times 14 \times \frac{100}{W}$$

where V_b = volume of 0.1 M sodium hydroxide used by blank, V_s = volume of 0.1 M sodium hydroxide used by sample, C = exact concentration of sodium hydroxide in mol/l, and W = weight of sample.

2.5.3 Determination of sulphur

Sulphur is determined by either the Schoeniger test, in which a sample is burnt in oxygen, or by one of a number of fusion methods in which a sample is ignited with an alkaline mixture. In the former, the combustion products are absorbed in alkali. In the latter, the melt is dissolved in water. In both cases the solution is acidified and the sulphate determined by precipitation as $BaSO_4$. The method given here is a fusion method.

1. Place in a platinum crucible (use nickel if the sample contains phosphate) 5 g of a mixture of magnesium oxide and sodium carbonate in the ratio 2:1 by weight. Add a sample containing about 0.05 g sulphur. Cover the sample with another 5–10 g reagent.
2. Heat on a hot bunsen flame for 15 min, then transfer to a furnace and heat at 800°C for 3 h. Allow to cool.
3. Transfer the melt to a beaker using 200 ml water to rinse the crucible.
4. Add bromine water and heat to 80°C to oxidise any sulphite.
5. Slowly add 5 M hydrochloric acid until the solution is acid. About 50 ml will be required. Do this and step 6 in a fume cupboard.
6. Heat to boiling and boil until all the bromine has been driven off, then slowly add 20 ml 10% barium chloride solution. Cover with a watch glass and stand on a steam bath until the precipitate has settled and the solution is clear.
7. Filter on a dry tared silica sintered crucible, wash the precipitate with water then with a few millilitres of ethanol, and dry under a vacuum. Dry in an oven at 105°C for 30 min.
8. Heat in a furnace at 600°C for 15 min, cool and weigh.

$$\% \text{ sulphur} = W' \times \frac{32}{233.4} \times \frac{100}{W}$$

where W' = weight of precipitate, and W = weight of sample.

Nitrogen and sulphur are more conveniently determined by means of an automatic elemental analyser, which burns the sample in a stream of oxygen and analyses the combustion products by gas chromatography. Such instruments are expensive, and certainly the capital outlay could not be justified by this application alone.

2.5.4 *Determination of phosphorus*

Alkyl and alkylether phosphates are most conveniently analysed by ^{31}P NMR spectroscopy, but if this technique is not available phosphorus can be identified and determined after destruction of the organic matter, by one of the many versions of the molybdenum blue test, as in the following method [8].

1. Place a sample containing 0.0025–0.0035 g phosphorus and transfer it to a micro-Kjeldahl flask. Clamp the flask so that the neck is inclined at about 45° and points towards the rear of a well-ventilated fume cupboard.
2. Add 5 ml concentrated sulphuric acid and 5 ml concentrated nitric acid (caution: both are corrosive) and heat slowly over a micro-burner until white fumes appear above the liquid.
3. Add dropwise by Pasteur pipette about 1 ml concentrated nitric acid and continue heating until the brown fumes disppear. Continue heating and adding nitric acid in 1 ml portions until the solution is clear and colourless. Allow to cool.
4. Cautiously add 25 ml water and two glass beads and boil gently over the micro-burner for 30 min to hydrolyse the condensed phosphates. Add more water if necessary to maintain the volume. Allow to cool.
5. Transfer quantitatively to a 100 ml volumetric flask. Add 10 M sodium hydroxide from a burette until the solution is just alkaline to phenolphthalein, then add 1 M sulphuric acid dropwise until the colour is just discharged. Dilute to volume. Dilute 10 ml of this solution to 100 ml. This is solution *S*.

Reagents

A. Sulphuric acid, 5 M.
B. Ammonium molybdate, 4%.
C. Ascorbic acid, 1.32 g in 75 ml water, freshly prepared.
D. Potassium antimonyl tartrate, 0.274%.

Mix 125 ml *A* with 37.5 ml *B*, then add the whole of *C* and 12.5 ml *D*. Prepare daily.

Calibration curve

1. Accurately weigh about 0.088 g potassium dihydrogen phosphate in water, transfer quantitatively to a 1 l volumetric flask and dilute to volume. Dilute 10 ml of this solution to 100 ml. The concentration of phosphorus is 22.79*W′* (approximately 2) μg/ml, where *W′* is the weight of KH_2PO_4.
2. Pipette 0, 5, 10, 15, 20 and 25 ml of the diluted KH_2PO_4 solution into six 50 ml volumetric flasks. Add 40, 35, 30, 25, 20 and 15 ml water respectively. When diluted to volume these solutions contain 0 and about 0.2, 0.4, 0.6, 0.8 and 1.0 μg/ml respectively.
3. Add 8 ml reagent to each flask, dilute to volume and mix.
4. Wait for 10 min, then read the absorbance of each solution against the blank in 1 cm cells at 882 nm.
5. Plot a graph of absorbance against the exact concentration of phosphorus in each solution. The plot should be linear.

Determination

1. Pipette 10 ml of solution *S* into a 50 ml volumetric flask and 10 ml water into another. Follow steps 3 and 4 of the calibration procedure.
2. Read the concentration of phosphorus from the calibration curve.

$$\%\text{phosphorus} = 0.50 \times P/W$$

where P = concentration of phosphorus found at step 2, and W = weight taken.

Provided it is present at a high enough level, phosphorus can also be determined by potentiometric titration with standard alkali, as in the method given below [9]. The titration depends on the formation of $H_2PO_4^-$ ions followed by precipitation of Ag_3PO_4 and titration of the liberated hydrogen ions. The equilibrium

$$H_2PO_4^- + 3Ag^+ \rightleftharpoons Ag_3PO_4 + 2H^+$$

moves steadily from left to right as the titration proceeds.

1. Follow steps 1–3 of the method for sulphur, using a nickel crucible and a sample containing 0.03–0.04 g phosphorus.
2. Slowly add 5 M nitric acid until the solution is just acid, then add a further 10 ml.
3. Heat to boiling, cover with a watch glass and stand on a steam bath for 1 h to ensure complete hydrolysis of the condensed phosphates. Cool to room temperature.
4. Adjust the pH to about 4.5 with 5 M sodium hydroxide, then add a few drops of 5 M nitric acid and adjust the pH to exactly 4.5 with 0.1 M sodium hydroxide.

5. Add 5 ml 5% ammonium nitrate solution followed by 50 ml 10% silver nitrate solution and stir.
6. Titrate potentiometrically with 0.1 M sodium hydroxide. The end point is at pH 4.8–4.9.

$$\% \text{ phosphorus} = V \times \frac{C}{1000} \times \frac{31}{2} \times \frac{100}{W}$$

where V = volume of 0.1 M sodium hydroxide, C = exact concentration of sodium hydroxide in mol/l, and W = weight of sample.

2.6 Spectroscopic identification

Detailed identification of surfactants, particularly in mixtures, is very difficult or impossible without spectroscopic help. Infrared and nuclear magnetic resonance spectroscopy and mass spectrometry are described in later chapters, but it is appropriate to mention them briefly here.

2.6.1 Infrared spectroscopy

For identification purposes mid-range infrared spectroscopy is of more general use than near-infrared. Both are used for quantitative work. Mid-range infrared is absorption spectroscopy in the frequency range 400 to 4000 cm^{-1}. The sample is usually presented in the form of a thin film trapped between two windows of rock salt, potassium bromide or silver chloride, since monatomic ions do not absorb in the infrared. Functional groups such as hydrocarbon chains, carboxy groups, polyoxyethylene chains and sulphate absorb within characteristic frequency bands. An infrared spectrum takes only a few minutes to obtain, and can yield a great deal of information about the types of group present, but it cannot normally tell whether two different groups are present in the same or different molecules. Simple, low-resolution instruments are adequate for 'finger-printing' work and are inexpensive.

2.6.2 Nuclear magnetic resonance (NMR) spectroscopy

NMR spectroscopy is radio-frequency spectroscopy in a powerful magnetic field. Certain atomic nuclei resonate in these conditions at frequencies that depend on their chemical environment; for example, a hydrogen atom in a CH_2 group in the middle of a hydrocarbon chain resonates at a slightly different frequency from a hydrogen atom in the terminal methyl group. The NMR spectrum permits the identification and at least the semi-quantitative determination of individual constituents of mixtures.

NMR can also determine lengths of hydrocarbon and ethylene oxide chains, and if two compounds are present it can measure their molar ratio. NMR and infrared spectroscopy should be regarded as complementary, not alternatives. Relatively inexpensive NMR spectrometers are available.

2.6.3 Mass spectrometry

Mass spectrometry is a technique in which ions of the substance under examination, or fragments of it, are accelerated through electric and magnetic fields to separate them according to their mass. Several different methods of ionisation can be employed, which bring about different degrees and modes of fragmentation. The obvious application is the determination of molecular weights, but knowledge of the way in which various structures fragment makes it possible to obtain a great deal of structural information, up to complete identification of single substances. Detergent alkylates from different sources often show small differences in their mass spectra, and it is sometimes possible to identify the source of the alkylate in a competitor's product by mass spectrometry. Mass spectrometers are very expensive.

These three techniques make the identification of surfactants a simple and straightforward matter in the majority of cases, although mixtures can present problems, and some degree of separation is usually necessary. The high cost of mass spectrometers puts that technique beyond all but the biggest research laboratories.

2.7 Separation/extraction techniques

Analysis of surfactant mixtures almost always requires separation of at least some components. The techniques outlined here are all useful and are all described more fully later.

2.7.1 Solvent extraction

Practical procedures for liquid–solid and liquid–liquid extraction are given in sections 3.1.1 and 3.1.2.

Liquid–solid extraction involves stirring or heating the sample with a solvent to dissolve selectively one component or one group of components. Its chief use is the extraction of surfactants from an inorganic matrix. The sample is dried and treated with a dry solvent, typically a low-molecular-weight alcohol. The solvent is then filtered and evaporated. For powders it is often most effective to do the extraction in a Soxhlet apparatus, but this is not very effective for spray-dried powders.

Liquid–liquid extraction in its simplest form involves shaking the sample in a separating funnel with two immiscible solvents, chosen so that the material to be extracted is much more soluble in solvent A than in solvent B, while the material from which it or they are to be separated is much more soluble in B than in A. The denser solvent is run off into a second separating funnel, and solvent B is re-extracted with more A. Several extractions may be necessary. The portions of solvent A are washed with solvent B. The extracted material is recovered by evaporating solvent A.

The liquid–liquid extraction procedure tends to be tedious, but if properly conducted it is very quantitative and it can be used to extract quite large quantities of material. Typical aqueous phases are water alone and mixtures of water with methanol, ethanol or propan-2-ol, while typical non-aqueous phases are petroleum ether or other hydrocarbons, diethyl ether, chloroform and butanols. It is used chiefly to extract fatty alcohols and ethoxylates, fatty acids, fatty amines and alkanolamides from mixtures containing more polar surfactants, but the possible range of uses is much wider than this.

2.7.2 Chromatography

Gas–liquid (GLC) and high-performance liquid (HPLC) chromatography are extremely useful techniques and are fully described in a later chapter. They are primarily used for quantitative analysis, GLC for more volatile and HPLC for less volatile substances. GLC can determine, for example, hydrocarbons, alcohols and esters, and HPLC can determine fatty acids and high-molecular-weight materials. Both can determine chain-length distributions of both hydrocarbon and ethylene oxide chains. It is frequently necessary to prepare derivatives of the materials to be separated by GLC, but this is not usually the case for HPLC. Neither is very useful as an aid to identification of unknowns.

Thin-layer chromatography can be used for identification, by comparison with chromatograms of reference materials, and this is now its chief use. It is, however, capable of only semi-quantitative analysis at best.

The use of ion chromatography for surfactant analysis has been reported [10], but in general it is not very useful in this field because of the strong hydrophobic interaction between the hydrocarbon tails of surfactants in general and the organic matrices of ion-exchange resins. The development of exchange media without organic matrices, e.g. ion-exchange silicas, may result in this problem being overcome, but it seems probable that any separations that could be achieved would be by chain length rather than by chemical class, i.e. it may be easier to separate, say, a C_{12} sulphate from a C_{14} sulphate than to separate a C_{12} sulphate from a C_{12} sulphonate.

2.7.3 *Ion exchange*

Simple ion exchange, i.e. the exchange of ions in solution for fixed ions in a solid matrix, is not ion chromatography. It does not involve differential migration down the column due to variations in the affinities of the solutes for the solid substrate; the ions either exchange and then stay where they are until eluted, or they pass through the column without interaction with it.

Ion exchange is the subject of chapter 4, and will not be discussed in any detail here. It is by far the most useful separation technique for dealing with appreciable quantities of surfactants, and can be used to separate the four main classes from each other. It can, however, achieve separation within classes to only a very limited extent. It is difficult to understand why advanced technology has not been introduced into this extremely useful technique until very recently. Now that process has started, it seems inevitable that the capabilities of ion exchange will increase very markedly during the next few years.

2.7.4 *Solid-phase extraction*

In solid-phase extraction (SPE) a solution of the sample is passed through a column of a sorbent, which retains the desired substance and possibly others. Unwanted materials can then be eluted and the desired substance selectively recovered. The available sorbents are mostly chemically modified silicas and include the following types.

Sorption type	*Sorbent functional groups*
Non-polar (reverse phase)	Alkyl—octadecyl, octyl, ethyl; cyclic—cyclohexyl; aromatic—phenyl
Polar (normal phase)	Cyanopropyl, diol, silica, amine
Ionic	Anion exchange—strongly basic (quaternary); weakly basic (primary, tertiary and mixed amine) Cation exchange—sulphonic acids; carboxylic acids
Covalent (vicinal diols)	Phenylboronic acid

These and other sorbents are supplied in prepacked columns. Analytichem (Varian Sample Preparation Products) Bond Elut, International Sorbent Technology (Jones Chromatography) Isolute and Waters Sep-Pak are all well-known names. The sorbents are also available in bulk, and a wide range of accessories is available, such as Luer stop-cocks, vacuum pumps, manifolds and regulators. The sorbents supplied by Varian and International Sorbent Technology appear to be identical, but the apparatus and accessories are not. The technique is very readily automated for handling large numbers of samples.

Buschmann *et al.* [11] have reported separation of anionic, cationic, nonionic and amphoteric surfactants with at least 95% recovery by using silica and alumina SPE cartridges.

Another interesting sorbent type is the Amberlite XAD range of resins. These are cross-linked polystyrene resins, like many ion-exchange resins, but without the ionic functional groups. XAD-1 to XAD-4 differ in specific surface and pore size. XAD-7 has carboxylic ester groups. They have been used for the analysis of nonionics [12–14] and cationics [15].

SPE might be described as HPLC without the technology, and at least for small quantities of material it offers advantages of speed, selectivity and convenience over liquid–liquid extraction.

2.7.5 Open-column chromatography

Open-column chromatography on alumina or silica has been extensively used in surfactant analysis. It is inexpensive but tedious and is now being superseded by HPLC.

Separations on alumina [16] include:

Ethoxylates, alkanolamides, anionics.
Amphoterics from other cationics.
Hydrocarbons, secondary alcohols, primary alcohols, waxes and fatty esters, lanolin, ethoxylates, alkanolamides, anionics.

Separations on silica are mostly in the nonionics field and include the following general separation [17]:

Hydrocarbons + fatty acid methyl esters, glycerides + fatty alcohols + fatty acids, fatty acid ethoxylates + alkylphenol ethoxylates, other ethoxylates, glycerol + high-molecular-weight polyethylene glycols.

Separations usually take a long time, use large volumes of solvents and often do not give clean separations. A description of the practical procedure is given in chapter 3. Some specific uses are described under the appropriate surfactants.

References

1. Christiansen, A. *Surfactant Science Series* vol. 12 (ed. Bluestein, B. R. and Hilton, L. H.), Marcel Dekker, 1982, pp. 1–70.
2. Lomax, E. *Recent Developments in the Analysis of Surfactants* (ed. Porter, M. R.), Elsevier, 1991, pp. 109–136.
3. ISO 2271: 1989 (E): Surface active agents—Detergents—Determination of anion-active

matter by direct manual or mechanical two-phase titration procedure. International Organization for Standardization, Geneva.
4. Rosen, M. J. and Goldsmith, H. A. *Systematic Analysis of Surface-Active Agents*, 2nd edn, Interscience, 1972.
5. Longman, G. F. *The Analysis of Detergents and Detergent Products*, Wiley, 1975, p. 155.
6. ISO 2869: 1973 (E): Surface active agents—Detergents—Anionic-active matter hydrolyzable under alkaline conditions—Determination of hydrolyzable and non-hydrolyzable anionic-active matter. International Organization for Standardization, Geneva.
7. ISO 2870: 1986 (E): Surface active agents—Detergents—Determination of anionic-active matter hydrolyzable and non-hydrolyzable under acid conditions. International Organization for Standardization, Geneva.
8. Murphy, J. and Riley, J. P. *Anal. Chim. Acta*, **27** (1962), 31.
9. Cullum, D. C. and Thomas, D. B. *Anal. Chim. Acta*, **24** (1961), 205–213.
10. Weiss, J. *J. Chromatog.*, **353** (1986), 303–307.
11. Buschmann, N., Kruse, A. and Schulz, R. *Comun. Jorn. Com. Esp. Deterg.*, **23** (1992), 317–322.
12. Leon-Gonzalez, M. E., Santos-Delgado, M. J. and Polo-Diez, L. M. *Analyst*, **115** (1990), 609–612.
13. Anielak, P. and Janio, K. *Tenside*, **27** (1990), 113–117.
14. Saito, T. and Hagiwara, K. *Fresenius Z. Anal. Chem.*, **312** (1982), 533–535.
15. Franke, L. P., Wijsbeek, J., Greving, J. E. and De Zeeuw, R. A. *Vet. Hum. Toxicol.*, **21** (Suppl., 1979), 193–196.
16. Longman, G. F., ref. 5, p. 51.
17. Rosen, M. J. *Anal. Chem.*, **35** (1963), 2074–2077.

3 Basic techniques

3.1 Introduction

This chapter deals with a collection of basic procedures, most of which have a range of applications. It presents a range of techniques that the analyst will use in various ways and combinations, and in some cases with different reagents, to attack problems. The methods described are all of a general nature, that is they can be applied to a range of materials. Methods specific for a particular surfactant or class of surfactants are described in later chapters.

In practice the distinction is not clear-cut, so that, for example, the potentiometric titration of ionic surfactants with surfactants of opposite charge is given in this chapter, because although it is a specific instance of potentiometric titration, it is not specific to a particular class of surfactants.

3.2 Extraction of surfactants

Most raw materials and many formulated products can be analysed without isolation of the surfactants. For some purposes, however, e.g. determination of molecular weight by titration, the surfactant must be isolated in a pure form. Liquid–solid extraction is useful for this purpose. It is also sometimes desirable to extract fatty acids, neutral fatty matter and/or weak fatty bases, either to determine them or to prevent interference with some other determination. Liquid–liquid extraction is appropriate for this. A special case is the analysis of structured bleaches, which may contain surfactants in strong hypochlorite. The bleach makes most analytical processes for the surfactants difficult or impossible, and the procedure given below is a useful first step.

3.2.1 Liquid–solid extraction

Surfactants can be separated from most inorganics by extraction with low-molecular-weight alcohols (up to and including butanols), and in some cases acetone or chloroform. Most inorganics are insoluble or sparingly soluble in all of these. Most inorganics are almost insoluble in propan-2-ol, and this solvent has a low enough boiling point to be easily evaporated off. For an unknown, it may be advantageous to extract sequentially with say propan-2-ol, acetone and chloroform. The solubility of inorganics decreases as the molecular weight of the alcohol increases. The solvent must be reasonably dry because low concentrations of water markedly increase the solubility of inorganics, but solvents as supplied by the manufacturer are normally suitable for use without any special treatment. Some non-surfactant organic substances may also be extracted.

The sample size will depend on its active content, if known, and on what further analyses are envisaged. For an initial extraction of an unknown, 5 g is suitable. Caution: organic solvents are mostly flammable or toxic or both, and appropriate precautions must be taken.

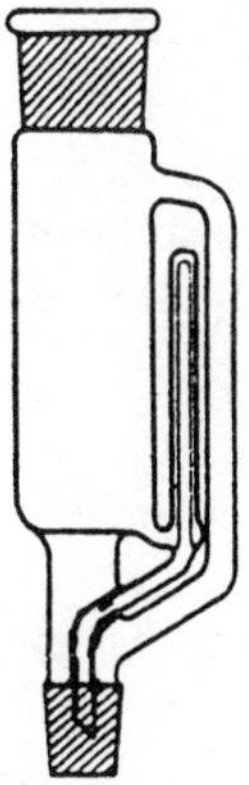

Figure 3.1 Soxhlet extractor (reproduced by kind permission of Merck Ltd).

Powders are best extracted in a Soxhlet apparatus, shown in Figure 3.1. It consists essentially of a cylindrical extraction chamber having an external bypass tube and a syphon tube running from the bottom to the top and down again, then re-entering the apparatus and discharging. In use, it is attached to a flask containing the solvent, and a reflux condenser. The sample is contained in a porous thimble, which is placed in the extraction chamber. When the solvent in the flask is boiling, the vapour flows up the bypass tube into the condenser. The condensate falls into the extraction chamber. When the chamber is full, the liquid syphons over and back into the flask. This constitutes one extraction cycle. The process is repeated until extraction is complete.

Procedure—1

1. Weigh a sample of the powder in a Soxhlet thimble, stand it in a beaker and dry to constant weight at 105°C.

2. Protect the sample with a plug of cotton wool and place the thimble in a Soxhlet extractor. Attach a tared receiving flask and a reflux condenser.
3. Slowly pour sufficient solvent (at least three times the volume of the extraction chamber) into the top of the condenser and allow it to syphon into the flask.
4. Heat to boiling using a heating mantle. Ignore the first extraction cycle but time the next two cycles and continue boiling for ten times as long, i.e. until about 20 cycles have occurred.
5. Remove the flask and replace it with another, add more solvent and continue for a further 10 cycles.
6. Evaporate the contents of both flasks to dryness, cool and weigh.

The purpose of the second extraction is to confirm that complete extraction was achieved by the first. The second flask should contain no solids, or at most a few milligrams. The hollow bead structure of spray-dried powders can make it difficult to obtain a complete extraction by the Soxhlet method, and the following procedure is then better.

Procedure—2

1. Weigh a sample into a 400 ml beaker and dry to constant weight at 105°C.
2. Add 200 ml solvent and heat to boiling on a steam bath or flameproof hotplate (caution: flammable or toxic vapour). Continue boiling for 5 min.
3. Allow the solids to settle and decant the supernatant through a filter, collecting it in a tared 800 ml beaker.
4. Repeat steps 2 and 3 twice more, using 100 ml solvent each time.
5. Dissolve the residue in the minimum quantity of hot water and evaporate to dryness.
6. Repeat step 2 twice more, using 100 ml solvent each time, and taking care to wet the whole of the dry residue thoroughly. Collect the filtrate in the same beaker as the main extract.
7. Evaporate to dryness, cool and weigh.

Pastes and most liquids are treated as follows.

Procedure—3

1. Weigh a sample in a round-bottomed flask or a beaker and dry first at 100°C until the sample appears dry, then to constant weight at 105°C, preferably in a vacuum oven. Do not apply the vacuum until the sample is fairly dry, otherwise it will probably spatter.
2. Add the extracting solvent and either boil under a reflux condenser or cover with a watch-glass and heat on a steam bath for 15 min.

3. Cool and carefully decant the solution into a tared beaker or flask, through a filter if suspended matter is present.
4. Repeat steps 2 and 3 at least twice more.
5. Evaporate to dryness, cool and weigh.

Compact solids such as soaps can be extracted by simply heating with the appropriate solvent.

3.2.2 Liquid–liquid extraction using separating funnels

In liquid–liquid extraction the sample is shaken with two immiscible liquids *A* and *B*, chosen so that the substance to be extracted is much more soluble in *A* than in *B*, while the substances from which it is to be separated are much more soluble in *B* than in *A*. At equilibrium the ratio of the concentrations of any component in the two solvents is approximately equal to the ratio of its solubilities in them. If a component is nine times more soluble in *A* than in *B*, at equilibrium about nine-tenths of it will be in *A* and one-tenth in *B*, if the volumes of *A* and *B* are equal. The extraction is usually repeated, sometimes several times, and the extracts are washed with the other solvent, either after being combined or sequentially.

The system may be envisaged as a column of separating funnels, each containing some of the lighter solvent, through which each portion of the heavier solvent passes in turn; this is so no matter which is the extracting solvent and which the extracted. At the end of the process one solvent will contain all of the desired substance, virtually free of all other material, and the other solvent will contain everything else. The solvent containing the desired substance may be evaporated off and the extract weighed and subjected to whatever other examination may be necessary. The procedure can vary in detail depending on the application.

Most chemists will already be familar with this technique, but two facts are not sufficiently appreciated: (a) it is more effective to extract with several small portions of solvent than with one portion of the same total volume; and (b) loss of the extracted substance is minimised by washing the extracts sequentially rather than after combining them. This is no more than good basic practice.

Nevertheless, many of the materials extracted in surfactant analysis are so insoluble in the solvent from which they are extracted that two extractions are sufficient, and the loss on washing is negligible even if the extracts are combined. An obvious example is the extraction of fatty acids with petroleum ether from acidified aqueous solution.

Extraction of neutral fatty matter (NFM). This is a procedure that is often used for extraction of unsulphated fatty alcohol or unsulphonated hydrocarbon, and it illustrates a basic technique that can be adapted to many other uses. It is assumed that the sample is either an alkyl sulphate or a hydrocarbon sulphonate. The neutral fatty matter from the latter will include materials other than unsulphonated hydrocarbon. It is possible to do the extraction in a number of ways, and a typical procedure is given here. Some versions use more or fewer extractions. It is not suitable for samples in which the NFM includes ethoxylated material.

The procedure uses aqueous ethanol to minimise emulsion formation, but a simple aqueous solution could be used if this were not a problem.

1. Dissolve the sample in 50% aqueous ethanol in a 500 ml separating funnel, make just alkaline to phenolphthalein with 1 M sodium hydroxide in 50% ethanol and adjust the total volume to about 100 ml with the same solvent.
2. Add 100 ml petroleum ether (caution: flammable vapour), stopper the separating funnel and shake thoroughly. Release the excess pressure by inverting the separating funnel and cautiously opening the stop-cock.
3. Allow the layers to separate and run the lower (aqueous ethanol) layer into another separating funnel.
4. Repeat steps 2 and 3 three more times.
5. Combine the petroleum ether extracts in the first separating funnel and wash with 50 ml 50% aqueous ethanol. Add the washings to the main aqueous ethanolic solution and keep for further analysis.
6. Evaporate the petroleum ether extract to dryness, dry and weigh.

Fatty alcohols and some hydrocarbons are relatively volatile and unnecessary heating must be avoided. A safe evaporation procedure is to evaporate on a steam bath to 5 ml or less, but not to dryness, take down to dryness under an air or nitrogen jet without heating, and then dry at room temperature in a vacuum desiccator, weighing every 15 min until the loss in weight is less than 2 mg [1].

Extraction of fatty acid from soap is similar, except that the solution is made acid at step 1. Extraction of fatty amine is as described, but if it is necessary to extract the NFM and the amine separately the extraction is done first from an alkaline medium (NFM and amine together), and the amine is then extracted with acidified aqueous ethanol, and re-extracted with petroleum ether after making the solution alkaline.

If all three were present, the NFM and amine could be extracted together as described, and the fatty acid extracted after acidifying the aqueous ethanol solution.

Extraction of surfactants from structured bleaches. Structured bleaches usually contain a high concentration of hypochlorite and low concentrations of one or more surfactants. Because of the aggressive nature of hypochlorite, the surfactant(s) cannot be determined in its presence. Chemical destruction of the hypochlorite, even with as mild a reagent as urea, which yields only nitrogen and carbon dioxide, results in extensive decomposition of the surfactants. The following procedure has been shown to recover the surfactants with only minimal damage. It uses both liquid–liquid and liquid–solid extraction. Because of the low concentrations of surfactants it may be necessary to use a large sample, e.g. 100 ml or 100 g.

1. Place a sample of the bleach in a separating funnel and dilute with four or five volumes of water.
2. Extract with three 100 ml portions of butan-1-ol or butan-2-ol (caution: irritant and flammable vapour). Discard the aqueous layer.
3. Wash the butanol extracts sequentially with 25 ml portions of 10% sodium sulphate solution until the washings remain colourless when treated with potasium iodide and hydrochloric acid.
4. Combine the butanol extracts and evaporate to dryness.
5. Dry the residue to constant weight at 105°C, then extract the surfactants by boiling with three 50 ml portions of dry propan-2-ol (caution: flammable) and decanting the solvent through a filter.
6. Evaporate to dryness, weigh the residue and analyse in whatever way is appropriate.

3.2.3 Liquid–liquid extraction using extraction columns

A recent innovation in liquid–liquid extraction is the use of extraction columns. These are simply glass or plastic tubes containing diatomaceous earth (kieselguhr). A measured volume of an aqueous solution of the sample is poured into the column and left to stand for 3–5 min, or mixed with the diatomaceous earth before packing the column. The organic extracting solvent is then poured in and is collected after flowing through the column under gravity. The advantages claimed are cleaner extracts and no emulsion formation.

Cross [2] determined neutral oil (NFM) in α-olefin sulphonates, alkyl sulphates, alkylbenzene sulphonates and alkylether sulphates by this technique. He reported good accuracy and reproducibility, and complete elimination of emulsion problems. His procedure was as follows. Celite 545 is a diatomaceous earth supplied by Johns Manville.

1. Weigh a 1 g sample in a 100 ml beaker. Add 2 ml 50% aqueous ethanol and mix until a clear solution is obtained.
2. Add 4 g Celite 545 and mix to a homogeneous semi-dry solid.
3. Transfer to a 19 mm × 400 mm glass column and pack down lightly with a flattened glass rod.

4. Add 5 ml 7% dichloromethane in petroleum ether and allow it to run into the bed. Then add a further 15 ml of the same solvent.
5. Collect the effluent in a tared flask.
6. Evaporate the solvent at 60°C under reduced pressure, cool and weigh.

Analytichem columns, supplied by Varian Sample Preparation Products (see section 2.7.4), are supplied prepacked and have sample capacities ranging from 0.3 to 300 ml. They are available in three basic varieties: buffered at pH 4.5 for extraction of acids, buffered at pH 9.0 for extraction of bases, and unbuffered. Inexpensive racks and collection tubes are available. The diatomaceous earth can be purchased in bulk, and it is easy to make one's own columns in plastic or glass.

The concept is an extension of partition chromatography, now superseded by HPLC. There would appear to be no reason why all, or at any rate most, liquid–liquid extraction processes should not be done in this way. Specific applications of liquid–liquid extraction are given in chapters 5, 6 and 7.

3.3 Acid–base titration

3.3.1 General principles

All acid–base titrations in aqueous or aqueous–alcoholic media consist of the neutralisation of hydrogen ions by hydroxide ions or vice versa, with the formation of water:

$$H^+ + OH^- \rightarrow H_2O$$

However, weak acids do not yield many hydrogen ions and weak bases do not yield many hydroxide ions, and titration of weak acids and bases or their salts with strong acids and bases may equally well be regarded as the addition or removal of a proton:

$$RCOOH + OH^- \rightarrow RCOO^- + H_2O$$

$$RCOO^- + H^+ \rightarrow RCOOH$$

$$RNH_3^+ + OH^- \rightarrow RNH_2 + H_2O$$

$$RNH_2 + H^+ \rightarrow RNH_3^+$$

It is therefore appropriate to speak of the protonated and deprotonated forms of weak acids and bases.

Hydrolysis of salts of weak acids and bases. In aqueous solution the end-point in titrations of strong acids with strong bases, or vice versa, occurs at or near pH 7. Salts of weak acids and bases, however, are hydrolysed in aqueous media, with the liberation of hydroxide or hydrogen ions respectively. The hydrolysis is an equilibrium state, not a continuing process.

$$RCOO^- + H_2O \rightleftharpoons RCOOH + OH^-$$

$$RNH_3^+ + H_2O \rightleftharpoons RNH_3OH\ (\rightarrow RNH_2 + H_2O) + H^+$$

Aqueous solutions of salts of weak acids with strong bases are alkaline, and aqueous solutions of salts of weak bases with strong acids are acidic. The end-point in the titration of a weak acid with a strong base is therefore displaced to a higher pH, while the end-point in a titration of a weak base with a strong acid is displaced to a lower pH.

The titration curve is also less steep for a weak acid or base, because more hydrogen or hydroxide ions are required to bring about a given change in pH when their concentration is greater than it is in water. This is a simple consequence of the logarithmic nature of the pH scale.

The weaker the acid or base, the greater the extent of the hydrolysis. In the case of long-chain carboxylic acids the effect is so great that the titration cannot be done in aqueous solution, and it is necessary to use alcohol as solvent, not merely because these compounds are insoluble in water but to suppress the hydrolysis. Free alkali in soaps must be titrated in alcoholic solution; any attempt to do it in aqueous solution simply results in precipitation of fatty acid, even though the solution remains alkaline.

Salts of weak acids can be titrated with strong acids. The hydroxide ions liberated by hydrolysis are neutralised, causing the equilibrium to shift to the right, and the process continues until the weak acid is completely protonated. Similarly, salts of weak bases can be titrated with strong bases. The neutralisation of the hydrogen ions causes the equilibrium to shift to the right, and the process continues until the weak base is fully deprotonated. Uses of these processes are described in section 3.2.3.

3.3.2 End-point detection

The end-point may be detected by the use of indicators or by using a pH meter (section 3.2.3). Most weak acids in alcoholic solution show a pH around 3 or 4, and their salts or deprotonated forms show a pH around 9 or 10. Most weak bases show a pH around 9 or 10, and their salts or protonated forms show a pH around 4. Actual values depend on concentration of the weak acid or base and the alcohol–water ratio. The fact that the protonated and deprotonated forms of acids and bases show similar pHs is fortuitous, and is due to the fact that their dissociation constants are fairly similar, to within an order of magnitude.

Suitable indicators for the low-pH end-point are bromocresol green and methyl red, or bromophenol blue and methyl orange, which change colour over a somewhat lower pH range. Suitable indicators for the high-pH end-point are phenolphthalein and thymolphthalein, the latter changing colour over a somewhat higher pH range. There are plenty of alternatives.

If an acid–base titration is done potentiometrically, the indicating electrode is a glass electrode, which responds to changes in the hydrogen-ion concentration. The procedure is described in section 3.3.2. The actual pH at the end-point, and the corresponding titration volume, is read from the titration curve, or printed out by the autotitrator. In some ways this is preferable to the use of indicators. It eliminates the need to guess the pH at the end-point, and is easily adapted for use in automatic analysis schemes.

Simple manual acid–base titrations with indicators are virtually free from bias and are capable of good reproducibility. Replicate titrations on the same-sized sample should agree within 0.10 ml or less, with normal care. With exceptional care it is sometimes possible to obtain replicates agreeing within 0.02 ml.

3.3.3 Determination of weak acids and bases and their salts

Weak acids and bases can be determined either alone or in the presence of strong acids and bases by titration between two pHs. It is immaterial whether they are present as free acid or base, or as a salt, or as a mixture of acid or base and salt.

Example 1. Analysis of a mixture of a fatty acid and a sulphonic acid

1. Dissolve the sample in ethanol and titrate with alcoholic alkali to bromophenol blue, or potentiometrically to the first point of inflection, at a pH around 3.
2. Continue titrating to phenolphthalein, or potentiometrically to the second point of inflection, at a pH around 9.

The volume of alkali to the first end point measures the sulphonic acid, but the fatty acid is still in the protonated (free acid) form, which is neutralised at the second end-point.

If the titration is done with indicators, it is preferable to titrate to phenolphthalein first, then add bromophenol blue and titrate to the low-pH end-point with alcoholic standard acid. The first titration measures both acids, and the second the fatty acid alone. This procedure makes the phenolphthalein end-point easier to see.

Example 2. Analysis of a mixture of a quaternary ammonium hydroxide and a fatty amine. Such a mixture might be present, for example, in the effluent from an anion-exchange column—see chapter 4.

1. Dissolve the sample in ethanol and titrate with alcoholic acid to phenolphthalein, or potentiometrically to the first point of inflection, at a pH around 9.
2. Continue titrating to bromophenol blue, or potentiometrically to the second point of inflection, at a pH around 4.

The volume of acid to the first end-point measures the quaternary hydroxide, but the amine is still in the deprotonated (free base) form, which is neutralised at the second end-point.

Comparison of the indicator and potentiometric procedures for specific cases may show that this choice of indicators is not always ideal.

Example 3. Determination of a weak acid or base present as a salt. The sample might contain the sodium salts of the two acids or the chloride and hydrochloride of the two bases.

1. Dissolve the sample in ethanol.
2. Either acidify by adding ethanolic hydrochloric acid or make alkaline by adding ethanolic sodium hydroxide.
3. In the first case titrate with alcoholic alkali as in Example 1. In the second case titrate with alcoholic acid as in Example 2.

Whichever procedure is used, the volume of titrant between the two end-points measures the weak acid or the weak base.

If the sample contains a partially neutralised weak acid or weak base, the proportions of free and neutralised material are determined by the following procedure.

1. Dissolve two samples in ethanol.
2. Titrate one with alcoholic alkali to phenolphthalein (or pH around 9), and the other with alcoholic acid to bromophenol blue (or pH around 3 or 4).

The titration with alkali measures either the free weak acid or the salt of the weak base. The titration with acid measures either the free weak base or the salt of the weak acid. For example, if this analysis was done on a sample of triethanolammonium lauryl sulphate containing excess triethanolamine, titration with acid to the low-pH end-point would measure the free triethanolamine, while titration with alkali to the high-pH end-point would measure the triethanolamine present as the lauryl sulphate and other salts.

Amphoterics, apart from the SS class, can be determined in this way, since they contain a weak acid group (WS), a weak base group (SW) or both (WW). There are, however, complicating factors, which are discussed in chapter 7.

3.4 Potentiometric titration

3.4.1 Principles

Potentiometric titration is done by following the potential difference between a pair of electrodes immersed in the solution. One of them, the indicating electrode, responds to changes in the concentration (strictly the activity) of one of the ions involved in the reaction, i.e. its potential changes as the concentration of that ion changes. The other, the reference electrode, displays a constant potential. The electrodes are connected to a suitable potentiometer, which measures the potential difference between them. The titrant is added in small increments, with continuous stirring, and the potentiometer reading (in millivolts) is recorded after each addition. A plot of millivolt reading (vertical axis) against volume of titrant (horizontal axis) is S-shaped, with the steepest part at the end-point.

Ideally the potential of the indicating electrode is given by the Nernst equation, which in its simplest form is

$$E = E^0 + 0.059 \log c$$

(for monovalent cations), where E is the potential of the indicating electrode in volts, E^0 is a constant and c is the concentration of the sensed ion. This means that the potential changes by 0.059 volt (59 millivolts) for any tenfold change in c. The S shape of the titration curve follows from this. It is not essential for the indicating electrode to show a Nernstian response for it to be useful as an end-point indicator.

Potentiometric titration can be used to determine any ion for which an indicating electrode can be devised. Apart from the familiar glass, silver–silver chloride and platinum electrodes, a wide range of ion-selective electrodes is now available, so that the scope for potentiometric titration is very wide. The only requirement of the reference electrode is that it must display a constant potential under the changing conditions during the titration. Calomel electrodes are usual, but silver–silver chloride can also be used, and there is no reason why indicating electrodes such as glass or platinum should not be used in some instances.

3.4.2 Applications

All of the following can be done potentiometrically.

Acid–base titrations, using a glass electrode.

Precipitation titrations, e.g. determination of halides by titration with silver nitrate, using a silver–silver chloride electrode.
Complexometric titrations, e.g. titration of barium with EDTA, using a divalent metal ion selective electrode.
Redox titrations, using a platinum electrode.
Titrations of anionic, cationic and amphoteric surfactants with surfactants of opposite charge, using a surfactant-sensitive electrode.
Titrations of cationic and amphoteric surfactants, and metal complexes of nonionic surfactants, with tetraphenylborate.

All of these find some application in surfactant analysis. The last but one is described more fully in section 3.6.

Manual potentiometric titration is straightforward. Most pH meters can be switched to display millivolts and can thus be used as potentiometers to monitor titrations other than acid–base. The basic procedure for all titrations is as follows.

1. Connect the indicating and reference electrodes to the correct sockets of the pH meter or potentiometer. Calomel electrodes are often used as the reference for titrations other than acid–base.
2. Place the solution to be titrated in a beaker and insert the electrodes, ensuring that the tips are immersed.
3. Stir continuously with either a magnetic or a propeller-type stirrer.
4. Add the titrant in small increments, allowing the indicated potential (or pH) to become constant after each addition. Record titrant volume and potential (or pH).
5. When the change in potential with each addition of titrant starts to increase, reduce the size of the increments, and proceed in single drops in the region of the end-point.
6. Stop titrating when the steepest point of the titration curve has clearly been passed, i.e. when the change in potential slows down.

It is, however, much simpler to use an autotitrator, which can usually be programmed to vary the increment size as the end-point is approached, and to print out the titration curve, the millivolt reading at the end-point and the titration volume, sometimes to three decimal places.

3.5 Methods for esters, amines, alcohols and unsaturated fatty materials

3.5.1 Determination of acid value, ester value and saponification value

These are traditional indices used chiefly in the analysis of oils and fats in the soap industry. They, or at least the methods used to determine them, do have application, however, in the analysis of surfactants which are esters. They are defined as follows.

Acid value. The number of milligrams of potassium hydroxide required to neutralise the free acidity in one gram of sample.

Ester value. The number of milligrams of potassium hydroxide required to saponify, i.e. turn into soap, the esters in one gram of sample.

Saponification value. The number of milligrams of potassium hydroxide required to saponify the fatty acids and esters in one gram of sample.

Procedure

1. Dissolve a convenient weight of sample, usually 2–3 g, in neutral ethanol in a round-bottomed flask. Industrial methylated spirit may be used in those countries in which it is permitted.
2. Add a few drops of 1% phenolphthalein and titrate with 0.5 M ethanolic potassium hydroxide.

$$\text{Acid value} = V_1 \times 28.05/W \text{ mg/g}$$

   where V_1 = volume of potassium hydroxide, and W = weight of sample.
3. Add a known excess, usually 25 or 50 ml, of 0.5 M ethanolic potassium hydroxide.
4. Boil under a reflux condenser for 1 h.
5. Cool and back-titrate with 0.5 M ethanolic hydrochloric acid.
6. Carry out a simultaneous control experiment using the same volume of alkali without a sample.

$$\text{Ester value} = (V_3 - V_2) \times C \times 56.1/W \text{ mg/g}$$

where V_3 = volume of hydrochloric acid required by the control, V_2 = volume of hydrochloric acid required by sample, C = exact concentration of hydrochloric acid in mol/l, and W = weight of sample.

$$\text{Saponification value} = \text{acid value} + \text{ester value}$$

Obviously the saponification value may be determined in one stage, without the preliminary neutralisation.

The method as written is traditional. The working medium must be alcoholic. The alkali may be sodium hydroxide and concentrations other than 0.5 mol/l may be used provided the boiling is continued for long enough to ensure complete saponification.

Some materials, e.g. beeswax and other materials containing very long-chain acids, need longer boiling. If in doubt, set up several samples and boil for 1, 2, 3, etc. hours. When two consecutive results agree within reasonable experimental error, saponification is complete.

3.5.2 *Determination of alkali value and base value*

There is some confusion in the literature about the meanings of these two terms. They are defined for the purpose of this section as follows, but there is not universal agreement about these definitions. The use of these two values in material specifications is largely confined to weak fatty bases.

Alkali value: the number of milligrams of potassium hydroxide equivalent to the free strong base in 1 g of sample.
Base value: the number of milligrams of potassium hydroxide equivalent to the free weak base in 1 g of sample.

They may be determined together, as follows.

1. Weigh a suitable sample, usually 2–3 g, in a conical flask.
2. Dissolve in 50 ml neutral ethanol.
3. Titrate to phenolphthalein with 0.5 M ethanolic hydrochloric acid.
4. Continue titrating with the same acid to bromophenol blue.

$$\text{Alkali value} = V_1 \times \frac{C}{1000} \times \frac{56\,100}{W}$$

$$\text{Base value} = V_2 \times \frac{C}{1000} \times \frac{56\,100}{W}$$

where V_1 = volume of 0.5 M acid (step 3), V_2 = volume of 0.5 M acid (step 4), C = exact concentration of 0.5 M acid, and W = weight of sample.

3.5.3 *Determination of hydroxyl value*

The hydroxyl value of a fatty alcohol or ethoxylate is defined as the number of milligrams of potassium hydroxide required to neutralise the acid required to esterify the hydroxyl groups in one gram of sample, i.e. the number of milligrams of potassium hydroxide that contains the same number of hydroxide ions as there are hydroxyl groups in one gram of sample.

Two methods are in general use, in which acetic anhydride and phthalic anhydride are used as the respective esterifying agents. ISO 4326 [3] and ASTM Method D 4252-89 [4] describe the acetic anhydride and phthalic anhydride methods respectively, and the reader is advised to refer to the appropriate standard method.

Both methods are applicable to ethylene oxide adducts of primary fatty alcohols, alkylphenols and fatty acids. They are also applicable to primary fatty alcohols, although this is not mentioned in either Standard.

According to ISO 4326, the acetic anhydride method is subject to interference from a large number of materials, including amines, amides, thiols, epoxides and some fatty acids and esters, and is not applicable to propylene oxide adducts. The phthalic anhydride method may be subject to the same constraints, but the ASTM method does not say so.

The analysis has three stages:

1. Esterification of the terminal hydroxyl groups with acetic or phthalic anhydride:

$$ROH + (CH_3CO)_2O \rightarrow CH_3COOR + CH_3COOH$$

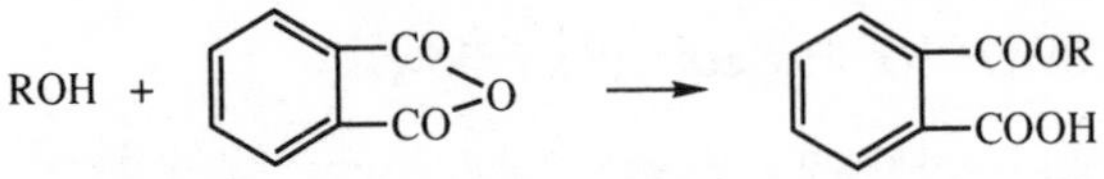

2. Hydrolysis of the excess reagent by adding water.
3. Titration of the liberated acid.

A blank determination is carried out, and the difference between the two titrations represents the anhydride used to esterify the hydroxyl groups.

Procedure: acetic anhydride method

Reagent. Dissolve one volume of acetic anhydride (caution: irritant vapour) in 10 volumes of pyridine (caution: toxic vapour).

1. Weigh a sample into a dry 250 ml conical flask with a ground-glass neck. Use approximately 380/HV, where HV is the expected hydroxyl value. The sample should preferably be dry, and its water content must not in any case exceed 1%.

Carry out the rest of the analysis in a well-ventilated fume hood.

2. Add by pipette or burette (do not pipette by mouth) 15 ml acetic anhydride reagent and swirl to mix.
3. Attach a reflux condenser and heat in a boiling water bath for 1 h.
4. Add 2 ml water via the condenser and continue heating for 5 min.
5. Cool to below 30°C and add 70 ml water via the condenser. Remove the condenser.
6. Add by pipette 25 ml 0.5 M sodium hydroxide. Titrate to phenolphthalein with the same solution.
7. Carry out a simultaneous blank experiment, replacing the sample with two or three drops of water.

The titration on the sample must be more than half the blank titration. If it is not, repeat the analysis with a smaller sample.

Procedure: phthalic anhydride method

Reagent. Dissolve 98 g of phthalic anhydride in 700 ml pyridine. Store in a dark bottle and discard after one week.

1. Weigh a sample into a dry 250 ml conical flask with a ground-glass neck. Use *M*/100(OH) g, where *M* is the molecular weight of the sample and (OH) the number of hydroxyl groups per molecule.

Carry out the rest of the analysis in a well-ventilated fume hood.

2. Add by pipette or burette (do not pipette by mouth) 25 ml phthalic anhydride reagent and swirl to mix.
3. Attach a reflux condenser and boil for 1 h.
4. Rinse the condenser with 25 ml pyridine and then with 25 ml water.
5. Titrate to phenolphthalein with 1.0 M sodium hydroxide.
6. Carry out a simultaneous blank experiment, replacing the sample with two or three drops of water.

The titration on the sample must be at least 75% of the blank titration. If it is not, repeat the analysis with a smaller sample.

$$\text{Hydroxyl value} = [(V_b - V_s) \times C \times 56.1/W] + \text{AV}$$

where V_b = volume of sodium hydroxide required by blank, V_s = volume of sodium hydroxide required by sample, C = exact concentration of sodium hydroxide in mol/l, W = weight of sample, and AV = acid value (positive) or alkali value (negative) of sample.

For the acetic anhydride method, the volume of sodium hydroxide includes the 25 ml added at step 6.

It may be more useful to calculate the percentage hydroxyl groups in the sample, which is given by:

$$\%\ \text{hydroxyl} = [V_b - V_s + (\text{AV} \times W/56.1C)] \times C \times 0.017 \times 100/W$$

3.5.4 Determination of iodine value

The iodine value of a fatty material is the amount of iodine, expressed as a percentage, required to saturate all the double bonds in the sample. It is not very reliable as a measure of the unsaturation in absolute terms, but is useful as an empirical parameter in setting specifications.

The method uses Wij's reagent, which is a 0.2 mol/l solution of iodine monochloride in glacial acetic acid, and which is most conveniently purchased ready-made. The reaction with an unsaturated compound is addition across the double bond. This is traditionally represented as the addition of a molecule of iodine monochloride, but as it is necessary to use at least a 100% excess of the reagent and the solution darkens during the reaction, it seems more likely that it is actually the addition of a molecule of chlorine.

An iodine flask is a conical stoppered flask having a deep well round the neck, into which a solution may be poured when the stopper is in place.

Procedure

1. Mix the sample and accurately weigh about 20/IV g, where IV is the expected iodine value, into an iodine flask or a 250 ml conical flask. It may be more convenient to weigh the sample into a small glass or plastic capsule and place this in the flask.
2. Add 10 ml chloroform (caution: toxic vapour) and swirl to dissolve the sample.
3. Add 20 ml glacial acetic acid.
4. Using a burette or an automatic pipette, add 20.0 ml Wij's reagent (caution: irritant and toxic vapour) and swirl to mix.
5. Stopper the flask and pour a few millilitres of 10% potassium iodide solution into the well of the neck.
6. Stand the flask in a dark cupboard for 30 min.
7. Momentarily remove the stopper, allowing the potassium iodide solution to fall into the flask, then replace it and shake the flask gently. If using an ordinary conical flask, add 10 ml potassium iodide solution, removing the stopper for the minimum possible time.
8. Titrate with 0.1 M sodium thiosulphate solution, shaking after each addition. The chloroform layer is pink during the titration and becomes colourless at the end-point.
9. Carry out a blank titration exactly as described but omitting the sample.

Optionally, a few drops of 1% starch solution may be used as indicator at step 8. The end-point is marked by the final disappearance of the blue colour from the aqueous layer.

The titration on the sample must be less than half the blank titration. If it is not, repeat the analysis on a smaller sample.

$$\text{Iodine value} = (V_b - V_s) \times \frac{C}{1000} \times 126.9 \times \frac{100}{W}$$

where V_b = volume of sodium thiosulphate required by blank, V_s = volume of sodium thiosulphate required by sample, C = exact concentration of sodium thiosulphate in mol/l, and W = weight of sample.

3.6 Two-phase titration of ionic surfactants with surfactants of opposite charge

3.6.1 Introduction

Anionic and cationic surfactants react together to form salts which are almost without exception insoluble or only sparingly soluble in water, but soluble in and therefore extractable by chloroform. This is the basis of two-phase titration of ionic surfactants with surfactants of opposite charge, which is probably the most widely used method in the whole field of surfactant analysis.

Detection of the end-point depends on the fact that anionic surfactants react with some cationic dyestuffs, and cationic surfactants with some anionic dyestuffs, to form salts which, while not particularly insoluble in water, can be extracted with chloroform. Over the last 50 years, many methods have been proposed which are all variants on the general idea of titrating an anionic with a cationic, or vice versa, in a chloroform–water system in the presence of an ionic dyestuff, the end-point being disclosed by the migration of colour from water to chloroform or vice versa. One of these variants [5] was developed at the behest of the Commission Internationale d'Analyses, and a critical review [6] was published two years later. These two papers repay close study.

The method has become the international standard method, and it is described in ISO 2271: 1989 [7], ISO 2871-1: 1988 [8] and ISO 2871-2: 1990 [9]. These methods are almost identical. ISO 2271 describes the titration of an anionic analyte with a standard solution of a cationic. ISO 2871-1 describes the determination of cationics with two alkyl chains by titration of a standard solution of sodium dodecyl sulphate with a solution of the cationic analyte. The cationic is dissolved in 20 ml of propan-2-ol and diluted to 1 l with water to give an approximately 0.003 M solution. ISO 2871-2 describes the determination of cationics with one alkyl chain by titration with sodium dodecyl sulphate, i.e. the exact converse of 2271. The titration can be done in either direction without difficulty in most cases.

The method uses a mixed indicator containing disulphine blue VN (Colour Index 42045, also known as Acid Blue 1), which is anionic, and dimidium bromide, which is a pink cationic dye.

The recommended cationic titrant is benzethonium chloride, which has the structure:

$$CH_3{-}\overset{\displaystyle CH_3}{\underset{\displaystyle CH_3}{\overset{|}{\underset{|}{C}}}}{-}CH_2{-}\overset{\displaystyle CH_3}{\underset{\displaystyle CH_3}{\overset{|}{\underset{|}{C}}}}{-}C_6H_4{-}O{-}(CH_2)_2{-}O{-}CH_2{-}\overset{\displaystyle CH_3}{\underset{\displaystyle CH_3}{\overset{|}{\underset{|}{N^+}}}}{-}CH_2{-}CH_6H_5\quad Cl^-$$

It is available as an accurately standardised solution containing 0.004 mol/l and as the pure crystalline monohydrate. It is marketed by Rohm and Haas under the brand name Hyamine 1622.

A solution containing 0.004 mol/l may be prepared by dissolving 1.8–1.9 g in water, diluting to 1000 ml and standardising by titration of an accurately standardised solution of pure sodium dodecyl sulphate. Two-phase titrations always use low concentrations (as low as 0.001 M) of both sample and titrant, to avoid problems of emulsification.

3.6.2 ISO 2271: Principle

A solution of the sample is placed in the titration vessel, mixed indicator and chloroform are added and the two-phase mixture is shaken or stirred. The pink cationic indicator dimidium bromide ('Dm^+Br^-') reacts with the anionic surfactant ('An^-') and the reaction product is extracted by the chloroform, colouring it pink.

$$Dm^+ + An^- \rightarrow DmAn \rightarrow \text{chloroform}$$

The water layer is yellow at this stage (Acid Blue 1 is also an acid–base indicator; it is yellow in acid solution and blue in alkaline solution).

The cationic titrant ('Cat^+') is added in small increments, with thorough agitation between additions. The following processes occur.

1. The benzethonium chloride reacts first with the anionic surfactant in the water layer, producing a colourless salt which is extracted by the chloroform.

$$Cat^+ + An^- \rightarrow CatAn \rightarrow \text{chloroform}$$

2. When most of the anionic surfactant in the water layer has been titrated, further increments of the cationic surfactant start to displace the cationic indicator from its salt with the anionic surfactant in the chloroform layer, and the pink colour starts to migrate back into the water layer.

$$Cat^+ \text{ (water)} + DmAn \text{ } (CHCl_3) \rightarrow CatAn \text{ } (CHCl_3) + Dm^+ \text{ (water)}$$

3. When all the cationic indicator has been displaced in this way, the chloroform layer is nearly colourless—actually a neutral grey colour. This is the end-point. The water layer is orange.
4. Further increments of cationic titrant react with the anionic indicator disulphine blue VN (Acid Blue 1, 'AB^-', to avoid confusion with <u>D</u>imidium <u>B</u>romide). The salt is extracted by the chloroform layer and colours it blue. This indicates that the end-point has been overshot.

$$Cat^+ + AB^- \rightarrow CatAB \rightarrow \text{chloroform}$$

These colour migrations are reversed when a cationic surfactant is titrated with an anionic. The titrant for this variation is 0.004 M sodium dodecyl sulphate.

3.6.3 ISO 2271: Procedure

Indicator: weigh 0.5 g disulphine blue VN and 0.5 g dimidium bromide in two 50 ml beakers. Add 25 ml hot 10% aqueous ethanol to each and stir until dissolved. Transfer both solutions to a 250 ml volumetric flask, dilute to volume and mix. This is the stock solution. Place 200 ml water and 20 ml stock solution in a 500 ml volumetric flask, add 20 ml 2.5 M sulphuric acid, dilute to volume and mix. This is the acid mixed indicator.

1. Take a sample containing 0.003–0.005 mol anion-active matter. As a rough guide, use 140/*A* g, where *A* is the expected active content, expressed as a percentage.
2. Dissolve the sample in water, dilute to 1 l and mix.
3. Transfer 25 ml by pipette to a titration vessel. This may be a 100 ml stoppered measuring cylinder or bottle, but it is much better to use the mechanical apparatus described in ISO 2271: 1989 (E), which is shown in Figure 3.2. Add 10 ml water, 15 ml chloroform and 10 ml acid mixed indicator.
4. Titrate with 0.004 M benzethonium chloride, shaking or stirring thoroughly after each addition. Proceed in increments of a single drop in the region of the end-point, shown by the disappearance of the pink colour from the chloroform layer, which is then grey. If the chloroform becomes blue, the end-point has been overshot.

$$\text{Anion-active content} = V \times \frac{C}{1000} \times M \times \frac{100}{W} \times \frac{1000}{25}$$

where V = volume of benzethonium chloride solution, C = exact concentration of benzethonium chloride in mol/l, M = molecular weight of the anionic active, and W = weight of sample.

The use of the mechanical apparatus with its semi-automatic dispensing of indicator and chloroform greatly reduces exposure of the analyst to toxic chloroform vapour, and the use of mechanical stirring also reduces operator fatigue, which in turn makes for better consistency in the results.

If the sample were diluted to some other volume V_1 ml instead of 1 l and an aliquot of V_2 ml were taken instead of 25 ml, the term 1000/25 would be replaced by V_1/V_2.

By performing the titration in acid and alkaline solution or before and after acid or alkaline hydrolysis it is possible to analyse some two- and three-component surfactant mixtures without prior separation. This is described in chapter 8. An alkaline mixed indicator is required for some

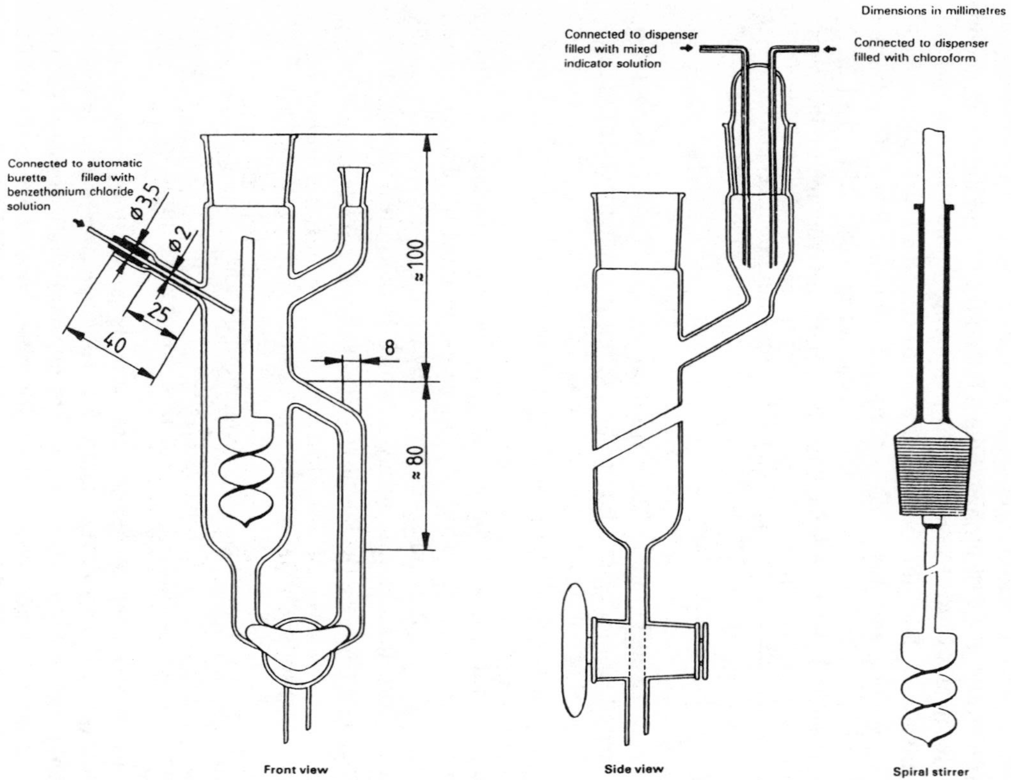

Figure 3.2 Mechanical two-phase titration apparatus (reproduced by kind permission of the International Organization for Standardization, Geneva).

purposes. It is prepared in the same way as the acid mixed indicator but adding 20 ml 5 M sodium hydroxide instead of the sulphuric acid.

This method is the accepted international standard and is well established in a very large number of laboratories. It should certainly be used wherever it is appropriate, but analysts should be aware that the original report [4] showed that the observed result is not stoichiometric, and should not regard it as the only acceptable procedure for two-phase titration. Some of the alternatives are discussed in sections 3.6.4 and 3.7.

3.6.4 Other variants of two-phase titration

Many variations have been published on the basic idea of two-phase titration, with colour transfer from one layer to the other to mark the end-point, and in view of the importance of this general technique it is desirable for the analyst to have some appreciation of the subtle but significant differences between the variants, only a few of which are discussed below.

The first two-phase titration method to come into widespread use was that of Epton [10], who used methylene blue as indicator. Methylene blue is cationic, and in the Epton method all the dye passes into the chloroform as its salt with the anionic at the beginning of the titration. In the region of the end-point it returns to the aqueous layer. Provided that the ratio of the volumes of aqueous and chloroform layers is 3:1 at the end-point, equal colour intensity in the two layers indicates equivalence between anionic present and cationic added. This is quite empirical and does not correspond with the completion of any chemical process or any clearly defined event. The more the volume ratio differs from 3:1, the greater the deviation of the observed result from true equivalence. Another difficulty is that the hues of the two layers are different, and matching them is very operator-dependent. Nevertheless, the method was the standard for many years and performed invaluable service to the surfactants industry. It is still in use in some laboratories and ASTM Method D 1681-83 [11] includes it, but ISO 2271 [7] and some of the methods described below are to be preferred.

Lew [12] used bromocresol green in alkaline solution, again taking the end-point as the point at which the two layers were of equal intensity. Bromocresol green is an anionic dye, and at this point most of the anionic and a substantial proportion of the indicator have been titrated, but it does not correspond with true stoichiometry (though that does not necessarily imply inaccuracy). Users report that the end-point is not very sharp and the solution is not clear at the end-point, which is no doubt due to the small amount of untitrated anionic remaining in the aqueous layer.

Battaglini *et al.* [13] recommended phenol red in alkaline solution for titration of α-sulphonated fatty esters, taking the first appearance of a pink colour in the chloroform layer as end-point, claiming as advantages that it makes the titration more responsive to lower-molecular-weight surfactants and that it avoids the need for colour matching. The basis for the first of these claims is unclear. The present author [14] has shown that when the first trace of colour appears in the chloroform layer a small amount of anionic remains untitrated, although the dye in question was bromophenol blue. The phenol red method does not seem to be very widely used outside the USA.

The present author [14] has also proposed the principle that the indicator should have the same charge as the species being titrated and that the end-point is the point at which the whole of the surfactant and the whole of the indicator have been titrated, i.e. when colour transfer to the chloroform layer is complete. Bromophenol blue is used for titration of anionics and methylene blue for titration of cationics. In both cases titration is continued until the top layer is clear and colourless. A measured 1 ml of indicator is used and a blank titration is done on the indicator alone, but the blank is appreciable only for bromophenol blue, because methylene blue is more intensely coloured and a much smaller quantity is used.

Many different cationics, at concentrations ranging from 0.001 to 0.005 M, have been used in this version of the titration, but 0.004 M benzethonium chloride is now the preferred titrant. The indicator solutions are 0.04% bromophenol blue and 0.004% methylene blue.

In the original version of this method, anionics were titrated in an approximately neutral solution (pH above 4.6), but it can be done equally satisfactorily in alkaline solution, where it is preferable to the ISO method for the titration of soaps; 10 ml 0.1 M sodium hydroxide is added for this variation.

At the end-point the solution becomes perfectly clear and the mode of separation of the two phases changes from the collapse of a network of hollow bubbles of chloroform to the settling out of 'solid' droplets. This is clearly visible, and a practised analyst can do the titration without an indicator, provided that it is roughly known how much titrant is needed.

An alternative approach is to use hydrophobic indicators, which remain in the chloroform layer throughout the titration and give a very sharp colour change. Eppert and Liebscher [15] used dimethyl yellow (cationic) in titrations with carbethoxypentadecyltrimethylammonium bromide (Septonex). The colour change in the chloroform layer is from pink to yellow and is very sharp. Tsubouchi and Matsuoka [16] used the potassium salt of the ethyl ester of tetrabromophenolphthalein (anionic) for titration with tetradecyldimethylbenzylammonium chloride (Zephiran) in a solution buffered at pH 6. The chloroform layer starts off yellow, becomes green near the end-point and turns sky blue when one drop of excess titrant is added.

Analysts should gain experience with all these variants.

3.7 Potentiometric titration with surfactants of opposite charge using a surfactant-sensitive electrode

3.7.1 Advantages of potentiometric titration

Surfactant-sensitive electrodes are commercially available, and they are also easily made in the laboratory. They can be used to detect the end-point in titrations of anionic and cationic surfactants with surfactants of opposite charge. They are used in exactly the same way as a glass electrode in acid–base titrations, or a silver–silver chloride electrode in titrations of chloride with silver nitrate. The main advantages of potentiometric as opposed to two-phase titration are:

1. It permits the use of much higher concentrations of titrant (0.02–0.04 M as against 0.004 M), resulting in sharper end-points and thus better reproducibility.
2. It is less subjective.
3. It eliminates the use of chloroform, whose vapour is toxic and may be carcinogenic.
4. It readily lends itself to automation.
5. It is much less fatiguing for the operator.

3.7.2 Construction and performance of surfactant-sensitive electrodes

The construction of surfactant-sensitive electrodes presents no difficulty, and three types are described here. The first type is a concentration cell, the second is a conductor coated with a membrane containing a surfactant salt, and the third is a conductor coated with a membrane not containing a surfactant salt. Many other electrodes have been described in the literature over the last 20 years or so, but most if not all of them are of one of these basic types.

Figure 3.3 shows the construction of the first type of electrode. A standard silver–silver chloride electrode is enclosed by a removable cap, the end of which is sealed by a plasticised PVC membrane. The membrane usually contains a salt of an anionic and a cationic surfactant, although a single surfactant of either charge, or even no surfactant at all, may be used. The electrode is filled with dilute sodium chloride solution, typically 0.01 mol/l, which may contain a surfactant, typically at 0.001 mol/l. The inclusion of the surfactant is said to improve selectivity, but the evidence for this is unconvincing.

Such an electrode can be made as follows. Capped silver–silver chloride electrodes may be obtained from most laboratory suppliers.

1. Accurately weigh 0.1063 g pure dry sodium dodecyl sulphate and 0.1719 g pure dry benzethonium chloride monohydrate, dissolve in tetrahydrofuran (caution: flammable; toxic vapour) and dilute to

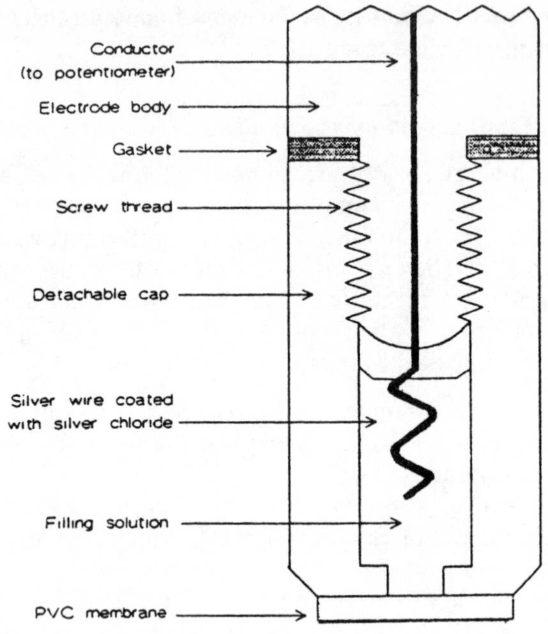

Figure 3.3 Surfactant-responsive electrode based on a silver–silver chloride electrode (reproduced by kind permission of the Society for Chemical Industry).

100 ml in a volumetric flask. Centrifuge to remove the precipitated sodium chloride. This yields a solution containing 0.25% benzethonium dodecyl sulphate (BDS).

2. Dissolve 0.25 g PVC powder and 0.75 g tritolyl phosphate in 4 ml of the BDS solution.
3. Place on a flat glass plate a glass ring of internal diameter 35–36 mm and having its circular ends ground flat. Pour the PVC/TTP solution into the ring, place a heavy weight on top of it to ensure a good seal between ring and plate, and leave for 24 h. This will produce a membrane approximately 1 mm thick.
4. Using a cork borer, cut discs of the membrane of a suitable size for the electrode cap.
5. Attach the membrane to the open end of the electrode cap using a 5% solution of PVC in tetrahydrofuran as adhesive. Ensure that the opening in the cap is completely sealed. Set the cap aside for a few hours for the adhesive to cure.
6. Meanwhile prepare a filling solution containing 0.0288 g sodium dodecyl sulphate and 0.0585 g sodium chloride in 100 ml water.
7. When the cap bearing the membrane is ready, fill it with filling

solution by means of a Pasteur pipette and screw it on to the electrode, which is then ready for use.

The various dimensions and quantities are not critical.

The type of electrode described above has no compelling advantage over any other type. It is sensitive to all types of ionic surfactant, and its selectivity is only between surfactant and non-surfactant ions. The inclusion of surfactants in the filling solution or in the membrane itself, and the chemical nature of such surfactants, has little effect on its performance as an end-point indicator. It would seem that almost any kind of electrode involving a membrane can be successfully used for this limited purpose. The performance is not significantly influenced by pH between pH 1 and pH 13.

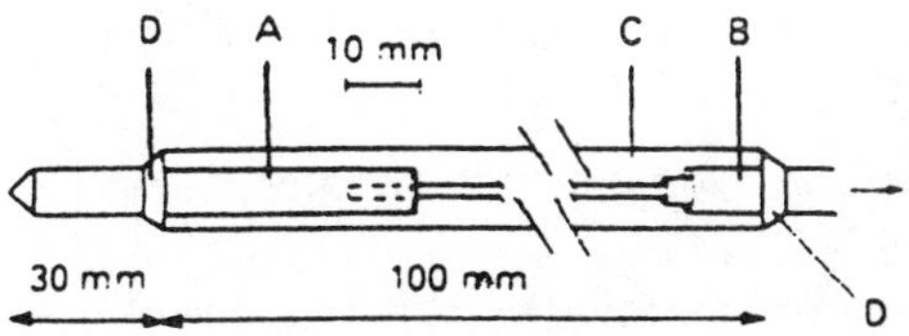

Figure 3.4 Surfactant-responsive electrode based on a graphite rod (reproduced by kind permission of the Royal Society of Chemistry).

Another type of electrode is shown in Figure 3.4. This design was proposed by Dowle *et al.* [17]. It is constructed as follows.

1. Drill a hole in the end of a spectroscopically pure graphite rod.
2. Insert a coaxial cable and coat the junction with conducting paint.
3. Fix the cable in place with epoxy resin.
4. Sheath the assembly in a glass tube, fixing with epoxy resin.
5. Immerse the tip in the coating solution to such a depth that the first 30 mm of the glass tube are coated.
6. Allow to dry for 30 min and repeat twice more.

Prepare the coating solution as follows.

1. Place 6.0 g tritolyl phosphate (TPP) and 75 ml tetrahydrofuran (THF) in a 150 ml beaker and stir magnetically.
2. Slowly add 4.0 g PVC powder (molecular weight 100 000; BDH) and continue stirring until a clear solution is obtained.
3. Add 0.01 g of the appropriate surfactant salt and stir until dissolved.

Prepare the surfactant salts *A* and *B* as follows.

1. Mix equal volumes of 0.001 M tetrabutylammonium hydroxide and sodium dodecyl sulphate to form salt *A*, which is cationic surfactant-selective.

2. Mix equal volumes of 0.001 M hexadecyltrimethylammonium bromide and sodium pentane-1-sulphonate to form salt *B*, which is anionic surfactant-selective.
3. Collect and wash the precipitates, wash with the minimum volume of water and dry at 80°C.

Electrodes containing salt *A* have been used as end-point indicators in titrations of cationics with dodecyl sulphate. Those containing salt *B* have been used as end-point indicators in titrations of anionics with benzethonium chloride. Both types give sharp end-points and good agreement with two-phase titration. They can be used for this purpose in solutions containing up to 20% ethanol, although their responses are then sub-Nernstian.

A third type of electrode is the very simple coated aluminium wire described by Vytřas [18], made in the laboratory from an aluminium wire coated in thick PVC, as follows.

1. Dissolve 0.09 g PVC and 0.2 ml 2,4-dinitrophenyl-*n*-octyl ether in 3 ml tetrahydrofuran.
2. Dip the aluminium wire into this solution, ensuring that the end of the PVC insulation is covered. Allow to dry.
3. Repeat step 2 several times.

The electrode is conditioned either by dipping it in a stirred solution of the analyte–titrant pair, or simply by doing a titration with it.

Vytřas and co-workers have reported successful titrations of large numbers of anionic, cationic, amphoteric and even nonionic surfactants using such electrodes. The paper cited [18] is a review with 38 references. The response appears to be sub-Nernstian, but when used as end-point indicators these electrodes exhibit potential jumps of up to 100 mV or more for very small increments of titrant.

With all types of surfactant-sensitive electrodes, high concentrations of electrolyte tend to depress the potential jump at the end-point, and some product types require separation of the surfactant from the product matrix.

3.7.3 Titration procedure

The procedure given here takes advantage of the opportunity to use higher concentrations of both sample and titrant than is possible in two-phase titration. This is not essential, and it is perfectly feasible to work with solutions down to 0.0001 M, although this is not recommended.

1. Weigh a sample containing 0.012–0.018 mol of the surfactant to be determined, dissolve in water and dilute to 1 l in a volumetric flask.

2. Pipette 20 ml into a 100 ml squat-form beaker, dilute to about 60 ml and add acid or alkali if necessary (see below), and a magnetic stirrer bar.
3. Connect the membrane electrode and a calomel reference electrode to a potentiometer switched to display millivolts, or better, an autotitrator.
4. Place the beaker on a magnetic stirrer and commence stirring.
5. Titrate anionics with 0.02 M benzethonium chloride and cationics with 0.02 M sodium dodecyl sulphate. Add the titrant in small increments and stir until the voltage reading becomes constant after each addition. Proceed in increments of one drop in the region of the end-point.
6. Plot a curve of millivolts (vertical axis) against volume of titrant (horizontal axis). Determine from the graph the volume corresponding to the point of inflection (i.e. the steepest point), which is the end-point.

Weak bases (amines and amine oxides) must be titrated in a medium containing not less than 0.01 mol/l H^+ and betaines in a medium containing not less than 0.1 mol/l H^+. Amphoterics must be titrated in a medium containing not less than 0.01 mol/l H^+ (cationic function) or not less than 0.01 mol/l OH^- (anionic function). These conditions are met by adding 10 ml 1.0 mol/l hydrochloric acid to solutions of betaines and 10 ml 0.1 mol/l hydrochloric acid or sodium hydroxide, as appropriate, in the other cases.

Failure to produce a smooth curve with a well-defined point of inflection is almost invariably due to a defect in the membrane or its attachment to the cap, or to a faulty electrical connection.

Reproducibility is excellent for most surfactants. The method is particularly good for betaines but is not recommended for fatty ester α-sulphonates, α-olefin sulphonates, soaps or sarcosinates. It is most useful for raw materials and products containing only low concentrations of non-surfactant electrolytes. Users should evaluate for themselves its usefulness for other product types.

3.7.4 Use of surfactant-sensitive electrodes with an autotitrator

If using an autotitrator, determine by trial and error the optimum instrument settings for the particular model. When using the method, make it standard practice to observe the following rules.

1. Always carry out one or more control titrations on a standard solution of sodium dodecyl sulphate or benzethonium chloride in each set of analyses.
2. Always programme the titrator to draw the titration curve, because autotitrators in general tend to find an end-point even when there isn't

one. The drawing of a smooth, steep curve demonstrates that the electrode and titrator are both functioning correctly.

3. After each titration wipe both electrodes gently with a paper tissue, place them in a dilute solution of a nonionic surfactant (e.g. 1% Nonidet LE or similar) and stir for a few seconds, to remove the sticky precipitate which might otherwise foul one or both.

If a large number of samples are to be analysed using an automatic sample changer, it is not practicable to wipe the electrodes after each titration. Instead, make every third sample a solution of a nonionic surfactant containing anionic or cationic surfactant at a very low level, say equivalent to 0.5 ml of titrant. The titrator must be able to find an end-point, otherwise it will titrate for ever, and this device keeps the electrodes clean whilst satisfying that requirement.

Some operators prefer to use a double junction reference electrode, to prevent contamination of the electrode by the surfactant solution and vice versa. This is basically a glass sheath with a frit at its lower end, which is filled with an electrolyte such as 1 M sodium nitrate and in which the calomel electrode is placed. An improvised version consists simply of a small beaker containing the electrolyte and the calomel electrode, and connected to the titration vessel via a piece of string soaked in the same solution.

3.8 Direct potentiometry: calculation of results

Chapter 6 includes a method that involves measuring the potential difference between a pair of electrodes, just as pH and hence hydrogen-ion concentration can be measured with a glass electrode and a calomel reference electrode. This has more applications than can be covered in this book. In such analyses there are basically two ways of translating a potential reading in millivolts into a percentage of an ingredient.

3.8.1 Calibration curves

The relationship between concentration of a component X and the millivolts recorded by an electrode sensitive to changes in that concentration is a logarithmic one. It is expressed by the Nernst equation, of which the simplest form is

$$E = E^0 + 59 \log c$$

where E is the electrode potential in millivolts, E^0 is a constant and c is the concentration.

It means that if c is changed by a factor of 10, E changes by

59 millivolts. This value, the change in potential per tenfold change in concentration, is the slope of the calibration curve. In practice, not many ion-sensitive electrodes show ideally Nernstian responses, and practical slopes are nearly always somewhat less than 59 mV. They do, however, have one very useful characteristic, namely that a plot of log concentration against millivolts is nearly always a straight line, at least over a concentration range of a few orders of magnitude.

A calibration curve is constructed by making several solutions of component X such that each is 10 times more (or less) concentrated than the preceding one, e.g. 10, 1, 0.1 and 0.01% or mg/l, inserting the electrode pair connected to a potentiometer and noting the millivolt reading when it has attained a constant value. Practical results might be, for example, 126, 70, 13 and −45 mV for this series. The changes in the readings are 56, 57 and 58 mV, so that the response in this case is nearly Nernstian. The mean slope is 57 mV.

A refinement is to make up the standard solutions of X in a solution of a formulation identical with the one in which X is to be determined, but without X, and at the concentration that will be used in practice. If the analyses are to be performed on a 1% solution of a product containing 10% X, the medium for the standard solutions should be a 0.9% solution of the basic formula minus X. This eliminates matrix effects.

Calibration curves can be used in two ways. The simpler way is to take a potential reading on a solution of the material to be analysed and to read off from the curve the corresponding concentration of X. This is satisfactory for many purposes.

A more sophisticated way is to evaluate E^0 and to calculate the concentration of X. E^0 is the observed value of E when $\log c = 0$, i.e. when $c = 1$. Obviously the actual value will depend on whether this is 1%, 1 mg/l or some other unit. In the above example concentration is expressed as a percentage, so $E^0 = 70$. We then have, for a solution of concentration c%,

$$E_c = 70 + 57 \log c, \text{ or } \log c = (E_c - 70)/57$$

Suppose an unknown solution gave a value of 84 mV for E_c. Then $\log c = (84 - 70)/57 = 0.246$, and $c = 1.76\%$.

This is a better way, because by doing several replicate measurements, very reliable values can be obtained for both E^0 and the slope, and calculation eliminates errors in reading off the concentration for a given potential.

3.8.2 Method of standard additions

It may be found that E^0 varies from one electrode to another, or with minor variations in the composition of the working medium. The labour of obtaining a result is minimised, and its accuracy maximised, by using the method of standard additions. It eliminates matrix effects and variations in E^0. Only the value of the slope is needed. The analysis is done as follows.

1. Make up two solutions of X in a medium as similar as possible to that in the test solution. Their concentrations must differ by a factor of 10 and must span the range within which the value to be measured is expected to lie. If this is unknown, do step 3 first.
2. Insert the electrode pair, connected to a potentiometer, into one of the solutions, stir continuously until the reading becomes constant and record it. Repeat with the other solution. The slope of the calibration curve, S, is the difference between the two readings, in millivolts.
3. Make up a solution of the sample to be analysed at an appropriate concentration. Take a fairly large measured volume, e.g. 50 or 100 ml, insert the electrodes and take a reading as in step 2.
4. Add a small measured volume of a solution of X, sufficient to cause an appreciable change in the concentration, say by a factor of at least 2. Take a reading as before.
5. Calculation. Refer to Figure 3.5. Then:

Let C be the concentration of X in the test solution and I the increase in concentration resulting from the standard addition. Let E_c be the reading on the test solution and E_i the reading on the solution after the standard addition. Let S be the slope of the calibration curve. S is positive for cations and negative for anions.

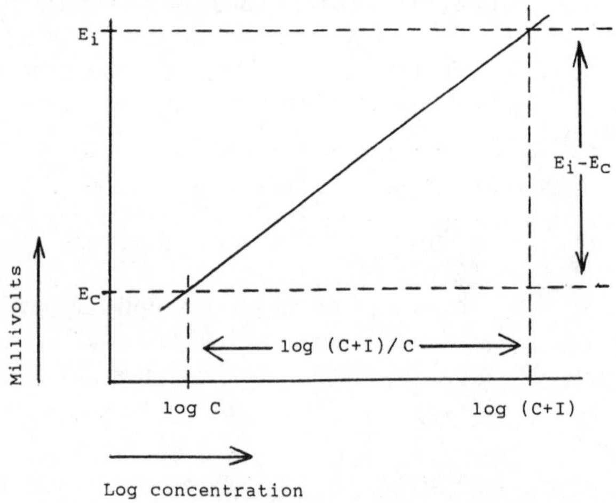

Figure 3.5 Relation between log concentration and millivolts before and after standard addition.

The vertical distance between the readings before and after the standard addition is the change in the millivolt reading. The horizontal distance between them is the difference between the logarithms of the concentrations before and after the standard addition, i.e. the logarithm of their ratio. The ratio of these two distances is the slope S.

$$\frac{E_i - E_c}{\log(C+I) - \log C} = S$$

Rearranging,

$$\log[(C+I)/C] = (E_i - E_c)/S$$

Taking antilogs,

$$(C+I)/C = \text{antilog}[(E_i - E_c)/S]$$

Let

$$\text{antilog}[(E_i - E_c)/S] = A$$

Then

$$C = I/(A-1)$$

For example, suppose the volume of the sample solution was 50 ml and that the standard addition was 1 ml of a 10% solution of X, so that $I = 0.2\%$. Suppose the millivolt readings before and after the addition were 20 and 56 mV and that the slope was 57 mV.

Then

$$A = \text{antilog}(56-20)/57 = \text{antilog}\,0.6316 = 4.28$$

Then

$$C = 0.2/3.28 = 0.061\%\ X$$

This is based on the assumption that the dilution caused by the standard addition is negligible. If it is not, redefine I as the increase in concentration that would have occurred if the same amount of X had been added without any increase in volume, and multiply A by the ratio, R, of the volumes after and before the standard addition. This changes the final equation to $C = I/(RA-1)$ and eliminates the dilution error. For example, if the volumes before and after the standard addition were 50 and 60 ml respectively, $R = 60/50 = 1.2$, and $C = I/(1.2A-1)$.

3.9 Open-column chromatography

Open-column chromatography usually employs either alumina or silica as sorbent and is gravity-driven. It does not require any specialised equipment and is therefore inexpensive, but it is tedious and does not always give clean separations.

Columns are glass or plastic tubes which may have a built-in frit to retain the sorbent, but burettes are often used. Solvent reservoirs can be custom-built, but separating funnels are a satisfactory alternative. The following is a general outline of the procedure.

1. Condition the sorbent if necessary, e.g. by heating to a high temperature, humidifying or both. Not all sorbents or analyses require conditioning.
2. Weigh the required amount of sorbent. This depends on the analysis required, but typically varies between 10 and 50 g.
3. Slurry the sorbent with whatever solvent is to be used as the first eluent. Partially fill the column with the same solvent. Transfer the slurry to the column a little at a time. Continue until all the sorbent has been transferred, running off surplus solvent to make room as required.
4. Protect the surface of the sorbent with a small plug of glass wool or a plastic frit. This is not essential.
5. Allow the solvent to drain almost to the surface of the sorbent.
6. Dissolve the sample, typically 0.5–2.0 g but sometimes as little as 0.01 g, in a little of the same solvent and transfer it quantitatively to the column.
7. Elute with one or more eluents in succession. Typical flow rates are 1–5 ml/min. If one is using an established procedure the compositions and volumes of successive eluents will be specified. Otherwise change the composition in small steps and collect 10 ml fractions. Evaporate each fraction to dryness and weigh the residue until no more material is eluted before each change in composition.
8. Continue until the analysis is complete.

Usually each eluent removes one class of compound from the column, but sometimes one eluent will remove two or more classes of compound at different rates, so that provided the column is large enough they will be separated. The following examples give an idea of typical eluent compositions and volumes, and the kind of separation that can be achieved.

It is often found that small amounts of the sorbent find their way into the collected eluates, and it is necessary to filter them off before evaporating and weighing.

Example 1 [19]

Sample: up to 1 g mixed nonionics dissolved in the minimum amount of petroleum ether.

Sorbent: alumina, Merck 1097.
Conditioning: heat to 800°C for 3 h and cool in a desiccator.
Column: 40 g alumina in petroleum ether in a 2 cm diameter column.
Elute at 3–5 ml/min, using 250 ml of each eluent.

Eluent	*Species eluted*
Petroleum ether	Hydrocarbons
Petroleum ether/diethyl ether, 9:1	Secondary alcohols
Petroleum ether/diethyl ether, 1:1	Primary straight-chain alcohols
Diethyl ether	Waxes and fatty esters
Chloroform	Alcohol ethoxylates
Chloroform/ethanol, 1:1	Alkanolamides

Example 2 [20]

Sample: 0.3 g mixed nonionics dissolved in chloroform.
Sorbent: Davison 922 silica gel, 200 mesh.
Conditioning: none.
Column: 10 g silica in chloroform in a 50 ml burette.
Elute with 70 ml of each solvent except where indicated. Flow rate is not specified, but it would be inadvisable to exceed 1 ml/min with a column of this diameter.

Eluent	*Species eluted*
Chloroform	Hydrocarbons, methyl esters of fatty acids
Chloroform/diethyl ether, 99:1	Glycerides, fatty alcohols, fatty acids
Chloroform/diethyl ether, 1:1	Fatty acid ethoxylates, alkylphenol ethoxylates
Chloroform/acetone, 1:1	Other ethoxylates with less than 10 EO
Chloroform/methanol, 19:1 and chloroform/methanol, 9:1	Ethoxylates with 10 or more EO
Chloroform/methanol, 2:1	Glycerol, high-molecular-weight polyethylene glycols

References

1. ASTM Method D 3673-89: Standard Test Methods for the Analysis of Alpha-Olefin Sulfonates. American Society for Testing and Materials, Philadelphia, USA.
2. Cross, C. K. *JAOCS*, **67** (1990), 142–143.

3. ISO 4326: 1980: Nonionic surface active agents—Polyethoxylated derivatives—Determination of hydroxyl value—Acetic anhydride method. International Organization for Standardization, Geneva.
4. ASTM Test Method D 4252-89: Standard Test Methods for Chemical Analysis of Alcohol Ethoxylates and Alkylphenol Ethoxylates. American Society for Testing and Materials, Philadelphia, USA.
5. Reid, V. W., Longman, G. E. and Heinerth, E. *Tenside*, **4** (1967), 292–304.
6. Reid, V. W., Longman, G. F. and Heinerth, E. *Tenside*, **5** (1968), 90–96.
7. ISO 2271: 1989: Surface active agents—Detergents—Determination of anionic-active matter by manual or mechanical direct two-phase titration procedure. International Organization for Standardization, Geneva.
8. ISO 2871-1: 1988: Surface active agents—Detergents—Determination of cationic active matter content—Part 1: High-molecular-mass cationic-active matter. International Organization for Standardization, Geneva.
9. ISO 2871-2: 1990: Surface active agents—Detergents—Determination of cationic-active matter content—Part 2: Cationic-active matter of low molecular mass (between 200 and 500). International Organization for Standardization, Geneva.
10. Epton, S. R. *Trans. Faraday Soc.*, **44** (1948), 226.
11. ASTM Method D 1681-83: Standard Test Method for Synthetic Anionic Active Ingredient in Detergents by Cationic Titration Procedure. American Society for Testing and Materials, Philadelphia, USA.
12. Lew, H. Y. *JAOCS*, **41** (1964), 297.
13. Battaglini, G. T., Larsen-Zobus, J. L. and Baker, T. G. *JAOCS*, **63** (1986), 1073–1077.
14. Cullum, D. C. *Proc. IIIrd International Congress on Surface Active Substances*, Cologne, 1960, vol. C, pp. 42–50.
15. Eppert, G. and Liebscher, G. *Z. Chem.*, **18** (1978), 188.
16. Tsubouchi, M. and Matsuoka, K. *JAOCS*, **56** (1979), 921–923.
17. Dowle, C. J., Cooksey, B. G., Ottaway, J. M. and Campbell, W. C. *Analyst*, **112** (1987), 1299–1302.
18. Vytřas, K. *Electroanalysis*, **3** (1991), 343–347.
19. Longman, G. F. *The Analysis of Detergents and Detergent Products*, John Wiley and Sons, 1975, p. 61.
20. Rosen, M. J. *Anal. Chem.*, **35** (1963), 2074–2077.

4 Ion exchange

4.1 Introduction

Ion exchange has been used to separate surfactants into the main classes for at least forty years, and it remains the most effective approach for this purpose, at least on the macro scale. It is one of the chemical analyst's most powerful tools in surfactant analysis, yet it is not much emphasised in any of the standard texts, and it seems to be poorly understood by analysts and analytical authors alike. The treatment in this chapter aims at giving the reader some insight into what goes on in ion exchange processes, and why, and how they can be adapted and combined to permit practical and useful separations. It aims more at practicality than at rigorous theoretical exposition. It is about what to do and how to do it, and does not include instructions for any specific analyses, apart from general deionisation.

Schmitt [1] gives a large number of literature references on ion exchange in surfactant analysis, all or most of which report specific applications of the principles set out here. Surprisingly, until recently the technology employed in ion exchange has developed hardly at all, and it is only in the last few years that publications have started to appear which describe the application to ion exchange of techniques long used in liquid chromatography. Good separations can certainly be achieved with very basic equipment, but the introduction of aspects of high technology will surely lead to greater convenience and effectiveness, and thereby to wider use.

4.2 What is ion exchange?

For the purpose of this book, ion exchange is a process whereby ions within the structure or on the surface of a solid matrix are displaced by ions of the same sign but of a different kind in a solution that passes through or over the matrix.

Solid ion-exchange media are materials whose structure is sufficiently porous to allow water and other solvents to flow through it, and which contain fixed ionic groups, either anionic or cationic. Electrical neutrality is maintained by the presence of mobile counterions. Under the right conditions, other ions of the same sign can displace these counterions.

This makes it possible to isolate surfactant ions of a particular class in one of two fundamentally different ways: to retain the unwanted ions in or on the ion-exchange medium, allowing only the wanted ions to pass through in the solution, or to retain the wanted ions and subsequently to recover them for further examination. The uptake of ions by an ion-exchange medium is called sorption, and the ions taken up from solution are said to be sorbed. Obviously nonionics cannot be sorbed by an ion exchanger (at any rate by ion exchange), and if they are present only the second of these processes is useful for their separation from ionic species.

The process is usually done in a column packed with the ion-exchange medium, down which the solution and washing solvent flow. It may be envisaged as a kind of molecular filtration, in which certain ions are retained by the 'filter'. It is not to be confused with ion chromatography, in which the ions migrate down the column at different rates and are thus separated into different species. In ion exchange as the term is used here, the ions are either retained by the column or are not, and there is little or no differential migration.

Retention of one kind of ion in exchange for another kind occurs because ion-exchange media show a clearly defined order of preference for ions of different species (see section 4.4.5 on Selectivity).

4.3 Ion-exchange media

4.3.1 Ion-exchange resins

Ion-exchange resins consist of an organic matrix, often but not necessarily cross-linked polystyrene, to which are attached ionic groups. In anion-exchange resins these fixed groups are cationic, e.g. —$N^+(CH_3)_3$, and their exchangeable anions may be, e.g., hydroxide or chloride. In cation-exchange resins the fixed ionic groups are anionic, e.g. —SO_3^-, and their exchangeable cations may be, e.g., hydrogen or sodium. The open structure of the three-dimensional polymeric network permits solvents such as water and low-molecular-weight alcohols to pass freely through it. Ions in solution may displace the counterions in the resin and be retained, the displaced ions being washed out of the column by the flowing solvent stream. The resin can be washed with pure solvent without removing the retained ions, and they may in turn be displaced by a suitably chosen electrolyte solution and so recovered. Resins are by far the most important ion-exchange medium in surfactant analysis, but this may not always be so.

4.3.2 Other ion-exchange media: membranes and silicas

Ion-exchange membranes are similar in structure to the resins, but only a very limited range is available, and they are highly cross-linked (see section 4.4.2), which makes them almost impermeable to the relatively large surfactant ions. Membranes of this type are of little practical use in this field.

Bio-Rad Laboratories market a novel type of ion-exchange membrane, which is perhaps better described as ion-exchange filters. They consist of a PTFE mesh (10%) containing dispersed beads of ion-exchange resin (90%), They are available in the form of discs of various sizes and as 12″ × 12″ sheets, and are used like filter paper. There is an anion exchanger (AG1-X8) supplied in the carbonate form and a cation exchanger (AG50-X8) supplied in the acid form, both with 8% cross-linking. There is also a chelating version. The theoretical exchange capacity (section 4.4.1) of the anion exchanger is 0.064 milliequivalents/cm^2, while that of the cation exchanger is 0.15 milliequivalents/cm^2. No applications in the analysis of surfactants have so far been reported. They may be useful for isolating small quantities of anionic and cationic surfactants.

Analytichem (Varian Sample Preparation Products, UK agent Jones Chromatography, Hengoed, Mid-Glamorgan) offer a range of ion-exchange silicas. These were mentioned in section 2.7.4. They are silicas with ionic groups grafted on to their structure, and having properties, including exchange capacities, similar to those of ion-exchange resins. A drawback is that they are destroyed by alkalis, but unlike resins they do not swell and shrink in response to changes in solvent composition. It is probable that the hydrophobic interaction with surfactants is less than it is for resins, and this may make them a better medium for genuine surfactant ion chromatography. These silicas may be bought in bulk, and in the form of pre-packed columns containing 50, 100 or 500 mg silica in plastic syringe bodies. International Sorbent Technology (UK agent Jones Chromatography) offer the same or closely similar silicas in bulk and in pre-packed columns containing from 50 mg to 10 g silica. A wide range of accessories is available, including vacuum manifolds into which the syringes can be plugged for vacuum filtration. They can be used for the rapid isolation of some surfactants, but there is little or no published information about this application. Varian Sample Preparation Products' *Handbook of Sorbent Extraction Technology* [2] is a comprehensive practical guide.

Other suppliers of SPE aids (e.g. Waters, Alltech) offer similar products.

4.4 Properties of ion-exchange resins

4.4.1 Exchange capacity

The exchange capacity of an ion-exchange resin is the number of ions a given quantity of resin is capable of taking up from solution. It obviously depends on the concentration of ionic groups in the resin. The maximum theoretical exchange capacity is equal to the concentration of ionic groups in the resin, but the capacity that can be achieved in practice depends on a number of factors and may be a great deal less than the theoretical maximum. Exchange capacities are expressed in milliequivalents per millilitre (meq/ml) of wet, i.e. fully swollen, resin, and range from about 0.5 to nearly 2.0 meq/ml. An equivalent is that weight of an anion or cation that carries unit ionic charge. For the majority of surfactants the equivalent weight is the same as the molecular weight, e.g. one equivalent of sodium dodecyl sulphate is 288.4 g.

4.4.2 Cross-linking

The polystyrene matrix used in many resins is cross-linked by including a proportion of divinylbenzene (DVB) in the polymerisation mix. The amount of DVB is often shown in the name of the resin by X followed by the percentage of DVB, as in Bio-Rad AG1-X4, which means 'Analytical Grade, resin type 1 with 4% cross-linking'. The degree of cross-linking varies from 2% to 16%. Low cross-linking results in a relatively low exchange capacity, high permeability to large ions such as surfactants, and a marked tendency for the resin, which is fully swollen when in equilibrium with pure water, to shrink in solvents containing a high concentration of electrolyte or organic solvent, e.g. methanol. High cross-linking results in a relatively high exchange capacity, lower permeability to large ions and a much reduced tendency to shrink. For surfactant separations, 4% cross-linking is the best compromise, and 2% is preferable to 8%. With 8% cross-linking, interaction with surfactant ions is largely confined to the outer layers. Although the theoretical exchange capacity of Bio-Rad AG1-X2, for example, is only half that of Bio-Rad AG1-X8, its practical capacity for large surfactant ions may well be greater.

The tendency of resins to swell and shrink in response to changes in solvent composition can have significant practical consequences. If a resin has been used in say a highly alcoholic solvent, and an attempt is made to flush the column with water, the swelling pressure can be great enough to split a glass column. Another effect is that ions that have been sorbed in a microregion of the resin in which the cross-linking is greater than the statistical average for the resin may become trapped if the resin shrinks because of a change in solvent composition. This accounts for at least some of the low recoveries of surfactants sometimes observed with highly cross-linked resins. High cross-linking certainly retards elution; MacDonald *et al.* [3] found that Dowex 50 with 8% cross-linking required more than three times as much acid eluent as the same resin with the same particle size but only 2% cross-linking.

4.4.3 Particle size

Resins have spherical particles, and are available in various size ranges. Resins for industrial applications are commonly 14–52 mesh, while those for laboratory use are more usually 50–100, 100–200 or 200–400 mesh. The mesh sizes are only nominal, and analytical grades have a narrower and more symmetrical size distribution than general-purpose resins of the same nominal mesh size. Small particles are more rapidly penetrated by solutes than large ones and therefore come to chemical equilibrium more quickly, but a column packed with small particles offers more resistance to solvent flow than one packed with large particles, and may require excessive pressure to give a useful flow rate. For open-column use, 50–100 mesh resins are satisfactory. Finer resins may be used in pumped columns, and are to be preferred because they require much smaller volumes of eluent; 100–200 mesh Dowex 50 with 2% cross-linking requires not much more than half as much 10% methanolic hydrochloric acid for complete elution as the same resin, 50–100 mesh [3].

4.4.4 Ionic groups: strong and weak acids and bases

The fixed anionic groups in cation-exchange resins are either carboxylic acid (weakly acidic) or sulphonic acid (strongly acidic) groups, or their salts. Weakly acidic resins are usually polyacrylates. The fixed cationic groups in anion-exchange resins are either tertiary amine (weakly basic) or quaternary ammonium (strongly basic) groups, or their salts. Quaternary anion exchangers are of Type 1 or Type 2. Type 1 resins contain $—N^+(CH_3)_3$ groups and Type 2 $—N^+(CH_3)_2C_2H_4OH$. The latter are less strongly basic, and in the hydroxide form are less likely to damage alkali-labile anions, but they have not found much application in surfactant analysis and are not considered in this chapter. The most important practical differences between weakly acidic or basic groups on the one hand and strongly acidic or basic groups on the other is in their relative affinities for hydrogen and hydroxide ions.

Weak acids are largely undissociated, so that when a hydrogen ion reacts with a salt of a weak acid, it displaces the cation and becomes covalently bound to the carboxyl group. If the weak acid happens to be an ion-exchange resin, the covalently bound hydrogen ions are not available for ion exchange to any significant extent. Thus any salt of a weakly acidic resin, when treated with a solution containing hydrogen ions, passes more or less completely to the free acid form, i.e. it prefers hydrogen ions to whatever other ions it was previously associated with. Thus, if the free acid is treated with any cation in the form of a salt, it stays in the acid form. For example, the equilibrium:

$$\text{Resin—COOK} + H^+ \rightleftharpoons \text{Resin—COOH} + K^+$$

lies almost entirely on the right. In other words, weakly acidic cation-exchangers retain hydrogen ions more strongly than all other cations. They are, however, readily neutralised by bases, because this is not an ion-exchange process; it is the neutralisation reaction $H^+ + OH^- \rightarrow H_2O$, which proceeds rapidly to completion.

In contrast, strong acids, including strongly acidic resins, are fully ionised and very readily exchange their hydrogen ions for other cations. The equilibrium:

$$\text{Resin—SO}_3\text{K} + \text{H}^+ \rightleftharpoons \text{Resin—SO}_3\text{H} + \text{K}^+$$

lies largely on the left. Strongly acidic cation exchangers retain hydrogen ions less strongly than any other cation except Li^+.

Weakly basic resins have very few exchangeable hydroxide ions; tertiary amines in the free base form exist predominantly as the amine, e.g. $—N(CH_3)_2$, and only to a very minor extent as the hydroxide of the protonated amine, e.g. $—NH^+(CH_3)_2 \cdot OH^-$. If a salt of a weakly basic resin is treated with an alkali, the hydroxide ions deprotonate the base, which becomes nonionic. For example, the equilibrium:

$$\text{Resin—NH}^+(\text{CH}_3)_2 \cdot \text{Cl}^- + \text{OH}^- \rightleftharpoons \text{Resin—N(CH}_3)_2 + \text{Cl}^- + \text{H}_2\text{O}$$

lies almost entirely on the right. The reaction of hydroxide ions with a salt of a weak base is the combination of the hydroxide ions with the protons of the amine salt to produce water, so that they both disappear from the system. Weakly basic anion exchangers may therefore be said to retain hydroxide ions more strongly than any other anion, although this is not strictly correct; rather, their salts react with hydroxide ions to give the free amine, which has no hydroxide ion to exchange. Neutralisation with acids is not an ion-exchange process but the addition of a proton ($—NR_2 + H^+ \rightarrow —NR_2H^+$), which proceeds rapidly to completion.

In contrast, strong (quaternary) bases exist only in the fully ionised form, e.g., $—N^+(CH_3)_3 \cdot OH^-$. The quaternary nitrogen cannot be deprotonated, and quaternary hydroxides retain their hydroxide ions less strongly than any other anion; the equilibrium

$$\text{Resin—N}^+(\text{CH}_3)_3\text{Cl}^- + \text{OH}^- \rightleftharpoons \text{Resin—N}^+(\text{CH}_3)_3\text{OH}^- + \text{Cl}^-$$

lies almost entirely on the left.

In summary, weak acids and bases prefer hydrogen and hydroxide ions respectively to all others, whereas strong acids and bases prefer any other ion to hydrogen and hydroxide. This is true only for ion-exchange processes involving salts; weak acids and bases are just as readily neutralised as any others by bases and acids respectively.

The distinction between strong and weak acids and bases is important when one intends to recover the exchanged ions (see section 4.5.4 on Elution and choice of eluent), and the same concept applies equally to the exchanged ions themselves: weak acids and bases in solution show little tendency to exchange with anions and cations in a resin.

4.4.5 Selectivity

When an ion-exchange resin is stirred with a solution of an electrolyte, an equilibrium is reached between the exchangeable ions in the resin (A^-, say) and the ions of the same sign in solution (B^-, say). This can be represented thus, for an anion exchanger:

$$\text{Resin}^+A^- + B^-(\text{soln}) \rightleftharpoons \text{Resin}^+B^- + A^-(\text{soln})$$

The equilibrium constant, which we will call K_{ex}, is given by

$$K_{ex} = \frac{[B_r^-][A_s^-]}{[A_r^-][B_s^-]}$$

in which square brackets denote concentration (strictly speaking activity) and the subscripts *r* and *s* refer to the resin phase and the solution phase respectively. K_{ex} is called the selectivity coefficient. If it has a value greater than 1, the resin retains ion B^- in preference to the ion A^-. If it has a value less than 1, the converse is true. The resin is said to be more selective for B^- than for A^- in the first case, and more selective for A^- than for B^- in the second.

To give some idea of the range of values that K_{ex} may have, for the strongly basic anion exchanger Bio-Rad AG1-X8 in the hydroxide form, it ranges from 1.6 for F^- through 22 for Cl^- and 50 for Br^- to 500 for benzenesulphonate. Cation exchangers are much less selective. For the strongly acidic cation exchanger Bio-Rad AG50W-X8 in the sulphonic acid form, K_{ex} ranges from 1.5 for Na^+ through 2.5 for K^+ and 2.7 for Cs^+ to 8.7 for Ba^{2+}.

Strongly and weakly acidic cation exchangers show broadly similar orders of selectivity for various ions, apart from the fact that H^+ lies at opposite ends of the ranges for weak and strong acids. Strongly and weakly basic anion exchangers also show broadly similar orders of selectivity, except that in this case OH^- lies at opposite ends of the ranges for weak and strong bases. Actual values of K_{ex} for specific ions may be very different for the weak and strong species, however. As a broad rule, resins are more selective for large ions than for small ones, and more selective for ions of higher valency than for those of lower valency, but there are many exceptions, and it is not possible without experiment to be sure whether a smaller ion of higher valency will be more or less selectively retained than a larger ion of lower valency.

4.4.6 *Selectivity: a practical example*

The practical significance of all this is best illustrated by an example. Consider a very practical instance, namely 20 ml Bio-Rad AG1-X8 in the chloride form, exchange capacity 1.33 meq/ml, stirred to equilibrium with 100 ml 1.0 mol/l NaOH solution. K_{ex} (Cl^-/OH^-) = 22 (see above), so K_{ex} (OH^-/Cl^-) = 1/22 = 0.045, which means that the resin strongly prefers Cl^- ions to OH^- ions. Using the exchange capacity as a fair approximation to the concentration of ions in the resin, the equation for K_{ex} can easily be solved. It shows that only 36.5% of the chloride in the resin has been replaced by hydroxide, i.e. an approximately fourfold excess of hydroxide ions has brought about only 36.5% conversion. This is very unlikely to be good enough, and this particular conversion is one that will have to be done quite often in practice.

The problem arose because the resin was stirred to equilibrium with the exchanging solution in a batch process, in which the displaced ions remained in contact with the resin. If the treatment is done by passing the solution down a column of the ion exchanger, the displaced chloride ions are carried away in the flowing solvent, so that equilibrium is never attained and the conversion proceeds to completion. Any ion can be completely displaced, even by an ion for which the resin is much less selective, if the conversion is done in a flowing system, because the displaced ions are removed from the contest. Conversion is complete when the concentration of the displaced ions in the column effluent falls to zero.

Almost all conversion processes from one ionic form to another are much better done in a column than batchwise. The only exceptions are:

1. Neutralisation of a weakly or strongly acidic resin with an alkali.
2. Neutralisation of a weakly or strongly basic resin with an acid.
3. Conversion of a salt of a weakly acidic resin to the free acid form by treatment with a strong acid.
4. Conversion of a salt of a weakly basic resin to the free base form by treatment with an alkali.
5. Deionisation using a mixed-bed resin.

4.4.7 *Mixed-bed resins*

A mixed-bed resin is a mixture of chemically equivalent amounts of a strongly acidic cation exchanger in the free acid form and a strongly basic anion exchanger in the hydroxide form. If such a mixture is stirred with a solution of any electrolyte Cat^+An^- until equilibrium is reached, the following processes occur:

$$\text{Resin—}SO_3^-H^+ + Cat^+ \rightleftharpoons \text{Resin—}SO_3^-Cat^+ + H^+$$

$$\text{Resin—N}^+(\text{CH}_3)_3\cdot\text{OH}^- + \text{An}^- \rightleftharpoons \text{Resin—N}^+(\text{CH}_3)_3\cdot\text{An}^- + \text{OH}^-$$

$$H^+ + OH^- \rightarrow H_2O$$

The hydrogen and hydroxide ions formed in the first two equilibria combine together to form water and are so removed from the system as fast as they are liberated. Both equilibria are thus displaced from left to right until all ions of either charge have been removed from the solution. This is the basis of a very general method for deionising solutions and isolating nonionic matter. It can be done completely effectively in a batchwise operation, but operators used to using columns will probably find it more convenient to do it in a column anyway.

4.5 Guidelines on using ion-exchange resins

4.5.1 Apparatus

The basic components of the apparatus necessary for ion exchange work are:

a solvent reservoir
a column to contain the resin
a means of controlling the flow, and
a receiving vessel for the effluent.

Completely satisfactory ion-exchange separations of surfactants into the four main classes can be accomplished using burettes as columns, capillary tubing and rubber bungs as connectors and separating funnels as solvent reservoirs, and no-one should be discouraged from using ion-exchange techniques by lack of custom-designed apparatus. But it is undoubtedly more convenient to use equipment designed for the purpose, and a wide range of columns, connectors, solvent reservoirs, pumps and meters is now available. A paper published in 1986 [3] described the use of HPLC technology for ion exchange, with a peristaltic pump to control the flow rate, a series of three columns that could be operated in series for the separation step and in parallel for the recovery step, switching valves to permit this and a computer to control the whole operation. No doubt the technology will continue to improve and the use of ion exchange will become progressively simpler and more effective. Meanwhile much successful 'low-tech' work can be and is being done.

A failure that sometimes happens with improvised apparatus is that the column is accidentally allowed to run dry. There is no way of retrieving the analysis; the only option is to start again. A simple device (Figure 4.1)

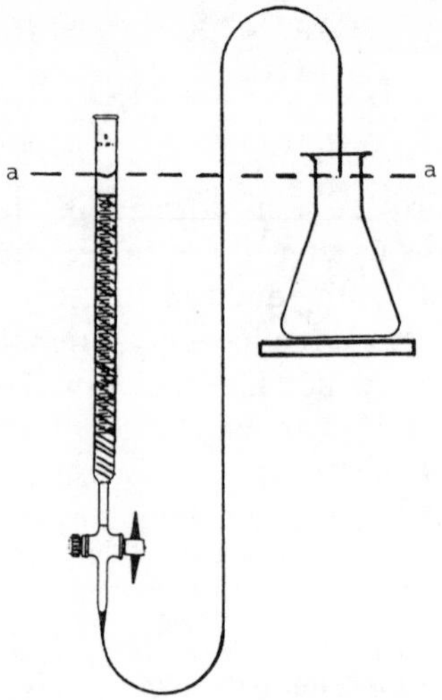

Figure 4.1 Device for preventing ion-exchange columns from running dry. Flow stops when the solvent reaches level a–a (adapted by kind permission of Merck Ltd).

eliminates this possibility. Assuming the improvised column to be a burette, attach a length of plastic capillary tubing securely to the jet, take it up above the level of the top of the resin bed, support it at the highest point, then clamp the end so that it points downwards with its open end above the level of the resin. The capillary tubing will then form an elongated letter S. Make sure it is not kinked. If the solvent in the reservoir connected to the top of the column runs out, the flow will then stop before the solvent level falls to the level of the resin, and no air will be admitted to the column.

4.5.2 Choice of resin: what retains what

The following are broad guidelines which should enable the analyst to devise a column or a combination of columns suitable for any particular separation. It should be remembered that ions of any non-surfactant species present in the sample will also interact with the resins. Each resin type retains the surfactant types shown in Table 4.1.

Examples of commercially available resins are given in Table 4.2. The mention of a resin here does not imply recommendation or endorsement of that resin in preference to any other resin of the same general type, nor any guarantee that it is suitable for any particular purpose.

The code '(ip)' which appears in two of the resin names stands for

Table 4.1 What retains what

Resin type	Surfactants retained
Strongly basic anion exchanger, chloride	Sulphates, sulphonates
Strongly basic anion exchanger, acetate	Sulphates, sulphonates, soaps, sarcosinates; not free carboxylic acids
Strongly basic anion exchanger, hydroxide	All anions, including soaps, sarcosinates and amphoterics with weakly basic nitrogen; all acids; not betaines
Weakly basic anion exchanger, hydrochloride	Sulphates, sulphonates
Weakly basic anion exchanger, free base	Only free acids, any kind
Strongly acidic cation exchanger, sodium salt	Quaternary cationics
Strongly acidic cation exchanger, free acid	All cations, including weak bases, amine oxides, betaines, amphoterics; all bases; not sulphobetaines with weakly basic nitrogen
Weakly acidic cation exchanger, sodium salts	Quaternary cationics
Weakly acidic cation exchanger, free acid form	Only free bases, any kind
Mixed-bed resins	All anions and cations

'isoporous'. This means that the pores in the resin matrix are more uniform in size than in resins not so designated.

4.5.3 Solvent and flow rate

The extent to which exchange occurs and the speed with which equilibrium is reached depend very much on the nature of the solvent. For the majority of surfactants, 80% methanol is a good solvent for both the sorption (uptake) and elution (recovery) steps. However, the rate of elution can be markedly affected by the concentration of alcohol in the eluent (see the end of section 4.5.4). Methanol has the added advantage that it is easily removed by evaporation. Ethanol and propan-2-ol (isopropanol) are also satisfactory, and some variation in concentration can be tolerated. Not all surfactants are readily soluble in methanol, and in such cases other solvents such as acetone or chloroform may be added. An example is the di(hydrogenated tallow)methylamine used in some soften-in-the-wash detergent powders, which requires the addition of up to 50% chloroform.

Fairly obviously, the sorption of ions is most efficient, i.e. most closely approaches the theoretical maximum, when each resin particle has time to take up as many ions as it is capable of retaining, i.e. when the flow is very slow. In practice, however, there is usually a time constraint on analyses, and a good compromise is to use a flow rate of 1 ml/min/cm^2 of column cross-section. This permits practical exchange capacities not

Table 4.2 Some commercially available ion-exchange resins

Resin type	Functional groups	Examples
Strongly basic anion exchanger	Quaternary ammonium	Bio-Rad AG-1 Dowex 1 Amberlite IRA 401 Zerolit FF(ip) Duolite A 113
Usually supplied as the chloride		
Weakly basic anion exchanger	Tertiary amine	Bio-Rad AG3 Dowex 3 Amberlyst A-21 Zerolit M(ip) Duolite A 378
Usually supplied as the free base		
Strongly acidic cation exchanger	Sulphonate	Bio-Rad AG50W Dowex 50 Amberlite IR 120 Zerolit 225 Duolite C 225
Usually supplied as the free acid		
Weakly acidic cation exchanger	Carboxylate	Bio-Rex 70 Amberlite CG 50 Zerolit 236 Duolite C 436
Usually supplied as the free acid		
Mixed-bed resins	Quaternary ammonium and sulphonate	Bio-Rad AG 501 Amberlite MB-1 Zerolit DM-F Duolite MB 5113
Supplied as hydroxide and free acid		

usually much less than half the theoretical maximum and often better than that, and recoveries within a manageable volume of eluent, i.e. displacing solution. Faster flow rates can be used, but it may then become necessary to use bigger columns. It is easy to find out by experiment. With basic apparatus, the flow rate can be controlled by adjusting the stop-cock of the burette (column) or by raising or lowering the solvent reservoir.

4.5.4 Elution and choice of eluent

Elution is the recovery of sorbed ions from the resin by washing with a displacing electrolyte solution. The reagent used for elution is called the eluent, and the effluent emerging from the column during elution is called the eluate. In principle any electrolyte will displace any ion, even when the selectivity coefficient is unfavourable, provided enough of it is used in a flowing system. In practice, factors to consider include the volume of eluent that will be required, the time it will take to pass through the column, the ease with which the eluted ions can be separated from the excess eluting electrolyte and whether the elution process restores the column to its original chemical form so that it can be re-used without further treatment other than a wash with solvent.

Sorbed ions that are salts of strong acids or bases can be eluted from strongly acidic and basic resins only with large volumes of eluent. Hydrochloric acid is effective for both. A molar solution is generally used, and 30 or more bed-volumes may be needed (one bed-volume is simply the apparent volume of the resin bed). Much smaller volumes of eluent will be needed if the separation can be done with a weakly acidic or basic resin, because elution with 1 M hydrochloric acid or 2 M ammonia rapidly converts the resin to the free acid or base. Since these are virtually nonionic, the sorbed ions 'fall off', so to speak, and complete elution can be achieved with five bed-volumes or even less. The eluate then contains all the exchangeable ions that were in the column. Similarly, if the sorbed species are anions of weak acids or cations of weak bases, the same eluents will render them virtually nonionic so that they are no longer bound to the resin.

Some eluates, e.g. acid eluates containing alkyl sulphate, may need to be neutralised before evaporation of the eluate to dryness to avoid decomposing the surfactant(s) they contain.

It must not be assumed that the residue obtained by evaporating any eluate contains only the desired surfactant ions and their counterions; in fact, this will not often be the case. The ions actually present depend on the initial form of the resin, the kind of non-surfactant ions present in the sample and the composition of the eluent itself. In the simple Voogt system [4], for example, the acid cation exchanger retains all the cations that were in the sample, the anion exchanger acetate retains strongly acidic surfactant anions, sulphonate hydrotropes, sulphate and possibly other organic and inorganic anions, and the anion exchanger hydroxide retains all other anions. Elution of the cations with hydrochloric acid recovers all the cations as their chlorides. Elution of the acetate column with hydrochloric acid recovers all the sorbed anions, including the acetate ions, as the corresponding acids. Elution of the hydroxide column with sodium hydroxide recovers all the other anions as their sodium salts, and the eluate also contains a large excess of the alkali. Ion exchange separates surfactant ions of different classes from one another, but not usually from non-surfactant ions having the same polarity. It is obviously an advantage to use a volatile eluting electrolyte if possible, but it will often be necessary to extract with a solvent or carry out some other process if isolation of pure surfactant is required.

Whatever eluent is used, it is desirable that it should use the same solvent as was used in the sorption step, to avoid problems with swelling or shrinking of the resin.

Solvent composition is important and can be crucial. MacDonald *et al.* [3] showed significant variations in eluent efficacy with concentration of methanol in both acid and alkaline eluents. 90% methanol was best for approximately 1 M hydrochloric acid and 80% methanol was best for 0.5 M potassium hydroxide in eluting from Dowex 1-X2 chloride and hydroxide respectively. Eluents must contain sufficient alcohol to suppress the hydrophobic interaction between surfactants and the resin matrix; the same workers found that no elution at all occurred when the eluent contained only 45% methanol.

4.5.5 Preparing a column

The preparation of an ion-exchange column consists of filling and, if necessary, conversion to another ionic form. It is assumed here that the column is a burette.

In a typical resin, the diameter of the biggest particles is at least twice that of the smallest, so that the biggest have at least eight times the volume of the smallest. The two main requirements of the filling operation are that these very unevenly sized particles must be packed reasonably homogeneously throughout the full depth of the column, and that no air bubbles must be admitted.

Filling. There are three cases to consider:

(a) the resin is a free acid or base which is to be converted to a salt by neutralisation;
(b) the resin is a salt which is to be converted to another ionic form by ion exchange; and
(c) the resin is to be used in the form in which it was supplied.

In case (a):

1. Half-fill a measuring cylinder with water and transfer a slight excess (say 10%) of resin into it, using a spatula.
2. Pour it into a beaker and add an excess of the appropriate acid or base, e.g. hydrochloric acid for a tertiary amine resin or sodium hydroxide for a sulphonic acid. It is convenient but not essential to use approximately molar reagents, in aqueous solution.
3. Stir for a few minutes; neutralisation is rapid, but the neutralising reagent needs a certain amount of time to penetrate to the innermost regions of the resin matrix.
4. Filter off the converted resin and resuspend it in the solvent to be used for the analysis.
5. Fill the column as described.

In case (b):

1. Measure out a slight excess of resin as before and fill the column as described, partially filling it with water rather than solvent.

In case (c):

1. Half-fill a measuring cylinder with the solvent to be used for the analysis and transfer into it the volume of resin required. Fill the column as described.

In all cases:

1. Fill the burette about two-thirds full with the solvent to be used.
2. Stir the resin suspension and transfer a small volume, say 0.5–1 ml, to the burette, using a Pasteur pipette. Allow the resin to settle before adding a further small quantity.
3. Continue in the same way, running off excess solvent to make room for more resin whenever necessary, until a column of the required size is obtained.
4. Push a small plug of glass wool or cotton wool gently down on to the resin bed to protect it from disturbance when in use.

In case (c) the column is then ready for use.

Conversion to another ionic form by ion exchange. The most frequent conversion other than neutralisation is that of a strongly basic anion exchanger from chloride to hydroxide. The ineffectiveness of batchwise conversion was shown in section 4.4.6.

1. After the column has been packed, using a suspension of the resin in water, attach a solvent reservoir. A large volume of reagent will be needed, so the reservoir should preferably have a capacity of about 50 bed-volumes. A separating funnel with the stem drawn out to a jet is satisfactory.
2. Insert into the neck of the burette a rubber bung carrying a short length of narrow-bore glass tubing. Connect this to the separating funnel with narrow-bore rubber or plastic tubing.
3. Fill the reservoir with approximately 1 M aqueous sodium hydroxide (or appropriate other reagent).
4. Allow the alkali to flow through the resin at about 2 ml/min/cm^2 of column cross-section. With a 10 ml burette, this will be about 1.0 ml/min.
5. When about 25 bed-volumes have passed through, collect a little of the effluent and test for the presence of chloride ions by acidifying with 1 M nitric acid and adding 1 ml 5% silver nitrate solution.
6. Continue the flow of alkali, testing for chloride every five bed-volumes, until the test solution remains clear.

Whichever conversion process (i.e. neutralisation or ion exchange) was used, attach a solvent reservoir filled with the solvent to be used for the analysis. Wash the resin with this, at the same flow rate as before, until the effluent is neutral to phenolphthalein (if alkali was used for the conversion) or methyl red (if acid was used). This should take only a few bed-volumes. The column is then ready for use.

4.5.6 Some examples of practical multicolumn systems

Voogt [4], as long ago as 1959, used a three-column apparatus to separate anionics, soaps, cationics and nonionics. This system is still in use. The first column contained a strongly acidic cation exchanger as the free sulphonic acid. This retained cationics and converted all anions to the corresponding acids. The second column contained a strongly basic anion exchanger acetate, to retain all surface-active sulphates and sulphonates. The third column contained the same resin as the hydroxide, to retain the fatty acids. The effluent contained the nonionics. The sorbed surfactants could all be recovered by elution with hydrochloric acid.

Draguez de Hault [5] used a three-column system, but only the first two columns were needed for the separation process. The first column contained a strongly acidic cation exchanger, as in the Voogt system, but the second contained a weakly basic anion exchanger, Amberlyst A-21, as the free base. The third column was a small column of the cation exchanger, and was used only to confirm complete sorption of the cations. The solvent was 50% propan-2-ol and the flow rate 5 ml/min. Tertiary amines and quaternaries were eluted from the cation exchanger with approximately 1 M hydrochloric acid, solvent not specified. Fatty acids were eluted from the anion exchanger with 96% ethanol saturated with carbon dioxide, and synthetic anionics with 0.3 M ammonium bicarbonate in 60% propan-2-ol. This system successfully separated a mixture of dodecylbenzene sulphonate, soap, tertiary amine, a quaternary ammonium salt and an ethoxylated nonylphenol.

MacDonald *et al.* [3] used a system similar to Voogt's, except that the anion exchanger acetate was replaced by the chloride. They recovered cationics and strongly acidic anionics by elution with 1 M hydrochloric acid in 90% methanol, and soaps with 0.5 M potassium hydroxide in 80% methanol. They reported successful analyses of four commercial detergents.

Milwidsky and Gabriel [6] used a three-column system in which the first column was a weakly basic anion exchanger hydrochloride, the second a strongly basic anion exchanger hydroxide and the third a strongly acidic cation exchanger as the free acid. They extracted soaps as fatty acids before passing the sample through the columns. The second column was stated to retain 'amphoterics', but in fact would retain only WW and SW types, betaines being retained on the third column with the cationics. Anionics were eluted from the first column with 3 M ethanolic ammonia, and amphoterics and cationics from the second and third columns with 1 M methanolic hydrochloric acid.

The same authors also described a simple method for determining the exchange capacity of columns, which consisted essentially of passing small measured amounts of a standard solution of a surfactant through the column, washing after each one, until the surfactant appeared in the effluent.

In the multicolumn systems it is better to retain soaps and acid-labile compounds on an anion exchanger before the solution passes through a strongly acid resin, which may catalyse esterification of fatty acids with the alcohol in the solvent and hydrolysis of such materials as alkyl sulphates.

Although ion exchange is a very powerful tool, there are some separations that are more conveniently done by liquid–liquid extraction, and for complex mixtures a combination of the two techniques may be the most effective approach.

4.5.7 Loading the sample, washing the column and elution

It is essential not to allow any air to enter the resin bed at any stage. The total amount of ionic matter in the sample applied to the column should not exceed half the theoretical exchange capacity of the column. If the resin has a capacity of, say, 1 meq/ml, and the column contains 10 ml of resin, the maximum sample size is 5 meq, or about 1.5–2.0 g, depending on the molecular weight of the surfactant(s) to be sorbed, or less if other ionic substances are present.

Loading the sample

1. Prepare a solution of the mixed surfactants in 80% methanol if possible, and preferably free from inorganics and other non-surfactant components, at a concentration of up to 5%.
2. Pipette a suitable volume of sample solution into the solvent reservoir.
3. Allow it to flow through the column at about 1 ml/min/cm^2, collecting the effluent in a beaker. Stop the flow when the surface of the solution just reaches the top of the resin bed.
4. With a clean pipette, transfer 10 ml solvent to the solvent reservoir, taking care to rinse the walls thoroughly. The pipette is used only to direct the flow, not to measure the volume. Allow the solvent to flow through the column until the surface again just reaches the top of the resin bed. Repeat the washing-in with a further 10 ml solvent.

Washing the column

5. Fill the solvent reservoir and allow the solvent to flow through the column, maintaining the same flow rate, until the effluent is free of dissolved matter. This should take no more than 5 bed-volumes.
6. Check by collecting one more bed-volume in a tared beaker and evaporating to dryness.

The effluent contains whatever components have not reacted with the resin, plus the displaced ions. Keep it for whatever examination may be appropriate.

Elution. Elution is necessary if the sorbed ions are required for examination, and were not removed just to facilitate analysis of the effluent. The volume of eluent required depends on both the resin and the nature of the sorbed ions; see later for guidelines.

7. Empty the solvent reservoir and refill it with the eluent. Allow this to flow through the column at the same rate as before, collecting the eluate in a beaker, until elution is complete.
8. When elution is thought to be complete, collect a further 1–5 bed-volumes in a tared beaker and evaporate to dryness. If the eluent is hydrochloric acid or ammonia, there should be no residue.

4.6 Practical applications of ion exchange

4.6.1 Introduction

The following sections describe the isolation of each of the four main classes of surfactant. It is not practicable to give precise instructions for every conceivable mixture, but most practical situations are covered. The isolation of nonionics (section 4.6.2) is described in a fair amount of detail, which is omitted from later sections. Amphoterics are classified as WW, WS, SW and SS, as defined in section 2.2.4.

In all ion exchange separations, the main precautions to be observed are:

—use sufficient resin
—pack columns correctly
—avoid excessive flow rates
—convert to other ionic forms in the correct manner
—avoid admitting air to the column
—confirm completeness of washing and elution by appropriate chemical tests

—neutralise acid or alkaline eluates before evaporating them to dryness, to avoid hydrolysis of constituents.

The behaviour of SW amphoterics is not definitely established by experiment, but it is difficult to imagine why they should not behave like other anionics in alkaline solution and like other zwitterions in acid solution. This behaviour would be analogous but opposite to that of WS amphoterics (betaines), which behave like other quaternary salts in acid solution and like other zwitterions in alkaline solution. Thus the fully deprotonated form would be expected to be retained by strongly basic anion exchanger chlorides, but not by weakly basic ion exchanger hydrochlorides, because the weakly basic nitrogen of the amphoteric would compete with the weakly basic nitrogen of the resin for the latter's proton. The protonated form would not be retained by the resin, being zwitterionic.

Sulphobetaines (SS amphoterics) do not interact with ion-exchange resins and appear with the nonionic fraction. Amphoterics in general are likely to contain anionic or cationic impurities which may turn up in different places from the amphoteric itself.

Phosphate esters present peculiar difficulties, because they may contain two or three of the possible esters (mono-, di- and triesters) in varying proportions. Very little information is available about their behaviour in ion-exchange systems, but it is certain that the triester will always appear in the nonionic fraction. The mono- and diesters are anions whose acid forms both contain a strongly acidic hydrogen ion, so one would expect them to behave like sulphates and sulphonates. There seems to be no published evidence for this, however. They are certainly retained by the hydroxide form of strongly basic anion exchangers.

The ability of the sodium salts of weakly acidic cation exchangers to retain quaternary ammonium salts has not been fully confirmed, and neither has the ability of the same resins in the acid form to retain weak bases, although there is some experimental evidence for both. The ability of free weakly acidic and basic resins to retain free bases and acids respectively is in any case of limited use, because once some of the resin has done this it is in salt form, and able to undergo ion exchange with other salts. For this reason such resins must be placed last in any multicolumn system.

4.6.2 Isolation of nonionics

This is the simplest of all ion-exchange separations, and it may be done equally well in a column or batchwise. The suggested ratio of resin volume to sample weight is based on an assumed exchange capacity of 0.75 meq/ml. The residue obtained at step 8 will include sulphobetaines (SS amphoterics).

1. Calculate how much mixed-bed resin will be required to remove all ionic matter from the test portion, and use three times this amount. As a guide, use 15 ml resin for every gram of solids in the test portion.
2. Measure out the resin as described earlier, and either pack it into a column or place it in a beaker with 100 ml solvent.
3. Prepare a solution of the sample in 80% methanol or other suitable solvent, and use a test portion containing about 0.5 g total nonionic matter.
4. If using a column, proceed as described in section 4.5.7.
5. If operating batchwise, pipette the test portion into the beaker containing the resin and stir mechanically or magnetically for 15 min.
6. Remove the stirrer and rinse it with 80% methanol.
7. Filter off the resin and wash it with at least three filter-funnels full of 80% methanol, allowing the solvent to drain thoroughly each time. Collect the filtrate in a tared beaker.
8. Evaporate the column effluent or the filtrate to dryness on a steam bath and dry to constant weight. Keep the nonionic matter for any further examination that may be necessary.

If necessary, confirm the completeness of the deionisation as follows. Dissolve the residue from step 8 in chloroform and dilute to volume in a volumetric flask. Test a portion of this solution for the presence of whatever ionic surfactant predominated in the sample, e.g. by adding water and acid mixed indicator and titrating with benzethonium chloride. If the nonionic matter is not completely free of ionic surfactants, repeat the analysis using twice as much resin.

4.6.3 *Isolation of anionics*

The most convenient procedure depends on which other classes are present. When nonionics and sulphobetaines are absent, the preferred method is to remove all other species and leave only the anionics in the column effluent. When nonionics are present, the anionics must be sorbed and eluted. Use 80% methanol or aqueous ethanol or propan-2-ol as solvent.

Nonionics and SW and SS amphoterics absent. Remove all cationics and amphoterics by retention on a strongly acidic cation exchanger, e.g. Bio-Rad AG 50W-X4 acid. The effluent contains all the anions, surfactant and otherwise, as the corresponding acids.

Nonionics and/or sulphobetaines present, all other amphoterics absent. Retain all anionics on a strongly basic anion exchanger, e.g. Bio-Rad AG 1-X4 hydroxide, and elute with at least 25 bed-volumes of

1 M methanolic hydrochloric acid. The effluent contains all the anions, surfactant and otherwise, as the corresponding acids.

Nonionics and WW or SW amphoterics present. The previous procedure would include all amphoterics with weakly basic nitrogen (WW and SW) in the anionic fraction. If a WW amphoteric is present the simplest procedure is to combine the two columns described, with the cation exchanger first. The WW amphoteric is retained on the cation exchanger and the anionics on the anion exchanger, from which they are eluted as described.

SW amphoterics, however, are not retained by the cation exchanger. In this case use a three-column system, as follows.

Column 1: strongly acidic cation exchanger, e.g. Bio-Rad AG50W-X4 acid. Retains all cationics and WW amphoterics. All anions emerge as the corresponding acids. SW amphoterics emerge as zwitterions.
Column 2: weakly basic anion exchanger. e.g. Bio-Rad AG 3-X4A hydrochloride. Retains sulphates and sulphonates (and very probably phosphate mono- and diesters). Zwitterionic SW amphoterics pass through. Elute with 2 M methanolic ammonia. A few bed-volumes will be required.
Column 3: strongly basic anion exchanger, e.g. Bio-Rad AG1-X4 hydroxide. Retains carboxylates and SW amphoterics. Elute with 1 M methanolic hydrochloric acid. A few bed-volumes will be required.

4.6.4 *Isolation of cationics*

As with anionics, the preferred method if nonionics are absent is to remove all other species and leave only the cationics in the column effluent. If nonionics are present, the cationics must be sorbed and eluted.

Nonionics and sulphobetaines (SS amphoterics) absent. Remove all anionics and WW and SW amphoterics by retention on a strongly basic anion exchanger, e.g. Bio-Rad AG 1-X4 hydroxide. The effluent contains all the cations, surfactant and otherwise, as the corresponding bases, and betaines in the zwitterionic form.

Nonionics and/or SS amphoterics present. Retain all cationics on a strongly acidic cation exchanger, e.g. Bio-Rad 50W-X4 acid, and elute with at least 25 bed-volumes of 1 M methanolic hydrochloric acid.

Nonionics and WW or WS amphoterics present. The previous procedure would include betaines and amphoterics with weakly basic nitrogen in the cationic fraction.

If a WW amphoteric is present, the simplest procedure is to combine the two columns described, with the anion exchanger first.

Betaines (WS), however, are not retained by the anion exchanger. If a betaine and/or weak bases (amines or amine oxides) are present, use a three-column system, as follows.

Column 1: strongly basic anion exchanger hydroxide. Retains all anionics and WW and SW amphoterics. All cations emerge as corresponding bases.

Column 2: strongly acidic cation exchanger, sodium salt. Retains quaternary cationics. Elute with at least 25 bed-volumes of 1 M methanolic hydrochloric acid.

It should be equally effective to use the sodium salt of a weakly acidic cation exchanger, e.g. Duolite 436, for this purpose. It is certainly worth trying it for individual cases, because elution with 1 M hydrochloric acid requires only a few bed-volumes.

Column 3: strongly acidic cation exchanger, free acid. Retains betaines (WS) and weakly basic cationics. Elute with 2 M methanolic ammonia. Only a few bed-volumes are required.

4.6.5 Isolation of amphoterics

WW amphoterics are cationic at low pH, anionic at high pH and zwitterionic at neutral pH.
WS amphoterics are cationic at low pH and zwitterionic at high pH.
SW amphoterics are anionic at high pH and zwitterionic at low pH.
SS amphoterics are zwitterionic at all pHs.

These four classes behave differently in ion-exchange systems, and the best method of isolation depends on which amphoteric and which other classes of surfactant are present. The behaviour of WW and WS amphoterics in ion-exchange systems is well characterised, at least for those with only one acid and one basic group, but that of SW amphoterics is not. They are believed to behave as described here, but there is little direct experimental evidence. SS amphoterics (sulphobetaines) do not interact with ion-exchange resins.

One might speculate about the anionic, cationic or zwitterionic character of amphoterics with more than one acid or basic group or with both strong and weak acid or basic functions (some of which are commercially available), but there appears to be no published information about their behaviour in ion-exchange columns. If confronted by one of these, it is a simple matter to pass samples through the columns of acidic and basic resins in the acid, base or salt form and see what happens. It should then be straightforward to work out a system for the isolation of the amphoteric. In any such experiment the pH must be adjusted to ensure that the amphoteric is in the appropriate ionic form (see section 4.7.4).

4.6.5.1 WW and SW amphoterics. Make up a solution in 80% methanol, if possible. If carboxylates are present, remove them by extraction with petroleum ether from acidified 50% methanol. Use the minimum possible amount of acid. Add 1½ volumes of methanol to give a final concentration of 80%. The procedure then depends on what other surfactants are present.

Anionics only present. Remove the anionics on a column of either a strongly basic anion exchanger chloride or a weakly basic anion exchanger hydrochloride. The effluent contains the amphoteric and the displaced chloride ions. For SW amphoterics ensure that the solution entering the column is acid, so that the amphoteric is in the zwitterionic form and will therefore not be retained.

Anionics absent and any other surfactant present. Retain the amphoterics on a column of a strongly basic anion exchanger hydroxide and allow cationics, betaines, sulphobetaines and nonionics to pass through. The acidity or otherwise of the solution is irrelevant since the resin will neutralise any acidity. WW amphoterics are rapidly eluted by 1 M methanolic hydrochloric acid, which protonates their carboxy groups and their basic nitrogen, making them cationic. SW amphoterics should be eluted by the same reagent, which protonates their basic nitrogen and renders them zwitterionic.

Anionics and any other surfactant present. Combine the two columns already described, in the order given. The first column retains sulphates and sulphonates, the second retains the amphoterics, and all other species pass through. Elute as before.

Separation of WW and SW amphoterics, anionics and nonionics only present. Use a three-column system, as follows.

Column 1: strongly acidic cation exchanger, acid form. Retains WW amphoterics. Elute with about 5 bed-volumes of 2 M methanolic ammonia, to deprotonate the amphoteric and make it anionic.
Column 2: strongly basic anion exchanger chloride, or weakly basic anion exchanger hydrochloride. Retains anionics.
Column 3: strongly basic anion exchanger hydroxide. Retains SW amphoterics. Elute with acid, as before, to neutralise the resin and render the amphoteric zwitterionic.

4.6.5.2 WS amphoterics (betaines). Again, the best procedure depends on what else is present.

Anionics and/or WW and SW amphoterics present, cationics, nonionics and sulphobetaines absent. Retain the anionics and other amphoterics on a strongly basic anion exchanger hydroxide. The effluent contains the betaine in the zwitterionic form plus non-surfactant cations as their hydroxides or free bases.

Anionics, nonionics and/or sulphobetaines present, cationics and other amphoterics absent. Retain the betaine on a strongly acidic cation exchanger, acid form. Elute with about 5 bed-volumes of 2 M methanolic ammonia or at least 25 bed-volumes of 1 M methanolic hydrochloric acid.

Anionics, quaternary salts, weak bases and nonionics present. Extract the weak bases with petroleum ether from alkaline 50% methanol or with diethyl ether from alkaline 33% methanol. Then add sufficient methanol to make the concentration 80% and proceed as follows.

Column 1: strongly acidic cation exchanger, sodium salt. Retains quaternary cationics.
Column 2: strongly acidic cation exchanger, acid form. Retains betaines. Elute with a few bed-volumes of 2 M methanolic ammonia.

The anionics and nonionics pass through.

4.6.6 The general case

Especially for the analysis of unknowns, a general method is required that will separate mixtures into at least the four main classes, and preferably with some subdivision. A four-column system is proposed here, which combines features of the established systems described by Voogt, Draguez de Hault and Milwidsky and Gabriel referred to earlier, with one additional column. It permits the separation into classes of any surfactant mixture. It is unlikely that a mixture containing all classes is ever going to be encountered in real life, but the proposed scheme is a general-purpose one for the separation of any mixture of unknown composition. One or more columns, however, will usually turn out to have served no purpose. If two or more members of any one class are present, they will obviously not be separated from each other in the primary separation. A four-column system must be pump-driven because the flow rate with gravity drive would be too slow.

The sample solution must be neutral when it is applied to the first column, to ensure that all amphoterics are in the zwitterionic form.

It must be understood that this scheme has not been validated in its entirety. It is, however, based on component parts that have either been completely validated or are at least supported by some experimental evidence. Matters not fully validated are the behaviour of SW amphoterics and the sorption of quaternary cationics on weakly acidic cation exchanger sodium salts. These should be confirmed by experiment before complete faith is placed in the scheme.

Column 1: weakly basic anion exchanger hydrochloride. Retains anionics (including phosphate mono- and di-esters?). Elute with ammonia to deprotonate and therefore deionise the resin.
Column 2: strongly basic anion exchanger hydroxide. Retains carboxylates and WW and SW amphoterics. Elute with hydrochloric acid. Carboxylates and amphoterics are eluted rapidly. Response of phosphate esters is uncertain, but if they were not retained by column 1 they would be retained by column 2 and also eluted rapidly. Quaternary salts are converted to the hydroxide, weakly basic cationics (amines, amine oxides) to the free base and betaines to the zwitterionic form. All of these pass through.
Column 3: Weakly acidic cation exchanger, sodium salt. Retains quaternary cationics. Weak bases and betaines pass through. Elute with hydrochloric acid, to protonate and therefore deionise the resin.
Column 4: Strongly acidic cation exchanger, free acid. Retains weak bases and betaines. Effluent contains nonionics, sulphobetaines (SS amphoterics) and phosphate triesters. Elute with ammonia, to deprotonate and deionise weak bases and to convert the betaine to the zwitterionic form.

The column 2 eluate contains carboxylates and WW and SW amphoterics. Soaps and sarcosinates can be separated from amphoterics and from each other by solvent extraction. Carboxylates and WW and SW amphoterics can be separated from each other by ion exchange, but it is necessary to evaporate the eluate to dryness and re-extract the surfactants.

The column 4 eluate contains WS betaines in the zwitterionic form and weakly basic cationics as the free bases. They can be separated by solvent extraction. Separation by ion exchange does not appear to be feasible.

4.7 Analysis of column effluents

4.7.1 Liberation of hydrogen and hydroxide ions

When any salt, whether organic or inorganic, is passed through a strongly acidic cation exchanger in the acid form, all the cations are exchanged for hydrogen ions, so that all the anions appear in the effluent in the form of the corresponding acids.

$$\text{Resin—SO}_3\text{H} + \text{CatCl} \rightarrow \text{Resin—SO}_3\text{Cat} + \text{HCl}$$

$$\text{Resin—SO}_3\text{H} + \text{CH}_3\text{COOCat} \rightarrow \text{Resin—SO}_3\text{Cat} + \text{CH}_3\text{COOH}$$

Similarly, when any salt is passed through a strongly basic anion exchanger hydroxide, all the anions are exchanged for hydroxide ions, so that all the cations appear in the effluent as the corresponding hydroxides or free weak bases.

$$\text{Resin—N(CH}_3)_3\text{OH} + \text{NaAn} \rightarrow \text{Resin—N(CH}_3)_3\text{An} + \text{NaOH}$$

$$\text{Resin—N(CH}_3)_3\text{OH} + \text{C}_{12}\text{H}_{25}\text{NH}_3\text{An} \rightarrow \text{Resin—N(CH}_3)_3\text{An} + \text{C}_{12}\text{H}_{25}\text{NH}_2 + \text{H}_2\text{O}$$

4.7.2 *Information to be obtained*

The effluents can be titrated with acid or alkali to obtain a measurement of the total anions or cations in the sample. Further, if both weak and strong acids or bases are present, they can be determined separately, because there are then two end-points. In the titration of acids in the effluents from a cation-exchange column, the end-point in the region of pH 3 or 4 corresponds with the neutralisation of all the strong acids, and the end-point in the region of pH 9 or 10 corresponds with the neutralisation of all the weak acids. In the titration of bases in the effluent from an anion-exchange column, the end-point in the region of 9 or 10 corresponds with the neutralisation of all the strong bases, and the end-point in the region of 3 or 4 corresponds with the neutralisation of all the weak bases.

4.7.3 *Titrating the effluents*

The titration must be done with alcoholic acid or alkali, and is best done potentiometrically with an autotitrator. Next best is manual potentiometric titration, but it is satisfactory to use indicators, methyl orange or bromophenol blue for the low-pH end-point and phenolphthalein for the high-pH end-point. The information obtained is often a useful cross-check on the mass balance in the analysis of an unknown, but it can also help in the analysis of WW amphoterics, which contain appreciable amounts of non-surfactant weak acids and bases.

If any constituent of the acid effluent is acid-labile, or if any constituent of the alkaline effluent is alkali-labile, collect the effluent in a measured excess of alkali in the first case or a measured excess of acid in the second case, and stir the contents of the receiving vessel continuously throughout the ion-exchange process.

Effluents should be kept after titration for analysis of whatever surfactants they contain.

4.7.4 *Amphoterics: loss of liberated ions*

WW, WS and SW amphoterics may not always yield the expected number of hydrogen or hydroxide ions. The circumstances in which this may occur are as follows. A WW amphoteric is represented for simplicity as NH^+——COO^-.

1. Acidic column, WW in zwitterionic or anionic (fully deprotonated) form, WS in zwitterionic form. The liberated proton may be captured by a carboxylate group and so be retained in the resin. For example,

$$\text{Resin—SO}_3\text{H} + \text{NH}^+\text{——COO}^- \rightarrow \text{Resin—SO}_3\text{NH——COO}^- + \text{H}^+$$

$$\text{NH}^+\text{——COO}^- + \text{H}^+ \rightarrow \text{NH}^+\text{——COOH}$$

   In the case of the WW in the anionic form, it is actually the accompanying cation that displaces a hydrogen ion from the resin, not the amphoteric itself. The end result is the same.

   The hydrogen ion cannot be lost when the amphoteric is in the fully protonated form.

2. Basic column, WW in zwitterionic or cationic (fully protonated) form, SW in zwitterionic form. The liberated hydroxide ion may capture the proton on a weakly basic nitrogen atom. For example:

$$\text{Resin—N(CH}_3)_3\text{OH} + {}^-\text{OOC——NH}^+ \rightarrow \text{Resin—N(CH}_3)_3\text{OOC——NH}^+ + \text{OH}^-$$

$$\text{NH}^+\text{——COO}^- + \text{OH}^- \rightarrow \text{N——COO}^- + \text{H}_2\text{O}$$

   In the case of the WW in the cationic form, it is actually the accompanying anion that displaces a hydroxide ion from the resin, not the amphoteric itself. The end result is the same.

   The hydroxide ion cannot be lost when the amphoteric is in the fully deprotonated form.

   The amphoteric molecule that captures the hydrogen or hydroxide ion may be the one that liberated it or another still in solution.

There seems to be no published information about the extent to which these losses of hydrogen and hydroxide ions occur. The analyst should be aware of them, and regard the results of the analysis of column effluents with some suspicion when an amphoteric is present. If the identity of the amphoteric is known, run control experiments with the amphoteric alone, to quantify exactly what happens.

An obvious way of guarding against such losses is to add a known excess of acid to the test solution before passing it through a cation-exchange resin, or a known excess of alkali before passing it through an anion exchanger. It is then necessary to run control experiments, by titrating further samples of the test solution with the same amounts of added acid and alkali, to allow for possible neutralisation of the added reagent by other components of the mixture.

References

1. Schmitt, T. M. *Analysis of Surfactants*, Surfactants Science Series Vol. 40, Marcel Dekker Inc., 1992, pp. 137–145.
2. *Handbook of Sorbent Extraction Technology*, Varian Sample Preparation Products, Catalog No. 1224-5005.
3. MacDonald, L. S., Cooksey, B. G., Ottaway, J. M. and Campbell, W. C. *Anal. Proc.*, **23** (1986), 448–451.
4. Voogt, P. *Recl. Trav. Chim. Pays-Bas Belg.*, **78** (1959), 899.
5. Draguez de Hault, E. *Proc. 2nd World Surfactants Congress*, 1988, Group C, pp. 67–91 (published by the Comité Européen des Agents de Surface et leurs Intermédiaires Organiques (CESIO), Paris).
6. Milwidsky, B. M. and Gabriel, D. M. *Detergent Analysis: A Handbook for Cost-Effective Quality Control*, 2nd edn, Micelle Press, 1982, pp. 270–273.

5 Analysis of anionics

5.1 Introduction

This chapter deals with the analysis of some individual anionic surfactants as raw materials and in some formulated products, or obtained as fractions separated by ion exchange or otherwise. It is assumed that materials obtained from unknowns will have been identified by spectroscopy or other means.

Anionic raw materials do not usually contain more than one surfactant species, and the determination of total active matter is usually straightforward, provided that the molecular weight is known. Anionics in formulated products can be determined without separation provided that no other ingredient interferes. Determination of mixtures with other surfactants, including mixtures of anionics, is covered in chapter 8.

Fractions obtained by anion exchange almost always contain other salts, as illustrated by the following examples.

1. Fractions obtained by removing cationics and amphoterics on a strongly acidic cation-exchange resin and neutralising the effluent contain salts of the anions of the cationics, and possibly anions of non-surfactant salts that were not removed by the initial extraction from the sample.
2. Fractions obtained by eluting from anion-exchange columns contain all the exchangeable anions from the resin. Hydroxide ions from a strongly basic resin eluted with hydrochloric acid emerge as water, and the excess acid is volatile, but if such an acidic eluate has been neutralised before evaporation it contains sodium or other metal chloride. Chloride ions from a weakly basic hydrochloride resin eluted with ammonia appear as ammonium chloride, which may be the chief constituent of the dried residue.
3. If the active has been converted to the free acid by passage through a strongly acidic cation exchanger, retained on a weak free base resin and then eluted with ammonia, the eluate contains the ammonium salts of any non-surfactant anions that were present in the sample.

The active matter can usually be isolated from such mixtures by drying and extracting with dry ethanol or propanol, but whether or not this is done, it will usually be necessary to determine the active by direct analysis. Section 5.2 gives some general methods, and the remainder of the chapter deals with methods for specific surfactant types.

All volumetric calculations required in this and subsequent chapters are performed by using the general equation given in section 1.4.1, and only those requiring some manipulation are written out in full.

5.2 General methods

5.2.1 Two-phase titration with benzethonium chloride

This is the most widely used method for total anion-active content, and was fully described in section 3.5. It can be used for raw materials, most formulated products, and without extraction for fractions from ion-exchange separations.

1. Weigh a sample containing about 1.5 g of anionic active. Dissolve in water, dilute to 1 l and mix.
2. Pipette 20 or 25 ml into a 100 ml stoppered measuring cylinder, a stoppered flask or the mechanical apparatus described in ISO 2271.
3. Add 10 ml water, 15 ml chloroform and 10 ml acid mixed indicator (disulphine blue VN and dimidium bromide).
4. Titrate with 0.004 M benzethonium chloride with thorough shaking or stirring after each addition until the pink colour is just discharged from the chloroform layer. If the chloroform layer becomes blue the end-point has been overshot.

If it is necessary to do the titration in alkaline solution, it may be easier to use the bromophenol blue method.

1. At step 3, replace the acid mixed indicator with 1.0 ml 0.04% bromophenol blue and 10 ml 0.1 M sodium hydroxide.
2. Titrate with 0.004 M benzethonium chloride with thorough shaking or stirring after each addition until the water layer is clear and colourless.
3. Carry out a blank titration on 25 ml water.

The blank titration should be 0.3–0.4 ml.

5.2.2 Potentiometric titration with benzethonium chloride

Construction of electrodes and experimental procedure were fully described in section 3.6. It is not recommended for fatty ester α-sulphonates or carboxylates. Whilst α-olefin sulphonates give good curves, it is uncertain which species is titrated. The method is satisfactory for many formulated products, but should be evaluated for individual applications. It is the method of choice when it is applicable. It is most conveniently done by autotitrator.

Indicating electrode: any of the electrode types described in section 3.6.2, and others, may be used. ASTM Method D 4251-89 [1] recommends the Orion 93-07 electrode, but an earlier version, D 4251-83, recommended the HNU ISE 20-31-00 nitrate electrode. Roth Scientific's RS 5000 electrode is also satisfactory. The Orion 93-05 fluoroborate electrode, which responds to cationics, can also be used.

Reference electrode: calomel or silver–silver chloride.

1. Weigh a sample containing about 3 g of anionic active. Dissolve in water, dilute to 250 ml and mix.
2. Pipette 20 ml into a 150 ml squat-form beaker.
3. Insert the electrodes and a magnetic stirrer bar or a propeller-type stirrer. Add enough water to ensure that the electrode tips are immersed.
4. Titrate slowly with 0.04 M benzethonium chloride, with continuous stirring. Allow the potential reading to become constant after each addition. Proceed in single drops in the region of the end point. Record volume and potential after each addition.
5. Plot a graph of titrant volume against millivolts. The end point is the steepest point of the curve.

5.2.3 Para-*toluidine precipitation/titration method* [2]

This is a general method for anionics, but its main use is for determination of molecular (equivalent) weight. The version given here assumes that the method will be used for that purpose. If the molecular weight is known, the drying and weighing step can be omitted, but it is still advisable to determine and correct for any extracted *p*-toluidine hydrochloride.

The anionic is precipitated as its *p*-toluidine salt, solvent-extracted, dried and weighed. It is then titrated in alcoholic solution with sodium hydroxide, which displaces the weak base.

The extracting solvent can be diethyl ether or carbon tetrachloride (tetrachloromethane). The former is advised in the method given here, because of its lower toxicity. ASTM Method D 4224-89 [3] is similar, but uses a 13.6% solution of the reagent, permits only carbon tetrachloride and directs that the initial extraction should be done by shaking for 10 min. It uses ortho-cresol red (*o*-cresol sulphonephthalein) as indicator, and adds the correction for extracted *p*-toluidine hydrochloride (step 10), which was not part of the original method.

1. Weigh a sample containing about 1 g of anionic into a 250 ml separating funnel.
2. Make just alkaline to phenolphthalein.

3. Extract with 50 ml diethyl ether (caution: flammable) or carbon tetrachloride (caution: toxic vapour). Add solid sodium sulphate a little at a time to break any emulsion. Discard the organic extract.
4. Add 100 ml 3.6% (0.25 M) *p*-toluidine hydrochloride and 50 ml of diethyl ether or carbon tetrachloride.
5. Shake until the solid phase has completely disappeared.
6. Allow the layers to separate and run the aqueous layer into a second separating funnel, or the carbon tetrachloride layer into a tared beaker. Take care to achieve a clean separation.
7. Repeat the extraction with a further 50 ml solvent, discarding the aqueous layer. Do not wash the organic extracts.
8. Evaporate the organic extracts to dryness in a tared beaker, dry in a vacuum oven for one hour at 50°C, cool and weigh.
9. Dissolve the residue in 100 ml hot ethanol and titrate to phenolphthalein with 0.1 M sodium hydroxide.
10. Cool, acidify with 1 M nitric acid and titrate potentiometrically with 0.05 M silver nitrate. Caution: do not use concentrated nitric acid, which reacts violently with ethanol.

The weight of *p*-toluidine hydrochloride in the residue (W_3) is given by:

$$V_2 \times \frac{C_2}{1000} \times 143.6 = W_3 \text{ g}$$

where V_2 = volume of 0.05 M silver nitrate (step 10) and C_2 = exact concentration of silver nitrate in mol/l.
Weight of *p*-toluidine salt of anionic in residue = $W_2 - W_3$ g, where W_2 = weight of residue (step 8).
Volume of sodium hydroxide consumed by *p*-toluidine salt of anionic = $V_1 - V_2C_2/C_1 = V_3$, where V_1 = volume of 0.1 M sodium hydroxide (step 9) and C_1 = exact concentration of sodium hydroxide in mol/l.
The molecular weight of the *p*-toluidine salt of the anionic, M, is given by:

$$1000(W_2 - W_3)/V_3C_1 = M$$

Molecular weight of surfactant in sample = $M + M_c - 108 = M_s$, where M_c = molecular or atomic weight of the original cation.

$$\%\text{ surfactant in sample} = V_3 \times \frac{C_1}{1000} \times M_s \times \frac{100}{W_1}$$

where W_1 = weight of sample (step 1).

5.2.4 Extraction of monosulphonic acids

Many monosulphonates can be extracted from 3 M hydrochloric acid with diethyl ether. Disulphonates and hydrotropes which are low-molecular-weight sulphonates are not extracted. In the case of α-olefin sulphonates the hydroxyalkane sulphonates are converted to sultones and extracted with the monosulphonic acid. The method is applicable to alkylbenzene sulphonates, mono- and dialkylsulphosuccinates and α-olefin sulphonates, but apparently not to alkane sulphonates.

1. Weigh a sample containing about 3 g active.
2. Dissolve in the minimum quantity of water. Transfer to a separating funnel, using sufficient water to bring the total volume to 75 ml.
3. Add 25 ml concentrated hydrochloric acid (see Note 1).
4. Extract with three 50 ml portions of diethyl ether. Keep the aqueous layer for extraction of disulphonic acids if necessary.
5. Combine the ether extracts and wash with two 25 ml portions of 3 M hydrochloric acid. Add the acid washes to the main aqueous extract.
6. Evaporate almost to dryness in a tared beaker (caution: flammable vapour) (see Note 2).
7. Add 25 ml diethyl ether and evaporate almost to dryness. Repeat (see Note 3).
8. Dissolve the residue in 50 ml water and titrate to phenolphthalein with 1 M sodium hydroxide.
9. Evaporate to dryness, dry to constant weight and weigh (see Note 4).

   Note 1. The volumes of water and acid are not critical provided that the final concentration of acid is not less than 3 M.

   Note 2. Stopping evaporation just short of dryness prevents decomposition of the highly acidic extracted monosulphonic acids.

   Note 3. The repeated evaporation with ether removes extracted hydrochloric acid. If necessary, determine chloride on the final residue and correct, but the correction will be negligible if the method has been carried out correctly.

   Note 4. The evaporation and weighing is to permit determination of the molecular weight. It can be omitted if this is already known.

5.2.5 Extraction of neutral oil

Neutral oil or neutral fatty matter is the nonionic fraction of an anionic surfactant. It is the fraction obtained by extraction under such conditions that the anionic itself is not extracted. It contains the unsulphated alcohol or unsulphonated hydrocarbon or other starting material, together with sultones, sulphones and possibly other trace constituents. This is an abbreviated version of the procedure given in section 3.1.2. The sample size may be varied. Other variations include the use of different volumes of the solvents or of other hydrocarbon extractants, or five extractions instead of four. The method is applicable to virtually any anionic except alkylether and alkylphenylether sulphates.

1. Weigh a 5 g sample, dissolve it in 50% ethanol and transfer to a separating funnel with the same solvent.
2. Make just alkaline to phenolphthalein with 1 M sodium hydroxide.
3. Extract with four 100 ml portions of petroleum ether.
4. Combine the petroleum ether extracts and wash with 100 ml of 50% ethanol.
5. Evaporate the petroleum ether extract almost to dryness in a tared beaker. Complete the evaporation under an air jet without heating. Dry to constant weight at room temperature in a vacuum desiccator.

The same basic procedure can be performed in an extraction column [4], using a much smaller sample.

5.3 Alkyl, alkylether and alkylphenylether sulphates

These materials are readily determined by two-phase titration (ISO 2271) or potentiometric titration with benzethonium chloride. On acid hydrolysis they yield the corresponding alcohol, a sulphate ion and a hydrogen ion, and this affords three additional approaches—determination of the increase in acidity, of the amount of fatty alcohol or ethoxylated alcohol liberated, and of the sulphate ion. The experimental procedure may be varied within limits; for example ISO 2870 [5] and ASTM D 1570-89 [6] differ with respect to the identity of the acid used, duration of boiling, choice of indicator and other details. The following procedures are similar to both of those standard methods. The choice of indicator is immaterial for these particular compounds, but is of crucial importance in some other cases.

5.3.1 Determination of acidity liberated on acid hydrolysis

1. Take 5 g of sample in a boiling flask, dissolve in 50 ml water, heating if necessary, and neutralise with 1.0 M hydrochloric acid or sodium hydroxide to phenolphthalein.
2. Add exactly 50 ml 1.0 M aqueous hydrochloric acid, heat on a steam bath under reflux for 30 min, then boil under reflux for a further 3 h.
3. Cool and back-titrate to phenolphthalein with 1.0 M sodium hydroxide.
4. Carry out a parallel experiment without the sample.

The difference between the two titrations measures the liberated hydrogen ions.

5.3.2 Extraction of fatty alcohol

1. Transfer the hydrolysed solution quantitatively to a separating funnel. If the solution contains unhydrolysed surfactants, e.g. alkylbenzene sulphonate, dilute with an equal volume of ethanol.
2. Extract three times with 50 ml petroleum ether.
3. Combine the petroleum ether extracts and wash with 25 ml water or 50% aqueous ethanol.
4. Re-extract the water or aqueous ethanol with 25 ml petroleum ether and add this to the main extract.
5. Transfer the extract quantitatively to a tared beaker and evaporate to a low volume on a steam bath (caution: flammable vapour). Complete evaporation under a jet of air without heating.
6. Add 15 ml dry acetone, swirl to dissolve and evaporate to a low volume on the steam bath (caution: flammable vapour). Complete evaporation under a jet of air without heating. Repeat twice more.
7. Dry to constant weight in a vacuum desiccator, or in a vacuum oven at a temperature not exceeding 50°C.
8. Determine the percentage of unsulphated fatty alcohol in the sample.

It is important to avoid unnecessary heating during the evaporation to avoid loss of lower-molecular-weight alcohols. The alcohol may be examined in any appropriate way, e.g. by gas–liquid chromatography to determine its chain length distribution and mean molecular weight.

$$\% \text{ active} = P \times \frac{(M_a + 79 + M_c)}{M_a} \times \frac{100}{W}$$

where P = weight of extracted fatty alcohol, corrected for unsulphated alcohol present before hydrolysis, M_a = molecular weight of fatty alcohol, M_c = atomic or molecular weight of the cation of the active, and W = weight of sample.

5.3.3 Extraction of alcohol and alkylphenol ethoxylates

If no other surfactants are present, transfer the hydrolysed solution quantitatively to a separating funnel. Proceed as described for fatty alcohols, replacing the petroleum ether with chloroform.

If other surfactants are present, proceed as follows.

1. Dilute the hydrolysed solution with four times its own volume of methanol.
2. Add 40 ml (apparent volume) of a mixed-bed ion-exchange resin (see section 4.4.7) previously washed in 80% methanol.
3. Stir for 15 min and filter into a tared beaker.

4. Wash the resin and the filter with several small volumes of 80% methanol, collecting them in the beaker.
5. Evaporate to a low volume on a steam bath (caution: flammable vapour).
6. Complete as described for fatty alcohols.

The calculation is as for alkyl sulphates.

5.3.4 Determination of sulphate ion

This can be done gravimetrically or volumetrically. In either case it is necessary to do a parallel determination on a sample that has not been hydrolysed, unless the hydrolysed sample had been rendered free of sulphate before hydrolysis, e.g. by extraction with dry ethanol or propan-2-ol.

Transfer the hydrolysed solution after extraction of the fatty alcohol or ethoxylate to a 500 ml volumetric flask, dilute to volume and mix. Use water for this dilution if the gravimetric procedure is to be followed, or acetone for the volumetric procedure. Continue as follows. The volumetric procedure is preferred if other surfactants are present.

A. Gravimetric procedure

1. Pipette 25 or 50 ml of the diluted solution into a 250 ml beaker.
2. If the hydrolysed solution has been neutralised, add 10 ml 1.0 M hydrochloric acid.
3. Heat to boiling and add 20 ml 10% barium chloride solution.
4. Cover with a watch glass and leave on a steam bath for 2 h.
5. Filter through a dry tared sintered silica crucible, ensuring that the precipitate is quantitatively transferred.
6. Wash the precipitate several times with water, ensuring that the whole of the inside of the crucible is washed.
7. Dry as far as possible under vacuum, then dry at 105°C for 30 min.
8. Heat in a muffle furnace at 600°C for 15 min, cool to room temperature and weigh.

$$\%\ \text{active} = P \times \frac{M}{233.4} \times \frac{100}{W} \times \frac{500}{V}$$

where P = weight of precipitate, M = molecular weight of active, W = weight of sample, and V = volume taken at step 1.

B. Volumetric procedure

1. Pipette 20 ml of the diluted solution into a 100 ml stoppered cylinder.
2. Dilute to 80 ml with acetone.

3. Add 1 ml of a 1% solution of dithizone in acetone.
4. Titrate with 0.05 M lead (II) nitrate solution. The colour changes from blue–green to vermilion at the end-point.

If the solution is not a clear blue–green initially, re-distil the acetone.

5.3.5 Extraction of active matter from alkylether and alkylphenylether sulphates

ISO 6843 [7] deals with the determination of the molecular weight of ethoxylated alcohol and alkylphenol sulphates in raw materials. It includes the following procedure for the extraction of active matter with a mixture of ethyl acetate and butan-1-ol in the ratio 9:1 by volume. This procedure is useful in itself and may be of more general use for extracting surfactants from aqueous solution.

1. Dissolve a sample containing up to 25 mmol, or about 10 g, of active matter in 50 ml 1.0 M sodium chloride in a separating funnel.
2. Extract with an equal volume of the ethyl acetate–butanol mixture.
3. Run the aqueous layer into a jacketed separating funnel maintained at 60°C.
4. Add 5 g solid sodium chloride and shake until dissolved.
5. Re-extract with 50 ml of the same mixed solvent.

Extraction of the active matter is nearly quantitative, and polyglycols, polyglycol sulphates and unsulphated ethoxylates are not extracted. The reader is strongly recommended to refer to the Standard for full details in context.

The Standard does not make clear why the use of a jacketed separating funnel is necessary, or how much inorganic matter is extracted. The purpose of the extraction is to isolate and determine the polyglycols, etc., mentioned above.

5.4 Monoalkanolamide and ethoxylated monoalkanolamide sulphates

5.4.1 Titration with benzethonium chloride and hydrolysis

These materials can be titrated with benzethonium chloride, either by ISO 2271 or potentiometrically.

The sulphate group is easily hydrolysed, but prolonged boiling with strong acid is necessary to hydrolyse the amide group. The products of acid hydrolysis are fatty acid, a monoalkanolamine or ethoxylated monoalkanolamine, a sulphate ion and a hydrogen ion:

$$RCONHCH_2CH_2OSO_3^- + 2H_2O \rightarrow RCOOH + NH_3^+CH_2CH_2OH + SO_4^{2-}$$

$$NH_3^+CH_2CH_2OH + OH^- \rightarrow NH_2CH_2CH_2OH + H_2O$$

However, direct titration of the liberated acidity is not really practicable, because of the high level of acid reagent. Also, the liberated hydrogen ion is neutralised by protonation of the alkanolamine, and titration to bromocresol green or methyl orange (ASTM method D 1570, ref. 6) would not measure it. On the other hand, titration to phenolphthalein (with alkali in alcoholic solution) would titrate not only that proton but the liberated fatty acid as well. It is therefore necessary to extract the fatty acid and to determine fatty acid and alkanolamine separately.

5.4.2 Determination of fatty acid and alkanolamine after acid hydrolysis

1. Weigh 5 g sample in a round-bottomed 250 ml flask.
2. Add 50 ml 6 M hydrochloric acid and heat on a steam bath for 30 min. Attach a reflux condenser and boil for at least 8 h.
3. Cool the hydrolysed solution and transfer it quantitatively to a 250 ml separating funnel.
4. Extract the fatty acid with three 100 ml portions of petroleum ether. Keep the aqueous layer.
5. Combine the petroleum ether extracts in a 500 ml separating funnel and wash with 50 ml water. Add the water to the main aqueous layer from step 4.
6. Evaporate the petroleum ether extract to dryness in a tared beaker (caution: flammable vapour), dry, cool and weigh.
7. If the molecular weight of the fatty acid is required, dissolve in neutral ethanol and titrate to phenolphthalein with 1.0 M ethanolic sodium hydroxide.
8. Evaporate the aqueous layer to a volume of about 10 ml, cool and add 100 ml ethanol.
9. Neutralise to phenolphthalein with 1.0 M ethanolic sodium hydroxide and titrate to bromophenol blue with 1.0 M ethanolic hydrochloric acid to determine the alkanolamine.

$$\%\ \text{fatty acid} = W_2 \times \frac{100}{W_1}$$

where W_2 = weight of fatty acid (step 6), and W_1 = weight of sample.

The molecular weights of the alkanolamines are 61 for monoethanolamine and 75 for monoisopropanolamine.

Sulphate may be determined if required on the titrated aqueous layer, following either of the procedures described in section 5.3.4.

5.5 Mono- and diglyceride sulphates

These materials yield fatty acid, glycerol, a sulphate ion and a hydrogen ion on acid hydrolysis, and soap and glyceryl sulphate on alkaline hydrolysis.

5.5.1 Determination of acidity liberated on acid hydrolysis

Use the same method as for alkyl sulphates etc. (section 5.3.1), but use methyl red or bromocresol green as indicator, to avoid titrating the liberated fatty acid. Alternatively, extract the fatty acid with petroleum ether before titrating. The identity of the indicator will then have a negligible effect on the result.

An alternative is to do duplicate hydrolyses and a blank, titrating one solution and the blank to bromocresol green and the other to phenolphthalein. For the titration to phenolphthalein, dilute the solution with four times its own volume of ethanol and use ethanolic alkali for the titration. If the three titrations are V_b (blank), V_{bg} and V_{pp}, $(V_{bg} - V_b)$ measures the liberated sulphuric acid and $(V_{pp} - V_{bg})$ the fatty acid.

5.5.2 Extraction of fatty acid after acid or alkaline hydrolysis

Proceed as for extraction of fatty alcohol (section 5.3.2). There is no need to avoid heating the extracted fatty acid as these are essentially involatile. The fatty acid may be dissolved in ethanol and titrated with alcoholic alkali to phenolphthalein or converted to its methyl ester and examined by GLC.

After alkaline hydrolysis, acidify with hydrochloric or sulphuric acid and then proceed as described.

5.5.3 Determination of sulphate

Proceed as for alkyl sulphates, section 5.3.4, using the aqueous layer from section 5.5.1, above.

If both sulphate and glycerol are to be determined, dilute the solution to a known volume and take suitable aliquots.

5.5.4 Determination of glycerol

Polyhydric alcohols with hydroxyl groups on three or more adjacent carbon atoms are oxidised by periodate, yielding aldehydes from the end groups and formic acid from the middle groups. One molecule of glycerol thus yields two molecules of formaldehyde and one molecule of formic acid.

$$CH_2OHCHOHCH_2OH + 2IO_4^- \rightarrow 2HCHO + HCOOH + 2IO_3^-$$

Glycerol can be determined by measuring the formaldehyde or formic acid produced or the periodate consumed. In this method the formic acid is measured. The weight of glycerol present must not exceed 0.4 g.

1. Transfer the aqueous layer from the extraction of fatty acid (or an aliquot) to a stoppered conical flask and neutralise to phenol red with 1.0 M sodium hydroxide.
2. Add about 1 ml 1 M hydrochloric acid and boil for 1 min (caution: flammable vapour will be given off) to expel carbon dioxide.
3. Cool and accurately neutralise with 0.1 M sodium hydroxide.
4. Add 50 ml 5% sodium periodate solution, stopper the flask, swirl to mix and place it in a dark cupboard for 30 min.
5. Add 5 ml ethane-1,2-diol (ethylene glycol), replace the stopper, swirl, and leave in the dark for a further 20 min.
6. Titrate with 0.1 M sodium hydroxide.

The amount of sodium periodate added is sufficient to oxidise just over 0.5 g of glycerol, which would give a titration of about 58 ml. If the titration exceeds 50 ml repeat the analysis on a smaller sample.

This method is similar to that described in BS 2621-5 (1979) [8], now superseded.

5.6 Alkylbenzene sulphonates

5.6.1 Desulphonation

Carbon–sulphur bonds are not hydrolysed by boiling with dilute mineral acids, but are broken by boiling with phosphoric acid of boiling point 215°C [9]. The sample size depends on the use to be made of the hydrocarbon, but typically 1–5 g is used. The hydrocarbon is distilled off and examined as required. Recovery of the hydrocarbon is not quantitative.

The phosphoric acid is prepared by boiling ordinary concentrated acid in an open beaker until the boiling point reaches 215°C.

1. Weigh a sample into a 500 ml distillation flask.
2. Add 200 ml phosphoric acid, boiling point 215°C.
3. Attach a Dean and Stark receiver for use with a heavy entrainer (Figure 5.1) and a reflux condenser.

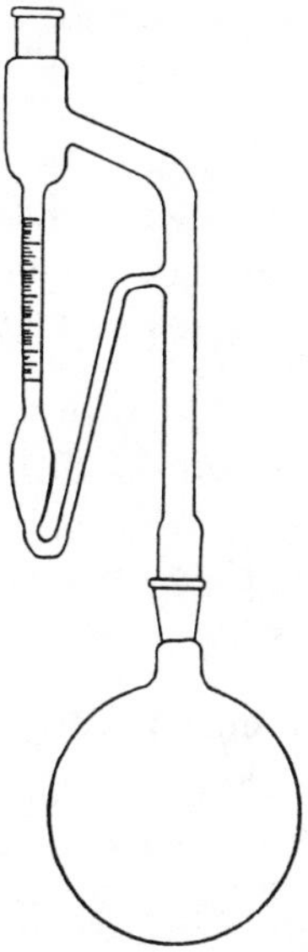

Figure 5.1 Dean and Stark distillation apparatus with trap for use with heavy entraining solvent.

4. Heat to boiling in a heating mantle and distil for 30 min. Read the volume of hydrocarbon in the receiver.
5. Continue distillation, reading the volume of hydrocarbon at 15-min intervals until two consecutive readings are equal.
6. Allow to cool, maintaining the flow of water through the condenser.
7. Transfer the contents of the receiver to a 100 ml separating funnel, using petroleum ether to rinse out the hydrocarbon, and extract with three 25 ml portions of petroleum ether (caution: flammable).
8. Combine the petroleum ether extracts and wash with 25 ml water.
9. Evaporate the petroleum ether almost to dryness. Complete the evaporation under a current of air without heating. Dry to constant weight in a vacuum desiccator without heating.

The hydrocarbon may be examined, for example, by gas chromatography.

5.6.2 Determination of sulphuric acid in alkylbenzene sulphonic acids

The method described by Sak-Bosnar *et al.* [10] depends on the fact that the second hydrogen ion of sulphuric acid is significantly weaker than the first, and than the hydrogen ion of alkylbenzene sulphonic acid (and others). It is therefore possible to determine sulphuric acid in alkylbenzene sulphonic acids by potentiometric titration in a non-aqueous medium.

1. Dissolve about 0.2 g of the mixed acids in 25 ml of a 1:1 mixture of acetonitrile (caution: toxic) and ethylene glycol.
2. Titrate potentiometrically through two end-points with 0.1 M triethylamine in methanol, with continuous stirring.

The mixed solvent was clearly the best of seven that were evaluated, and tolerates up to 10% water with only slight deterioration in the sharpness of the end-points.

ASTM Method D 4711-89 [11] gives a long-established method which depends on the same principle.

1. Dissolve about 0.5 g of the mixed acids in 100 ml methanol.
2. Titrate potentiometrically through two end-points with 0.1 M cyclohexylamine, with continuous stirring.

In both methods the titration to the first end-point measures the alkylbenzene sulphonic acid plus one hydrogen ion of the sulphuric acid. The additional titration to the second end-point measures the second hydrogen ion of the sulphuric acid.

There are several variations on this idea, dating back at least to 1959. Solvents include ethanol and 50% methanol–benzene. Tetrabutylammonium hydroxide has been used as titrant. None of these variations appears to offer any advantage over the two methods quoted.

It would be advantageous to use larger samples for materials with low sulphuric acid contents.

5.6.3 Determination of short-chain sulphonates

It is sometimes necessary to determine sodium (or potassium) toluene, xylene or cumene sulphonate present as a hydrotrope in products containing alkylbenzene sulphonates. This can be done by removing the alkylbenzene sulphonate and determining the short-chain sulphonate by ultraviolet spectroscopy. BS 3762, Part 3.10 [12], describes a method for toluene and xylene sulphonates. A similar method can also be used for cumene sulphonate.

1. Extract the alkylbenzene sulphonic acid as described in section 5.2.4. The aqueous extracts contain the short-chain sulphonate.
2. Combine all the aqueous extracts and washes and evaporate to dryness.
3. Dissolve the residue in water and neutralise to pH 4.5 with 2.5 M sodium hydroxide.

4. Evaporate to dryness, add 5 ml acetone and evaporate to dryness (caution: flammable vapour). Repeat.
5. Add 50 ml 50% aqueous ethanol and heat on a steam bath for 10 min.
6. Filter through a glass sinter and wash the filter and insoluble matter with 50% aqueous ethanol.
7. Evaporate the filtrate to dryness, dissolve the residue in water and dilute to 100 ml.
8. Measure the absorbance in 1 cm silica cells, using a pure short-chain sulphonate to calibrate. A suitable concentration is 0.02–0.04%.

Absorption maxima are found at the following wavelengths: toluene sulphonate: 220, 256, 262 and 268 nm; xylene sulphonate: 220, 272 and 278 nm; cumene sulphonate: 231 and 260 nm.

The method can therefore be used to differentiate between short-chain sulphonates as well as to determine them.

An alternative method of separation is to sorb all the anions on a strongly basic anion exchanger hydroxide and to elute the short-chain sulphonates with at least 30 bed-volumes of 1 M aqueous hydrochloric acid. If soap is present, elute the fatty acids first with at least 12 bed-volumes of 1 M ethanolic acetic acid, followed by a wash with 5 bed-volumes of water.

5.7 Alkane sulphonates

Alkane sulphonates contain both mono- and disulphonates. The monosulphonates can be determined by direct titration with benzethonium chloride according to ISO 2271, and this is the procedure recommended in ISO 6121 [13], but see the comment at the end of section 5.7.3.

This section describes methods for total alkane sulphonates, extraction and determination of mean molecular weight of alkane monosulphonates and separation and determination of alkane mono- and disulphonic acids.

5.7.1 Determination of total alkane sulphonate

In this method the non-surfactant matter, chiefly sodium sulphate, is precipitated by adding an organic solvent and removed by filtration. The procedure is similar to that described in ISO 6122 [14].

1. Take a sample containing 0.5–1.0 g total active matter.
2. Dissolve in 15 ml 20% sodium sulphate solution.
3. Add 150 ml of a mixture of three parts by volume of butan-1-ol and two parts of acetone. Stir thoroughly and allow to settle.
4. Filter. Wash the filter and residue with the same mixed solvent.
5. Evaporate to dryness, dry at 105°C for 1 h, cool and weigh.

6. Determine the chloride content of the residue by potentiometric titration with silver nitrate and correct the observed weight.

$$\% \text{ total alkane sulphonate} = 100W_2/W_1$$

where W_2 = corrected weight of residue (step 6), and W_1 = weight of sample.

5.7.2 Determination of mean molecular weight of alkane monosulphonates

Alkane monosulphonates cannot be extracted from an acid medium by a simple separating-funnel technique because they are not sufficiently soluble in hydrocarbon solvents, the distribution coefficient between pentane and aqueous alcohol being only 0.08–0.09. Extraction with diethyl ether, which is effective for other monosulphonates, has apparently not been reported, but may be practicable.

The method given here was originally described by Kupfer *et al.* [15] and has subsequently been adopted as an International Standard [16].

1. Weigh a sample containing 0.6–0.8 g alkane monosulphonate in a 150 ml beaker.
2. Dissolve in 50 ml 50% v/v ethanol and add 20 ml concentrated hydrochloric acid, S.G. 1.18 g/ml.
3. Transfer the solution quantitatively to a 300 ml liquid–liquid extractor (Figure 5.2), using 30 ml 50% ethanol to effect the transfer.
4. Fit the extractor to a 500 ml round-bottomed flask containing about 400 ml petroleum ether, boiling range 40–60°C, and attach a reflux condenser.
5. Heat the petroleum ether to boiling on a water bath and extract for 5 h at a distillation rate of about 1.5 l/h.
6. Evaporate the extract to dryness on a steam bath under a stream of nitrogen or in a rotary evaporator.
7. Dissolve the residue in 50 ml 95% v/v ethanol and pass the solution through a column containing a strongly acidic cation-exchange resin with 2% cross-linking, in the free acid form, in 95% ethanol. Wash the column with 100 ml 95% ethanol.
8. Titrate the combined effluent to bromophenol blue with carbonate-free 0.1 M sodium hydroxide (see Note 1).
9. Evaporate to dryness as in step 6.
10. Dry for 1 h in a vacuum oven at 120°C, cool and weigh.
11. Repeat step 10 until two successive weighings differ by not more than 1 mg.
12. Calculate the mean molecular weight.

Note 1: preparation of carbonate-free 0.1 M sodium hydroxide: weigh 8.1 g sodium hydroxide pellets into a 250 ml conical flask fitted with a

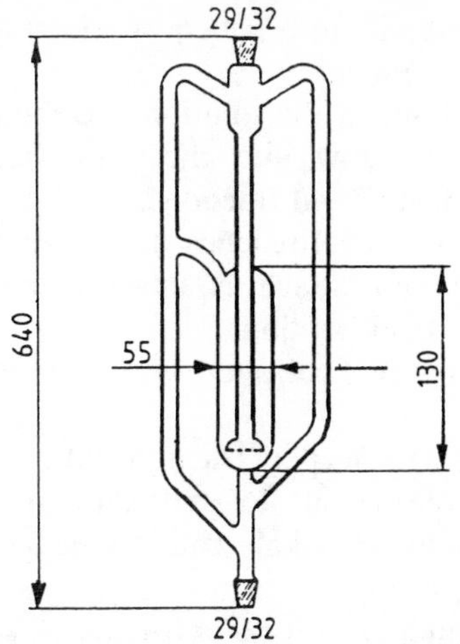

Figure 5.2 Continuous liquid–liquid extractor; dimensions are in millimetres (reproduced by kind permission of the International Organization for Standardization, Geneva).

soda-lime guard tube. Dissolve in 100 ml 95% ethanol. Allow to cool. When the insoluble matter has settled out, pipette (caution: caustic) 10 ml of the clear supernatant into a 200 ml volumetric flask, dilute to volume with freshly boiled and cooled water and mix. Standardise against potassium hydrogen phthalate using phenolphthalein.

5.7.3 Separation and determination of alkane mono- and disulphonates

The first two methods described here both involve sorption of the sample on a column followed by successive elution of the disulphonates and monosulphonates, in that order.

The first method is that of Scoggins and Miller [17]. The mixed acids are adsorbed on a cross-linked polystyrene resin and eluted separately. The authors successfully separated and determined four different mono-/disulphonic acid pairs and one mono-/dicarboxylic acid pair, so the method appears to be very widely applicable. The composition of the aqueous eluent and the volume required to elute the disulphonic acid must be determined by experiment. The volume of methanol required to elute the monosulphonic acid is 100–150 ml.

The method as described is for dodecane and tetradecane mono- and disulphonic acids. Minor variations are necessary for the other species studied by the authors.

1. Prepare a 25 cm × 1.2 cm bed of Amberlite XAD-2 resin in a glass column. The dimensions are not critical. Wash the resin with 5–6 bed-volumes of water.

2. Dissolve a sample containing 2–3 meq of mixed alkanesulphonic acids in water and dilute to 100 ml.
3. Pipette a suitable volume of the solution into the column (see Note 1).
4. Allow the solution to flow into the resin bed at 1.5–2.0 ml/min, collecting the effluent in 10-ml fractions.
5. As soon as the solution level reaches the top of the resin bed, fill the column with water and attach a solvent reservoir full of water. Continue collecting 10 ml fractions.
6. Titrate each fraction as it is collected with sodium hydroxide (see Note 2).
7. When the alkali titration per fraction falls to zero (after 10–12 fractions), change the eluent to methanol and continue collecting 10-ml fractions. Analyse by alkali titration, or by dilution and titration with benzethonium chloride.
8. When elution is complete (after 5–6 fractions), flush the column with water. It is then ready for re-use.
   Note 1. The authors do not define a 'suitable volume', but 50 or 100 ml would appear to be reasonable.
   Note 2. The authors do not specify either the concentration of the alkali or the indicator to be used, but 0.01 M and bromocresol green would be appropriate.

The above method is applicable to the sulphonic acids. Neutral sulphonates may be converted to the acids by passing them through a strongly acidic cation exchanger. The first of the two columns used in the following method is effective.

A more convenient method is separation by ion exchange [18]. The sample is extracted with ethanol to remove sulphate. The solution is passed through a strongly acidic cation-exchange resin to convert the sulphonates to sulphonic acids, and then through a weakly basic anion exchanger in the free base form, on which mono-, di- and any polysulphonates are retained. The di- and polysulphonates are eluted first with a wholly or largely aqueous eluent, and the monosulphonates with an eluent in which the solvent is 60% propanol. The recommended ion exchangers are Bio-Rad AG50W-X8 and Sephadex DEAE A 25, a diethylaminoethyl substituted cellulose. Bio-Rad AG50W-X4 and a dimethylamino-substituted weakly basic resin might offer advantages.

1. Set up two columns in series, the upper one (external diameter 1.7 cm, depth of resin 10 cm) containing the cation exchanger and the lower containing 6–8 meq of the anion exchanger in the free base form, both in ethanol.
2. Dissolve a sample of the alkanesulphonate in ethanol, and if it is not completely soluble, filter and wash with ethanol.
3. Pass the solution through the columns, discarding the effluent.

4. Wash the columns with three 30 ml portions of ethanol.
5. Separate the columns and wash the anion exchanger with 50 ml water.
6. Elute the anion exchanger with 1 M ammonium bicarbonate solution in water or 30% methanol. The composition of the solvent and the volume required vary with the identity of the sulphonate (see Note 1).
7. Elute with 0.3 M ammonium bicarbonate in 60% propanol (see Note 2).
8. Evaporate the eluates from steps 6 and 7 separately to dryness and heat until free from ammonium bicarbonate.
9. Determine the disulphonate content of the first eluate and the monosulphonate content of the second (see Note 3).

   Note 1. For octadecanedisulphonate use 30% methanol. For lower homologues use water. Use 100 ml eluent for dodecanedisulphonate. For higher homologues use 500 ml.

   Note 2. Use 350 ml eluent for dodecanemonosulphonate. For higher homologues use 500 ml.

   Note 3. For determination of the recovered mono- and disulphonic acids the authors recommended passing the pure ammonium salts obtained at step 8 through the cation exchanger and titrating with potassium hydroxide to bromocresol purple. The monosulphonates but not the disulphonates can also be determined by titration with benzethonium chloride.

A simpler method is butanol extraction of the aqueous layer from the extraction of alkane monosulphonates (section 5.6.2).

1. Quantitatively transfer the acidic aqueous ethanolic layer from the liquid–liquid extractor to a separating funnel, using water to effect the transfer.
2. Extract with three 50 ml portions of butan-1-ol.
3. Combine the butan-1-ol extracts and wash with 50 ml 3 M hydrochloric acid.
4. Evaporate the butan-1-ol extracts almost to dryness. Add 5 ml diethyl ether and evaporate to dryness under an air jet without heating. Repeat.
5. Dissolve the residue in 50 ml neutral ethanol and titrate with 0.1 M sodium hydroxide.
6. Evaporate to dryness in a tared beaker, dry for 1 h at 105°C, cool and weigh.
7. Dissolve the residue in water, acidify with 1 M nitric acid and titrate the chloride potentiometrically with 0.1 M silver nitrate.
8. Correct the weight of the residue (step 6) for the chloride content and calculate the sodium alkane disulphonate content of the sample.

ISO 6121 states without qualification that the monosulphonate content of mixed alkane mono- and disulphonates can be determined by two-phase titration with benzethonium chloride (ISO 2271). This is in agreement with the findings of Scoggins and Miller [17], who state that disulphonates do not titrate by ASTM Method D 1681-62, which uses the Epton procedure (section 3.5.4) with cetyltrimethylammonium bromide as titrant and methylene blue as indicator, but they refer to disulphonates as 'hyamine inactive', so it is not clear which titrant they actually used.

Kupfer *et al.* [15] state that pure disulphonates do not titrate, but that they do influence the results of titrations on monosulphonates, their consumption of titrant approaching 1 mol/mol at low disulphonate contents. These authors titrated with dodecyldimethylbenzylammonium chloride, but did not specify indicator and procedure.

Lew [19] has shown that 2-octadecylbutane-1,4-disulphonate is titrated by benzethonium chloride using the Epton (methylene blue) procedure, consuming 2 mol/mol.

The evidence is apparently inconclusive, but the truth is probably that whether disulphonates can be titrated with quaternary ammonium surfactant titrants or not, and, if they can, the number of mol of titrant they consume per mol, depend primarily on their molecular weight, and perhaps to a minor extent on the titrant, the indicator and the titration procedure. If in doubt, check by experiment.

5.8 Alpha-olefin sulphonates

5.8.1 Composition

Commercial α-olefin sulphonates are complex materials, containing the following constituents in appreciable proportions:

2,3-olefin sulphonate	$R{-}CH{=}CH{-}CH_2{-}SO_3^-$
3,4-olefin sulphonate	$R{-}CH{=}CH{-}CH_2{-}CH_2{-}SO_3^-$
3-hydroxyalkane sulphonate	$R{-}CHOH{-}CH_2{-}CH_2{-}SO_3^-$
4-hydroxyalkane sulphonate	$R{-}CHOH{-}CH_2{-}CH_2{-}CH_2{-}SO_3^-$
1,3-sultone	$R{-}CH{-}CH_2{-}CH_2$ (CH bonded to O, CH_2 bonded to SO_2, O—SO_2)
1,4-sultone	$R{-}CH{-}CH_2{-}CH_2{-}CH_2$ (CH bonded to O, CH_2 bonded to SO_2, O—SO_2)
'Disulphonate' (sulphatosulphonate)	$R{-}CH_2{-}CH(OSO_3^-){-}CH_2{-}SO_3^-$

Typically the total olefin sulphonate content is 60–70%, the hydroxyalkane sulphonate content 25–30% and the disulphonate content up to 5% or more. The 2,3-olefin sulphonate predominates over the 3,4 and the 3-sultone over the 4. The sulphonic acids produced by SO_3 sulphonation of α-olefins contain up to 30% sultones and no hydroxyalkane sulphonate, but subsequent alkaline hydrolysis converts the former to the latter, and the neutralised product contains little sultone. There may also be a little alkane sulphonate, from saturated hydrocarbon present as impurity in the parent olefin, and a little 2-hydroxyalkane sulphonate.

5.8.2 Titration with benzethonium chloride

Two-phase titration of α-olefin sulphonates by the ISO 2271 procedure is straightforward, and the end-point is usually but not always fairly sharp. The bromophenol blue procedure (section 3.5.4) is rather better, but may give significantly higher results. Potentiometric titration with benzethonium chloride yields good-quality curves, but with at least some samples the result does not agree with that found by either of these two-phase methods. The magnitude of the disagreement can be quite serious, up to 5% relative or even more. The three techniques may measure different things, and there is good evidence showing that the sulphatosulphonate ('disulphonate') is not titrated. There is therefore an element of uncertainty in defining the species titrated and determining its/their mean molecular weight.

In setting a raw-material specification it may be adequate to specify the active content in millimol per 100 g, determined by an arbitrarily chosen titration method. In other cases it is necessary to determine at least olefin sulphonate and hydroxyalkane sulphonate separately.

5.8.3 Determination of total sulphonate

This procedure measures the total matter soluble in propan-2-ol and its apparent mean molecular weight. It is assumed that the product is the sodium salt; if it is the free mixed acids, it must first be neutralised to phenolphthalein with sodium hydroxide. The results give no information about the composition of the propan-2-ol solubles, but the method could be useful for quality control of raw materials.

1. Dry a 5 g sample in a 250 ml beaker, add 100 ml dry propan-2-ol, cover with a watch-glass and boil gently on a steam bath for 30 min.
2. Cool and filter into a tared beaker. Wash the filter with propan-2-ol, collecting the washings in the same beaker as the filtrate.
3. Evaporate to dryness on a steam bath (caution: flammable vapour), dry at 105°C for 1 h, cool and weigh. Calculate the percentage of propan-2-ol-soluble matter.

4. Dissolve the dried residue in 50 ml 80% methanol.
5. Pass the solution through a column containing 40 ml of a strongly acidic cation-exchange resin with 4% cross-linking, in the free acid form, in 80% methanol.
6. Collect the effluent and wash the column with five bed-volumes of 80% methanol, and then with a further 3 bed-volumes, collecting the second washings in a separate receiver.
7. Titrate effluent and washings with 1.0 M sodium hydroxide to bromocresol green. Calculate the mean molecular weight.

The titration on the second washings (step 6) should be no more than a few drops.

An alternative procedure is to pass the 80% methanol solution of the propanol solubles through a weakly basic anion exchanger hydrochloride, elute with ammonia, evaporate to dryness and dry to constant weight, then to determine the NH_4^+ and chloride contents of the residue.

5.8.4 Determination of alkene monosulphonates and hydroxyalkane sulphonates

Alkene monosulphonates may be extracted as the sulphonic acids from strongly acid solution and determined by titration with alkali. Under these conditions the hydroxyalkane sulphonates are converted back to the sultones from which they were derived, and although these are also extracted, they are not titrated by alkali. Alternatively the alkene sulphonates in the extract may be titrated with benzethonium chloride. The hydroxyalkane sulphonates may be determined by alkaline hydrolysis followed either by measurement of the alkali consumed or by a second titration with benzethonium chloride. The following method is adapted from those of Ranky and Battaglini [20] and Martinsson and Nilsson [21]. The alkali used for hydrolysis of the sultones is alcoholic simply to prevent excessive foaming.

1. Dissolve a sample containing about 1–2 g active matter in 50 ml water in a 250 ml separating funnel.
2. Add 50 ml 6 M hydrochloric acid and extract with three 50 ml portions of diethyl ether (caution: flammable).
3. Combine the ether extracts and wash with 25 ml 3 M hydrochloric acid.
4. Evaporate the ether extract to about 10 ml.
5. Add 50 ml ethanol and titrate to phenolphthalein with 0.1 M ethanolic sodium hydroxide. Calculate the percentage of alkene sulphonate.

Then either

6. Add 50 ml 0.1 M ethanolic sodium hydroxide, attach a reflux condenser and boil for 1 h.

7. Cool and titrate with 0.1 M hydrochloric acid.
8. Carry out a blank experiment on 50 ml of the 0.1 M alkali.

$$\% \text{ hydroxyalkane sulphonate} = (V_b - V_s) \times \frac{C}{1000} \times M \times \frac{100}{W}$$

where V_b = volume of 0.1 M acid used by blank (step 8), V_s = volume of 0.1 M acid used by sample, C = exact concentration of acid in mol/l, M = molecular weight of hydroxyalkane sulphonate, and W = weight of sample.

Or

6. Transfer to a 250 ml volumetric flask, dilute to volume with water and mix.
7. Dilute 50 ml to 250 ml in a second volumetric flask.
8. Titrate 20 ml of this solution with 0.004 M benzethonium chloride according to ISO 2271.
9. Pipette a second 50 ml portion of the solution from step 1 into a 250 ml round-bottomed flask.
10. Add 50 ml 0.1 M ethanolic sodium hydroxide, attach a reflux condenser and boil for 1 h.
11. Cool and neutralise to phenolphthalein with 0.1 M hydrochloric acid.
12. Transfer to a 250 ml volumetric flask, dilute to volume with water and mix.
13. Pipette 20 ml of this solution into a titration vessel, add 20 ml water and titrate with 0,004 M benzethonium chloride.

$$\% \text{ alkene sulphonate} = V_1 \times \frac{C}{1000} \times M_1 \times \frac{100}{W} \times \frac{250}{20}$$

$$\% \text{ hydroxyalkane sulphonate} = (V_2 - V_1) \times \frac{C}{1000} \times M_2 \times \frac{100}{W} \times \frac{250}{20}$$

where V_1 = volume of 0.004 M benzethonium chloride (step 8), V_2 = volume of 0.004 M benzethonium chloride (step 13), C = exact concentration of benzethonium chloride, M_1 = molecular weight of alkene sulphonate, M_2 = molecular weight of hydroxyalkane sulphonate, and W = weight of sample.

As a check, the sum of the alkene sulphonates and hydroxyalkane sulphonates in mol/l may be determined on the original sample by titration with benzethonium chloride. The result includes any 2-hydroxyalkane sulphonate, which is not converted to the sultone by the acid treatment described, and so may differ slightly from the sum of the alkene sulphonates and hydroxyalkane sulphonates determined separately.

Total hydroxysulphonates may be determined by measuring the hydroxyl value as described in section 3.4.3.

5.8.5 Determination of disulphonate

1. Extract the monosulphonates as described in section 5.8.4, steps 1–3. Keep the aqueous layers.
2. Extract the dilsulphonate with butan-1-ol and complete the determination as described in the final part of section 5.7.3.

If desired the sulphatosulphonate can be acid-hydrolysed in the same way as an alkyl sulphate and determined by two-phase titration of the resulting hydroxyalkane sulphonate with benzethonium chloride. Any true disulphonates present are not measured.

5.8.6 Determination of sultones

The non-detergent organic matter ('neutral oil', 'unsulphonated matter') obtained by petroleum ether extraction from aqueous ethanol in separating funnels or extraction columns contains sultones, but Martinsson and Nilsson [21] advocate 48 h continuous liquid–liquid extraction with petroleum ether from 50% aqueous ethanol. The sultones plus all other nonionic matter are more easily obtained by deionisation using a mixed-bed resin, but as there is some danger that strongly basic anion exchangers may hydrolyse sultones, it is better to use a strongly acidic cation exchanger in the acid form followed by a weakly basic anion exchanger in the free base form. This is the preferred method.

Sultones at 500 p.p.m. or more may be determined by the following method [21]. At lower concentrations they may be determined by liquid chromatography or colorimetry.

1. Extract the non-detergent organic matter from a 5 g sample.
2. Evaporate the extract to dryness (caution: flammable vapour) in a 100 ml round-bottomed flask.
3. Add 10 ml 0.1 M potassium hydroxide, fit a reflux condenser and boil for 1 h.
4. Neutralise to phenolphthalein with 0.1 M hydrochloric acid.
5. Titrate with 0.004 M benzethonium chloride according to ISO 2271.

Sultones at concentrations from 10 to 500 p.p.m. may be determined as follows [22].

Colour reagent: dissolve 0.35 g methylene blue in 200 ml water in a 1 l volumetric flask. Slowly and with continuous swirling add 6.5 ml concentrated sulphuric acid. Dilute to volume and mix.

1. Evaporate the hydrolysed solution from step 3 of the previous procedure to dryness.
2. Dissolve in water, transfer to a separating funnel and dilute to 85 ml.
3. Add 15 ml 0.1 M sulphuric acid, 5 ml colour reagent and exactly 15 ml chloroform.
4. Shake for 2 min.
5. Allow the layers to separate and run the chloroform into a 50 ml volumetric flask. Dilute to volume with chloroform and mix.
6. Measure the absorbance at 642 nm in a 1 cm cell.

Determine the concentration of sultone in the solution and hence in the sample by reference to a calibration curve prepared using pure sultones. These can be extracted from a sultone-rich α-olefin sulphonate by deionisation, or any of the suggested methods, and standardised by alkaline hydrolysis and benzethonium chloride titration.

5.9 Oil-soluble sulphonates

Alkylnaphthalene sulphonates, petroleum sulphonates and other high-molecular-weight sulphonates, especially salts of alkaline-earth metals, may all be determined by two-phase titration, using ISO 2271 or similar methods, after conversion to the sulphonic acids by ion exchange. Any strongly acidic cation exchanger is effective, but the solvent may be a problem. Fialko and Balashova [22] advocate benzene and 96% ethanol, 7:3. The subsequent determination may be done by titration with a base or a quaternary ammonium salt. Markó-Monostory and Börzsönyi [23] used potentiometric titration with tetrabutylammonium hydroxide or two-phase titration with Septonex (carboxypentadecyltrimethylammonium bromide) using dimethyl yellow (0.002% in 0.1 M hydrochloric acid containing 2.5% ammonium chloride) as indicator and a 1:1 mixture of chloroform and 1,2-dibromomethane as solvent. The colour change is from pink to yellow and the colour remains in the organic phase throughout, thus overcoming one of the problems experienced with very dark-coloured samples.

5.10 Acyl isethionates

5.10.1 Determination of active content

The active content of acyl isethionates in raw materials and products is determined either by titration with benzethonium chloride or by saponification. The fatty acid may be extracted after acidification for molecular-weight determination, etc. The only likely surfactant impurity is free fatty acid or soap, which is determined by extraction with petroleum ether from acid solution.

5.10.2 Determination of sodium isethionate

Unreacted sodium isethionate is the only other impurity likely to be present in significant amounts. It is determined by determining total anions by ion-exchange conversion to the corresponding acids and titration with alkali, then determining the active, soap and chloride contents and finding the isethionate by difference.

1. Prepare an ion-exchange column containing 40 meq of a strongly acidic cation exchanger hydroxide in 80% methanol.
2. Dissolve a sample containing about 5 g active in 50 ml 80% methanol and neutralise to phenolphthalein with methanolic sodium hydroxide.
3. Transfer the solution to the column and allow it to flow through, collecting the effluent in a 500 ml volumetric flask. Wash with five bed-volumes of water, collecting the effluent in the same flask. Dilute to volume and mix.
4. Wash the column with a further three bed-volumes of water.
5. Titrate 100 ml of the solution obtained at step 3 to bromocresol green with 0.1 M sodium hydroxide.
6. Titrate the effluent obtained at step 4 with the same titrant.
7. Determine the active content of the sample in mol %.
8. Determine the soap or fatty acid content of the sample in mol %.
9. Determine the chloride content of the sample in mol %.

From the titration at step 5 calculate the total anion content in mol %:

$$\% \text{ sodium isethionate} = 148([\text{An}^-] - [AI] - [S] - [\text{Cl}^-])$$

where square brackets denote concentration in the sample in mol %, An^- means total anions, AI means acyl isethionate and S means soap.

The titration at step 6 should be no more than two or three drops, and if so it may be ignored. If it is more than that but less than 1.0 ml, include it in the calculation as shown. If it is more than 1.0 ml repeat the analysis using slower flow rates.

5.11 Fatty ester α-sulphonates

5.11.1 Composition

Fatty ester α-sulphonates normally contain all of the following species. It is assumed here that the sample is the sodium salt, but the sulphonic acid can be analysed in the same way. R′ is usually methyl but may be ethyl, propyl or butyl.

Sulphonated ester	$RCH(SO_3Na)COOR'$
Sulphonated carboxylate salt (or carboxylic acid)	$RCH(SO_3Na)COONa$ (−COOH)
Soap (or fatty acid)	RCH_2COONa (−COOH)
Unsulphonated fatty ester	RCH_2COOR'
Low-MW sodium alkyl sulphate (R′ = methyl to butyl)	$R'OSO_3Na$

The unsulphonated ester and the fatty acid can be extracted with petroleum ether. The low-molecular-weight alkyl sulphate is alkali-hydrolysable, which affords a method for its determination if required.

When the sulphonated ester is alkali-hydrolysed, not only does the ester group hydrolyse, as expected, but the sulphonate group is slowly removed as well. Saponification is therefore not a useful way of determining the main active. Acid hydrolysis destroys the ester group but has no effect on the sulphonate group.

There are several approaches to the complete analysis of raw materials. The following range of methods gives the analyst a degree of choice. There are similar schemes based on the same principles, e.g. that of Battaglini *et al.* [24]. The analysis of formulated products probably calls for only the determination of the active and the chief impurity, i.e. the sulphonated carboxylate salt.

5.11.2 Determination of unsulphonated ester and soap

1. Weigh a 5 g sample in a beaker, dissolve in 50 ml neutral ethanol and titrate to phenolphthalein with 0.1 M sodium hydroxide. Transfer to a separating funnel, using 50 ml water.
2. Extract with three 50 ml portions of petroleum ether (caution: flammable). Keep the lower layer.
3. Combine the petroleum ether extracts and wash with 50 ml water. Add the water to the lower layer from step 2.
4. Evaporate the petroleum ether extract almost to dryness on a steam bath. Complete the evaporation under an air jet without heating. Add 5 ml acetone and repeat the evaporation. Dry to constant weight in a vacuum desiccator at room temperature. The residue is the unsulphated ester.
5. Acidify the combined lower layers from steps 2 and 3 with 1 M

hydrochloric acid and extract with three 50 ml portions of petroleum ether.
6. Combine the petroleum ether extracts and wash with 50 ml water.
7. Evaporate the petroleum ether extracts to a low volume in a tared beaker.
8. Add 50 ml neutral ethanol and titrate to phenolphthalein with 0.1 M ethanolic sodium hydroxide.
9. Evaporate to dryness (caution: flammable vapour), dry for 1 h at 105°C, cool and weigh. The residue is the sodium soap.

If the molecular weight of the fatty acid is known, steps 2, 3 and 4 may be omitted. The residue at step 9 then contains the ester (and therefore must not be heated) and the sodium soap, and the latter can be found by calculation from the titration at step 8.

5.11.3 Determination of sulphonated surfactant species

Titration with benzethonium chloride in acid solution measures the sulphonated ester plus the sulphonated carboxylic acid, only the sulphonate group of the latter being titrated. In alkaline solution the titration measures the sulphonated ester plus twice the sulphonated carboxylate, both the sulphonate and the carboxylate group being titrated. Alpha-sulphonated esters give poor potentiometric titration curves, and two-phase titration is strongly preferred.

If the titration in alkaline solution is done without pretreatment, the soap will be titrated as well. The following method extracts the fatty acid to eliminate this complication.

1. Dissolve a sample containing 1.0–1.5 g active in 50 ml water and transfer to a separating funnel using 50 ml ethanol.
2. Just acidify with 1 M hydrochloric acid and extract with three 50 ml portions of petroleum ether. Collect the lower layer in a 1 l volumetric flask.
3. Combine the petroleum ether extracts and wash with 50 ml water. Collect the washings in the 1 l flask. Discard the petroleum ether extracts, unless they are required for the determination of ester and fatty acid.
4. Neutralise the contents of the volumetric flask to phenolphthalein with 1 M sodium hydroxide, dilute to volume and mix.
5. Pipette a 20 ml portion into a titration vessel, add acid mixed indicator and chloroform and titrate with 0.004 M benzethonium chloride according to ISO 2271.
6. Pipette a second 20 ml portion into a titration vessel, add 10 ml 0.1 M sodium hydroxide, 1.0 ml 0.04% bromophenol blue solution and 25 ml chloroform, and titrate with 0.004 M benzethonium chloride until the

upper layer is clear and colourless. Carry out a blank titration on 25 ml water.

The titration at step 5, V_1, measures the sulphonated ester plus the sulphonate group of the carboxylate salt. The titration at step 6, V_2, measures the sulphonated ester plus both the sulphonate and the carboxylate group of the salt. Hence $2V_1 - V_2$ measures the α-sulphonated ester and $V_2 - V_1$ measures the α-sulphonated carboxylate salt.

5.11.4 Determination of α-sulphonated ester

The ester cannot be determined by saponification because this slowly removes the sulphonate group as well as hydrolysing the ester group. It can be determined by acid hydrolysis followed by measurement of the α-sulphonated carboxylate salt produced, either by potentiometric titration of the weak acid or by two-phase titration with benzethonium chloride. Both procedures measure the α-sulphonated ester plus the unsulphonated ester. If the latter is present at a significant level, it can be determined (section 5.11.2) and corrected for.

1. Weigh a sample containing about 5 g active in a round-bottomed flask.
2. Add 25 ml 3 M aqueous hydrochloric acid and heat on a steam bath for 30 min. Attach a reflux condenser and boil for 2 h. Cool.

Then either (procedure 1):

3. Add 100 ml neutral ethanol and either titrate potentiometrically with 0.5 M ethanolic sodium hydroxide through end-points at pH about 3.0 and 7.7, or make just alkaline to phenolphthalein with ethanolic 3 M sodium hydroxide and titrate with 0.5 M ethanolic hydrochloric acid, first to phenolphthalein and then to bromophenol blue.
4. Carry out the procedure described in section 5.11.5, using a 5 g sample.

Or (procedure 2):

3. Transfer the hydrolysed solution to a 500 ml volumetric flask. Rinse the boiling flask with 30 ml 3 M sodium hydroxide and add this to the volumetric flask. Complete the transfer, dilute to volume and mix.
4. Dilute 25 ml of this solution to 500 ml in another volumetric flask and mix.
5. Pipette 25 ml of the final dilution into a titration vessel and titrate by the bromophenol blue method as described in section 5.11.3, step 6.
6. Take a second 5 g sample, dilute as in steps 3 and 4 and titrate as in step 5.

$$\%\ \alpha\text{-sulphonated ester (procedure 1)} = \left[\frac{V_1 C_1}{1000 W_1} - \frac{V_2 C_2}{1000 W_2}\right] \times 100M$$

where V_1 = volume of titrant used between end-points at step 3, C_1 = exact concentration of this titrant in mol/l, V_2 = volume of titrant used between end-points at step 5 of the procedure of section 5.11.5, C_2 = exact concentration of this titrant in mol/l, M = molecular weight of α-sulphonated ester, W_1 = weight of sample taken at step 1, and W_2 = weight of sample used for procedure of section 5.11.5.

% α-sulphonated ester (procedure 2)

$$= \left[\frac{V_1}{W_1} - \frac{V_2}{W_2} \right] \times \frac{C}{1000} \times \frac{500}{25} \times \frac{500}{25} \times 100M$$

where V_1 = volume of titrant used at step 5, V_2 = volume of titrant used at step 6, C = exact concentration of this titrant in mol/l, M = molecular weight of α-sulphonated ester, W_1 = weight of sample taken at step 1, and W_2 = weight of sample taken at step 6.

The analyst may wish to use both procedures, using the solutions obtained in the acid–base titrations for the benzethonium chloride titrations.

5.11.5 Determination of total unesterified carboxylate

The combined sulphonated carboxylic acid and fatty acid or their salts are determined by potentiometric acid–base titration, after oxidation of sulphite/sulphur dioxide.

1. Weigh about 7.5 g of the well-mixed sample in a 200 ml beaker or autotitration vessel.
2. Add 20 ml ethanol and stir until dissolved. If the sample is the free sulphonic acid, add about 25 ml carbonate-free 1 M ethanolic sodium hydroxide and stir.
3. Add about 3 ml 30% hydrogen peroxide.
4. Add 1 M ethanolic hydrochloric acid or carbonate-free 1 M ethanolic sodium hydroxide to adjust the pH to between 1.5 and 2.0.
5. Titrate potentiometrically with carbonate-free 0.2 M sodium hydroxide. There are end-points around pH 3 and 7.7.

The titration between the end-points measures the total unesterified carboxylate groups.

The fatty acid can be determined separately and corrected for. This then provides an alternative route to the sulphonated carboxylate salt.

5.11.6 Determination of low-molecular-weight alkyl sulphate

Battaglini *et al.* [24] showed that the methyl sulphate ion, $CH_3OSO_3^-$, was almost completely hydrolysed by boiling for 30 min in 0.5 M potassium hydroxide in ethylene glycol. They did not report results on ethyl, propyl or butyl sulphate, and results on any of these would need to be confirmed by the technique of boiling two samples for different lengths of time, since the sulphate group is increasingly resistant to alkaline hydrolysis as the chain length increases. The ester group of the active must also be at least partially hydrolysed, and a correction is made for this. The version given here uses acid–base titration on two sampless, one hydrolysed and one not hydrolysed, to determine the amount of alkali consumed by the low-molecular-weight alkyl sulphate.

1. Weigh a sample containing about 3 g active in a round-bottomed flask.
2. Add exactly 50 ml 0.5 M potassium hydroxide in ethylene glycol, attach a reflux condenser and boil for 30 min. Run a blank experiment alongside.
3. Cool and titrate first to phenolphthalein and then to bromophenol blue with 0.5 M hydrochloric acid.
4. Carry out a parallel experiment on a similar sample but omitting the boiling.

The amount of alkali needed to hydrolyse the low-molecular-weight alkyl sulphate (LMWAS) and saponify the sulphonated and unsulphonated esters in 1 g of sample, A, is given by:

$$\frac{V_q}{W_2} - \frac{V_p}{W_1} = A,$$

where V_p = the acid titration to phenolphthalein on the hydrolysed sample, V_q = the acid titration to phenolphthalein on the unhydrolysed sample, W_1 = weight of hydrolysed sample (step 1), and W_2 = weight of unhydrolysed sample (step 4).

The amount of alkali needed to saponify the sulphonated and unsulphonated esters in 1 g of sample is equal to the amount of acid needed to protonate the —COO^- groups produced by saponification of the esters, B, is given by:

$$\frac{V_r}{W_1} - \frac{V_s}{W_2} = B$$

where V_r = the acid titration to bromophenol blue on the hydrolysed sample, V_s = acid titration to bromophenol blue on the unhydrolysed sample, and W_1 and W_2 are the sample weights as before.

The amount of alkali used in hydrolysing the LMWAS = $A - B$, and:

$$\% \text{ LMWAS} = (A - B) \times \frac{C}{1000} \times 100M$$

where M = molecular weight of LMWAS, and C = exact concentration of the 0.5 M hydrochloric acid in mol/l. The molecular weight of sodium methyl sulphate is 134.

The volume of alkali consumed by the LMWAS in a sample of the size used is small; if the sample contains 85% active and 5% sodium methyl sulphate, the volume used is 2.6 ml. It is necessary to use a large excess of alkali to allow for the amount used in saponifying the esters.

Alternatively, the correction for esters can be done by titrating the hydrolysed and unhydrolysed samples with benzethonium chloride in alkaline solution. The bromophenol blue method is recommended.

5.12 Mono- and dialkylsulphosuccinates

5.12.1 Composition

Sulphosuccinate surfactants are of two main types, namely monoesters of C_{12}–C_{18} alcohols and diesters of C_6–C_8 alcohols. Monoesters of ethoxylated alcohols and monoalkanolamides also exist. Both types may be synthesised by addition of sodium sulphite or ammonium bisulphite across the double bond of a mono- or diester of maleic acid. In the latter case the product may contain small quantities of mono- or dialkyl aspartate.

$NH_3^+CH_2COOR$
$\quad\quad |$
$\quad CH_2COO^-$

Monoalkyl aspartate

NH_2CH_2COOR
$\quad\quad |$
$\quad CH_2COOR'$

Dialkyl aspartate

Monoalkylaspartates are amphoteric and dialkylaspartates are weakly basic. They therefore behave as cationics in acid solution, and interfere in two-phase titrations if the carbon chains are long enough.

Monoesters may contain low concentrations of diester, and vice versa. Small amounts of unesterified fatty alcohols are usually present. They are determined by extraction with petroleum ether. Particular care must be taken to avoid losses due to the volatility of low-molecular-weight alcohols, and evaporation of extracts must be completed under an air jet without heating.

5.12.2 Two-phase titration of anion-active matter and aspartate

Mono- and dialkylsulphosuccinate may be determined by two-phase titration with benzethonium chloride. Materials made by the sodium bisulphite route can be titrated in acid solution and give a good end-point. Materials made by the ammonium bisulphite route contain aspartate, and are better titrated in alkaline solution.

Aspartate and anion-active content can be determined simultaneously by the following procedure, which uses two different concentrations of benzethonium chloride.

1. Dissolve a 3–4 g sample in water, transfer quantitatively to a 250 ml volumetric flask, dilute to volume and mix.
2. Pipette 25 ml into a 100 ml beaker. Insert a surfactant-sensitive electrode and a reference electrode. Add 10 ml M hydrochloric acid. If necessary add water until the electrode tips are immersed.
3. Titrate potentiometrically with 0.04 M benzethonium chloride with continuous stirring.
4. After the end-point has been reached, add 5 ml 0.004 M sodium dodecyl sulphate.
5. Titrate potentiometrically with 0.004 M benzethonium chloride to just past the end-point. Add 10 ml 2 M sodium hydroxide and continue titrating with 0.004 M benzethonium chloride until the end-point is again reached.

$$\% \text{ anion-active matter} = \frac{(V_1C_1 + V_2C_2)}{1000} \times M_1 \times \frac{100}{W} \times \frac{250}{25}$$

$$\% \text{ aspartate} = V_2 \times \frac{C_2}{1000} \times M_2 \times \frac{100}{W} \times \frac{250}{25}$$

where V_1 = volume of 0.04 M benzethonium chloride, C_1 = exact concentration of 0.04 M benzethonium chloride in mol/l, V_2 = volume of 0.004 M benzethonium chloride added between the end-points in acid and alkaline solution, C_2 = exact concentration of 0.004 M benzethonium chloride in mol/l, W = weight of sample, and M_1 and M_2 are the molecular weights of the alkylsulphosuccinate and aspartate respectively. If the aspartate content is 1% of the alkylsulphosuccinate content, V_2 will be one-tenth of V_1.

5.12.3 Determination of monoalkyl sulphosuccinate, alkaline hydrolysis and extraction of fatty alcohols

The carboxylate group of monoalkylsulphosuccinates is determined by acid–base titration. Both mono- and dialkyl sulphosuccinates are readily hydrolysed by alkalis, and the hydrolysis and subsequent extraction of fatty alcohols are performed by standard procedures. It is essential to stop the evaporation at step 6 short of dryness because the fatty alcohols, particularly those of low molecular weight, are quite volatile.

1. Weigh about 2.5 g sample in a round-bottomed flask and dissolve in 50 ml ethanol.
2. Make the solution acid to bromophenol blue with ethanolic 1 M hydrochloric acid.
3. Titrate with 0.5 M ethanolic sodium hydroxide, first to bromophenol blue and then to phenolphthalein. The volume of titrant between the end-points measures unesterified carboxyl groups plus aspartate, if present.
4. Add 50.0 ml 0.5 M ethanolic sodium hydroxide, attach a reflux condenser and boil for 1 h. Run a blank experiment alongside on 50 ml ethanol and 50.0 ml alkali.
5. Cool and titrate with 0.5 M ethanolic hydrochloric acid to the phenolphthalein end-point.
6. Evaporate to less than 10 ml on a steam bath (caution: flammable vapour). Avoid evaporating to dryness.
7. Add 50 ml water and 50 ml diethyl ether (caution: flammable), stopper the flask and shake.
8. Transfer the mixture quantitatively to a 250 ml separating funnel.
9. Shake the separating funnel vigorously and allow the layers to separate. Run off the lower layer and extract the aqueous layer with two more 50 ml portions of diethyl ether.
10. Combine the ether extracts, wash with 50 ml water and evaporate to a low volume on a steam bath (caution: flammable vapour). Complete the evaporation under a jet of air without heating.
11. Dry with 10 ml acetone, completing the evaporation without heating. Dry to constant weight in a vacuum desiccator at room temperature.

The alkali consumed measures total ester, which includes monoester, diester and aspartate, if present.

The fatty alcohols may be used, e.g., for determination of chain-length distribution by gas chromatography.

5.12.4 Isolation and determination of mono- and dialkyl aspartates

This is an alternative to the procedure given in section 5.12.2. The mono- and dialkylsulphosuccinates are sorbed on a strongly basic anion exchanger chloride. The eluate contains the aspartate and free fatty alcohols. If desired, the fatty alcohols can be removed by extraction with petroleum ether from acid solution. The aspartate is determined by titration with sodium dodecyl sulphate.

1. Prepare a column containing about 40 meq of a strongly basic anion exchanger chloride in 80% methanol.
2. Dissolve a sample containing about 5 g active in 100 ml 80% methanol.
3. Pass the solution through the column at a flow rate not exceeding 2 ml/min/cm^2. Wash the column with three bed-volumes of 80% methanol. Collect the effluent in a beaker.
4. Evaporate the effluent to dryness.
5. Dissolve the residue in 10 ml 0.1 M hydrochloric acid.
6. Optional: transfer to a separating funnel. Rinse the beaker with petroleum ether and add this to the separating funnel. Extract twice with petroleum ether. Discard the petroleum ether extracts.
7. Quantitatively transfer the hydrochloric acid solution to a two-phase titration vessel.
8. Titrate with 0.004 M sodium dodecyl sulphate using acid mixed indicator (cf. ISO 2871-2).

One mol of aspartate reacts with 1 mol of mono- or dialkylsulphosuccinate when the latter is titrated with benzethonium chloride in acid solution. One mol of aspartate consumes 1 mol of acid when titrated, and/or 2 mol of alkali during alkaline hydrolysis.

5.12.5 Separation of mono- and dialkylsulphosuccinates

This is not very often required, but is straightforward. Dialkylsulphosuccinates are soluble in propanol. Monoalkylsulphosuccinates are not. The method of separation is obvious.

5.13 Sulphonated amides

In principle a large number of sulphonated amide surfactants can exist, but only acyl taurates and methyltaurates and sulphosuccinamates are of practical interest.

All may be determined by two-phase or potentiometric titration with benzethonium chloride. Titration of sulphosuccinamates in acid solution measures the sulphonate group alone (but see section 5.13.3), and titration in alkaline medium measures the carboxylate group as well (cf. section 5.11.3). Sulphosuccinamates, possessing a carboxylate group, can also be determined by potentiometric acid–base titration.

Possible surface-active impurities are soap in the taurine derivatives and fatty amine in sulphosuccinamates. They are easily determined by extraction with petroleum ether or by two-phase titration in acid and alkaline solution. Soap may also be determined by potentiometric acid–base titration.

5.13.1 Determination of fatty acid, taurine and methyltaurine

The amide is acid-hydrolysed and the hydrolysis products are separated and determined. Amides resist hydrolysis and require prolonged boiling at high temperature with a high concentration of acid. The fatty acid is extracted and the amino acid determined by acid titration. The same procedure can also be used for sarcosinates.

1. Place 5 g sample in a pressure vessel with 50 ml 6 M hydrochloric or 5 M sulphuric acid, seal and heat in an oven at 160–180°C for 6 h. Allow to cool to room temperature before cautiously opening the pressure vessel.
2. Transfer the solution quantitatively to a separating funnel and extract the fatty acid with petroleum ether. Keep the aqueous layer.
3. Evaporate the petroleum ether extract to dryness. Dry, cool and weigh the fatty acid.
4. Neutralise the aqueous layer to phenolphthalein with 5 M alkali, then make just alkaline. (Caution: the solution will become hot during neutralisation. Add the alkali a little at a time, with continuous cooling.)
5. Titrate potentiometrically with 0.5 M hydrochloric acid to an end-point at a pH around 9, then titrate with the same acid to a pH around 3. Alternatively, titrate first to phenolphthalein and then to bromophenol blue.

The molecular weight of the fatty acid may be determined by titration or gas chromatography. Its iodine value may also be determined, since acyl taurates and methyltaurates are often oleic acid derivatives.

Taurine or methyltaurine consumes one hydrogen ion per mol, but sarcosine consumes two:

$$CH_3NHC_2H_4SO_3^- + H^+ \rightarrow CH_3NH_2^+C_2H_4SO_3^-$$

$$CH_3NHCH_2COO^- + 2H^+ \rightarrow CH_3NH_2^+CH_2COOH$$

5.13.2 Determination of sulphosuccinic acid and fatty amine from alkylsulphosuccinamates

1. Acid hydrolyse as in section 5.13.1.
2. After transferring the solution to a separating funnel, neutralise to phenolphthalein with 5 M sodium hydroxide, then add 10 ml excess. (Caution: the solution will become hot during neutralisation. Add the alkali a little at a time, with continuous cooling.)
3. Extract the fatty amine with petroleum ether. Keep the aqueous layer.
4. Evaporate the petroleum ether extract to dryness, cool and weigh the fatty amine. If required, dissolve the fatty amine in ethanol and titrate

to bromocresol green with alcoholic hydrochloric acid to determine its molecular weight, or analyse by gas chromatography.

5. Neutralise the aqueous layer to phenolphthalein and make just alkaline.
6. Complete as in section 5.13.1, step 5. The sulphosuccinate ion requires two protons.

5.13.3 Determination of soap and free fatty amine

Three approaches are possible, which are briefly described here. Free fatty amine in sulphosuccinamates cannot be determined by potentiometric titration because the carboxylate group is also titrated.

Solvent extraction

1. Dissolve a 5 g sample in 50% aqueous ethanol.
2. For extraction of fatty acid, acidify with 1 M hydrochloric acid. For extraction of fatty amine, make alkaline with 1 M sodium hydroxide.
3. Extract with three 50 ml portions of petroleum ether. Wash the combined extracts with 50 ml 50% aqueous ethanol, evaporate to dryness, cool and weigh.

Two-phase titration

1. Dissolve about 1.5 g sample in water, dilute to 1 l and mix.
2. Titrate two 25 ml portions with 0.004 M benzethonium chloride, one portion using acid mixed indicator and the other using bromophenol blue in alkaline solution.

The titration in acid solution measures acyl taurates or methyltaurates. The titration in alkaline solution measures the same actives plus soap.

Alkyl sulphosuccinamates are slightly more complicated. The titration in acid solution measures the sulphonate group minus the free fatty amine. The titration in alkaline solution measures the sulphonate group plus the carboxylate group, which is equal to twice the sulphonate group. Half of this titration minus the first titration gives the amine.

Potentiometric titration

1. Dissolve a 5 g sample in 100 ml ethanol.
2. Make acid to bromophenol blue with 1 M hydrochloric acid.
3. Titrate potentiometrically with 0.1 M ethanolic sodium hydroxide through end-points at low and high pH.

5.14 Carboxylates

5.14.1 Titration with benzethonium chloride

Soaps, sarcosinates and alkylether carboxylates (ethoxycarboxylates, polyethylene glycol monoesters) give poor curves in potentiometric titrations with benzethonium chloride, and ISO 2271 in alkaline solution does not always work very well for these compounds, particularly those of shorter chain length. The bromophenol blue method works well provided that the solution is distinctly alkaline; it is necessary to add 5 ml 0.1 M sodium hydroxide before titrating.

5.14.2 Solvent extraction

Fatty acids and acyl sarcosines may be extracted with petroleum or diethyl ether from an acidified aqueous solution if no other surfactants are present, or with petroleum ether from acidified 50% ethanol if they are. The extract is evaporated and the residue weighed. Alkylether carboxylic acids cannot be quantitatively extracted with petroleum ether. They can be extracted from aqueous solution with chloroform, but they are best determined by two-phase titration with benzethonium chloride in akaline solution (bromophenol blue method) or direct potentiometric acid–base titration.

Fatty acids can be separated from acyl sarcosines by extraction from 70% propanol with petroleum ether. The fatty acids are extracted. It is necessary to do at least four extractions and three washes with 70% propanol.

5.14.3 Acid–base titration

Carboxylic acids and their salts can be determined by potentiometric titration between end-points at low and high pHs. It does not matter whether an acid solution is titrated with alkali or an alkaline solution with acid. The titration must be done in alcoholic solution.

Soaps and sarcosinates can be determined in aqueous solution by titrating to bromocresol green or bromophenol blue with hydrochloric or sulphuric acid in the presence of petroleum ether or diethyl ether. The solution is shaken after each addition of titrant to extract the liberated fatty acid or acyl sarcosine. Alkylether acids can probably be determined in the same way if chloroform is used as extracting solvent.

5.14.4 Determination of soaps in fatty products

Soaps are strongly basic in glacial acetic acid solution, and may be titrated with hydrogen bromide in the same solvent. Zarins *et al.* [25] determined soaps in various types of fatty products by this method.

1. Dissolve a 2 g sample in 15 ml glacial acetic acid and 5 ml benzene.
2. Titrate with 0.1 M hydrogen bromide in glacial acetic acid, using crystal violet as indicator. The colour change is from violet to blue–green.

Both the titration flask and the burette must be protected from atmospheric moisture with drying tubes. The titrant enters the flask via a Luer tip and a PTFE tube passing through a neoprene stopper. During the titration the solution is stirred with a magnetic stirrer.

5.15 Alkyl and alkylether phosphates (phosphate esters)

5.15.1 Composition and approaches

Commercial phosphate esters are usually mixtures of the free acids, although they may be sodium or triethanolamine salts. Monoesters nearly always contain some diester, and diesters both mono- and triester, with the nominal product predominating. They cannot be hydrolysed by either acids or alkalis at normal temperatures, but acid hydrolysis with strong acid at high temperature may be effective.

Although there is no obvious reason why the mono- and diesters should not be titratable with benzethonium chloride, difficulties due to flocculation have been reported in titrating them according to ISO 2271. There is clearly scope for experimentation, e.g. with lower concentrations, increasing the pH and including some ethanol in the solvent, or by adding excess benzethonium chloride and back-titrating with sodium dodecyl sulphate. Potentiometric titration with benzethonium chloride has not been reported in the literature. The monoester has one weakly acidic and one strongly acidic hydrogen ion, and may behave differently at high and low pH.

Mixtures of mono- and diester and phosphoric acid can be determined by alkali titration.

Analysis of raw materials can be done in several ways, but analysis of formulated products is more difficult.

5.15.2 Determination of total organic phosphate

The sample is ashed to produce chiefly phosphorus pentoxide; this is hydrolysed to orthophosphoric acid, which is titrated with alkali.

1. Dry a sample containing 0.5–1.0 g phosphate esters in a beaker.
2. Add 100 ml dry propan-2-ol, cover with a watch-glass and boil on a steam bath for 30 min, filter, wash and evaporate to dryness in a nickel crucible. Do not use a platinum crucible.
3. Heat gently over a flame, periodically attempting to ignite the vapour. When the vapour ignites, remove the flame and allow the sample to burn. If it goes out, reheat and reignite. When the sample is completely carbonised heat strongly until no further change is apparent, then heat in a furnace at 600°C until a white ash is obtained.
4. Dissolve the ash in a few ml of 0.5 M sulphuric acid, transfer to a round-bottomed flask using a total of not more than 20 ml of the acid, and boil under reflux for 30 min to hydrolyse condensed phosphates.
5. Cool and transfer quantitatively to a 100 ml beaker. Insert pH electrodes and add 1 M sodium hydroxide until the pH is about 3.5. Titrate potentiometrically with 0.1 M sodium hydroxide through two end-points to a pH of about 10. Determine the volume used between the end-points, which are around pH 4.5 and pH 9.2. Alternatively, transfer the acid solution to a conical flask, make just alkaline to phenolphthalein with *c.* 1 M sodium hydroxide and neutralise carefully with 0.1 M hydrochloric acid. Titrate to bromocresol green with the same acid.

The titration at step 5 involves the loss or gain of one hydrogen ion per phosphorus atom.

5.15.3 Conversion to alkyl or alkylether phosphoric acids

The phosphate esters are converted to the free acid form, if they are present as salts, and determined by alkali titration. Monoesters have one weakly acidic and one strongly acidic proton. Diesters are strongly acidic.

Sokolov *et al.* [26] have reported successful analysis of a mixture of pure sodium dodecyl sulphate and pure sodium dodecyl phosphate (mono- or diester not specified) by this method, and Ivanov *et al.* [27] analysed alkyl phosphates by titration of the column effluent with alkali to bromocresol green.

1. Dry a 1 g sample and extract the organic matter by boiling with dry propan-2-ol as in section 5.13.1.
2. Cool and dilute with one-third volume of water.
3. Pass through a column of a strongly acidic cation exchanger in the free acid form and wash the column free of solute, using propan-2-ol or ethanol (about three bed-volumes). Collect the effluent and washings in the same beaker.
4. Titrate potentiometrically with 0.1 mol/l alcoholic alkali through two end-points. The titration to the first end-point, around pH 4, measures

the sum of mono- and diesters. The titration between that and the second end-point, around pH 9, measures the monoester.

5.15.4 Alkali titration

This method was described by Kurosaki *et al.* [28]. Their account lacks detail and the calculations are difficult to follow, but this appears to be a correct version.

1. Weigh a sample containing about 7.5 g of mixed mono- and diester in the free acid form. Dissolve in 100 ml 65% ethanol, transfer to a 250 ml volumetric flask, dilute to volume with the same solvent and mix.
2. Titrate a 100 ml portion potentiometrically with 0.5 M sodium or potassium hydroxide through two end-points.
3. Titrate a second 100 ml portion to the first of these end-points. Add excess silver nitrate (amount not specified; 10 ml of a 10% solution should be sufficient) and titrate through another end point at about pH 6.

At step 2, first end point:
$$ROPO(OH)_2 \rightarrow ROP(OH)O_2^-$$
$$(RO)_2PO(OH) \rightarrow (RO)_2PO_2^-$$
$$H_3PO_4 \rightarrow H_2PO_4^-$$

At step 2, second end point:
$$ROPO(OH)O^- \rightarrow ROPO_3^{2-}$$
$$H_2PO_4^- \rightarrow HPO_4^{2-}$$

At step 3, second end point:
$$ROPO(OH)_2 \rightarrow ROPO_3Ag_2$$
$$(RO)_2PO(OH) \rightarrow (RO)_2PO_2Ag$$
$$H_3PO_4 \rightarrow Ag_3PO_4$$

All the hydrogen ions replaced by silver ions are titrated.

If V_1 is the titration to the first end-point, V_2 the additional titration to the second end-point and V_3 the additional titration to the end-point at step 3, the three titrated species are measured by the following volumes.

Monoester: $2V_2 - V_3$ ml
Diester: $V_1 - V_2$ ml
Phosphoric acid: $V_3 - V_2$ ml

In each case 1 ml titrant represents 0.5 millimol.

5.15.5 Degradation using mixed anhydride reagent

Tsuji and Konishi [29] described the preparation of the acetate of the alcohols in phosphate esters, and the diacetate of the ethylene glycol derived from the ethylene oxide in alkylether phosphates, with subsequent examination by gas chromatography. Their method permits characterisation of both the fatty alcohols and the ethylene oxide chains. They used a reagent consisting of the mixed anhydrides of acetic and *p*-toluenesulphonic acids. The reaction is said to be as follows. Alkyl phosphates behave in the same way.

$$RO(C_2H_4O)_nPO_3^{2-} \rightarrow ROOCCH_3 + nCH_3COOCH_2CH_2OOCCH_3 + H_3PO_4$$

The authors did not say how quantitative this is, but it is potentially an attractive approach, because all three products would be easy to determine by chemical means.

Reagent. Place 120 g *p*-toluenesulphonic acid in a 300 ml round-bottomed flask. Slowly and with stirring, add 80 g acetic anhydride. Attach a reflux condenser and heat at 120°C for 30 min on an oil bath. Cool and store in a brown bottle. The reagent will keep for 1 month.

Procedure

1. Place 0.1 g sample in a 20 ml flask and add 2 g reagent.
2. Heat under reflux at 120°C on an oil bath for 2 h.
3. Cool and transfer to a 100 ml separating funnel.
4. Extract with three or four 20 ml portions of diethyl ether.
5. Combine the ether extracts and wash with several portions of water until the washings are no longer acid to methyl orange.
6. Dry over anhydrous sodium sulphate and evaporate to dryness (caution: flammable vapour).

The authors then analysed the mixed acetate esters by gas chromatography. It would also be possible to determine the phosphoric acid in the water washings, after suitable dilution, by a colorimetric method. There are many published methods, e.g. [30] (see also section 2.5.4), which depend on the formation of molybdophosphoric acid and its subsequent reduction to molybdenum blue.

5.15.6 Determination of phosphate esters in formulated products

The following approaches are all likely to be effective.

1. If no other phosphate is present, determine the total phosphate content of both product and raw material.
2. If no other salt of a weak acid is present, determine the monoester content of both product and raw material by cation exchange and potentiometric titration. If no weak base is present, the cation exchange can be omitted.

3. Determine the triester content, if any, of both product and raw material. The triester is isolated in the nonionic fraction. If other nonionic matter is present, determine the triester from its total phosphate content. This is likely to be less useful for monoesters than for diesters.
4. Analyse by ^{31}P NMR. This is the simplest and best approach, but the apparatus is expensive.

References

1. ASTM Method D 4251-89: Standard Test Method for Active Matter in Anionic Surfactants by Potentiometric Titration. American Society for Testing and Materials, Philadelphia, USA.
2. Marron, T. U. and Schifferli, J. *Ind. Eng. Chem., Anal. Edn,* **18** (1946), 49.
3. ASTM Method D 4224-89: Standard Test Method for Anionic Determination by Para-toluidine Hydrochloride. American Society for Testing and Materials, Philadelphia, USA.
4. Cross, C. K. *JAOCS*, **67** (1990), 142–143.
5. ISO 2870: 1986: Surface active agents—Detergents—Determination of anion-active matter hydrolyzable and non-hydrolyzable under acid conditions. International Organization for Standardization, Geneva.
6. ASTM Method D 1570-89: Standard Test Methods for Sampling and Chemical Analysis of Fatty Alkyl Sulfates. American Society for Testing and Materials, Philadelphia, USA.
7. ISO 6843: 1988: Surface active agents—Sulfated ethoxylated alcohols and alkylphenols—Estimation of the mean relative molecular mass. International Organization for Standardization, Geneva.
8. BS 2621-5, 1979: Specifications for Glycerol. British Standards Institution, Milton Keynes, UK.
9. Knight, J. D. and House, R. *JAOCS*, **36** (1959), 195.
10. Sak-Bosnar, M., Zelenka, Lj. and Budimir, M. *Tenside*, **29** (1992), 289–291.
11. ASTM Method D4711-89: Standard Test Method for Sulfonic and Sulfuric Acids in Alkylbenzene Sulphonic Acids. American Society for Testing and Materials, Philadelphia, USA.
12. BS 3762: Section 3.10: 1989: Analysis of formulated detergents. Method for determination of short-chain alkylbenzene sulphonates. British Standards Institution, Milton Keynes, UK.
13. ISO 6121: 1988: Surface active agents—Technical alkane sulfonates—Determination of alkane monosulfonates content by direct two-phase titration. International Organization for Standardization, Geneva.
14. ISO 6122: 1978: Surface active agents—Technical alkane sulphonates—Determination of total alkane sulphonates content. International Organization for Standardization, Geneva.
15. Kupfer, K., Jainz, J. and Kelker, H. *Tenside*, **6**(1) (1969), 15–20.
16. ISO 6845: 1989: Surface active agents—Technical alkane sulfonates—Determination of the mean relative molecular mass of the alkane monosulfonates and the alkane monosulphonate content. International Organization for Standardization, Geneva.
17. Scoggins, M. W. and Miller, J. W. *Anal. Chem.*, **40**(7) (1968), 1155–1157.
18. Mutter, M. *Tenside*, **5**(5/6) (1968), 138–140.
19. Lew, H. Y. *JAOCS*, **49** (1972), 665–670.
20. Ranky, W. O. and Battaglini, G. T. *Soap Chem. Spec.*, **43**(4) (1968), 36–39 and 78–86.
21. Martinsson, E. and Nilsson, K. *Tenside*, **11**(5) (1974), 249–251.
22. Fialko, M. M. and Balashova, A. N. *Khim. Tekhnol. Topl. Masel*, No. 7 (1973), 23.
23. Markó-Monostory, B. and Börzsönyi, S. *Tenside*, **22**(5) (1985), 265–268.

24. Battaglini, G. T., Larsen-Zobus, J. L. and Baker, T. G. *JAOCS*, **63**(8) (1986), 1073–1077.
25. Zarins, Z. M., White, J. L. and Feuge, R. O. *JAOCS*, **60** (1983), 1109–1111.
26. Sokolov, V. P., Sapelnikova, T. V. and Onishchenko, P. P. *Zavod. Lab.*, **50** (1984), 7–9.
27. Ivanov, V. N., Chaplanova, A. M. and Malakova, T. N. *Otkritiya Izobret.*, **22** (1988), 148–149.
28. Kurosaki, T., Wakatsuki, J., Imamura, T., Matsunaga, A., Furugaki, H. and Sassa, Y. *Comun. Jorn. Com. Esp. Deterg.*, **19** (1988), 191–205.
29. Tsuji, K. and Konishi, K. *JAOCS*, **52** (1975), 106–109.
30. Murphy, J. and Riley, J. P. *Anal. Chim. Acta*, **27** (1962), 31–36.

6 Analysis of nonionics

6.1 Introduction

This chapter deals with the analysis of individual nonionic surfactants. Analysis of raw materials includes not only the determination of the active content, but also of impurities and by-products, and determination of oxyalkylene chain length. Any nonionic can be determined in formulated products by deionisation, which was described in section 4.6.2 and is not further discussed here. Nonionic fractions isolated in this way may have more than one principal active, and some of the methods given here are applicable to the analysis of such mixtures. There are also a few procedures for the determination of nonionics in formulated products.

6.2 Ethoxylated alcohols, alkylphenols and fatty acids

6.2.1 Composition

All ethoxylates contain a range of ethylene oxide (EO) chain lengths, because the product of the addition of one molecule of ethylene oxide is always an alcohol and therefore capable of adding on another molecule of ethylene oxide. Alcohol ethoxylates always contain a disproportionate amount of unreacted alcohol, with the nominal product predominating among the ethoxylated species, although not very markedly. The exact proportions depend on the synthetic process, but a typical dodecanol-3EO contains about 22% unreacted dodecanol and only about 16% of the 3EO compound. In contrast, a typical nonylphenol-3EO contains no unreacted phenol and about 27% of the 3EO adduct [1]. Both species contain minor amounts of polyethylene glycol formed from traces of water in the starting material.

Ethoxylated carboxylic acids may be regarded as monoesters of polyethylene glycol. If made by ethoxylation of the acid, all the carboxylate groups become ethoxylated before further addition of ethylene oxide occurs, so that there is no unreacted acid in the product. However, transesterification can occur between two molecules of the ethoxylate, with production of a molecule of diester and a molecule of polyethylene glycol:

$$2RCO(OC_2H_4)_nOH \rightarrow RCO(OC_2H_4)_nOCOR + H(OC_2H_4)_nOH$$

Such materials therefore usually contain free polyethylene glycol and diester in equimolar proportions. The spread of EO chain lengths is similar to that in alkylphenol ethoxylates. If they are made by esterification of polyethylene glycol, the spread of EO chain lengths obviously depends on the composition of the starting material, and the product always contains unreacted free fatty acid as well as diester and polyethylene glycol [2].

Hydrocarbon and ethylene oxide chain-length distributions are determined by gas or liquid chromatography.

6.2.2 Determination by two-phase titration

Ethoxylates with six or more ethylene oxide units form pseudocationic complexes with many metals, and the tetraphenylborates of these are insoluble or sparingly soluble in water. Several volumetric methods for the determination of ethoxylates are based on this. Some of them are two-phase procedures.

Tsubouchi and Tanaka [3] described the following method, in which the sodium–nonionic tetraphenylborate is extracted into an organic solvent and titrated with a quaternary surfactant, which displaces the sodium–nonionic complex. Anionics do not interfere chemically, but cause emulsification problems and are better removed. Potassium, ammonium, calcium, chloride and sulphate ions do not interfere because the complex is extracted. The aqueous layer remains colourless throughout the titration.

The buffer solution used at step 4 is made from a solution containing 0.3 mol/l sodium dihydrogen phosphate and 0.05 mol/l sodium borate by adding 6 M sodium hydroxide until the pH is 9.0, although the actual pH is not critical between pH 8.5 and 9.5.

1. Place 10 ml of a solution of the sample containing 0.2–2.0 mg nonionic and less than 10^{-4} mol/l anionic surfactant in a separating funnel.
2. Add 2 ml 6 M sodium hydroxide, 5 ml 0.002 M sodium tetraphenylborate (TPB) and 10 ml 1,2-dichloroethane (caution: toxic vapour). Shake vigorously. Allow the layers to separate and filter the dichloroethane through a filter paper.
3. Transfer 5 ml of the filtrate to a stoppered titration vessel.
4. Add 5 ml buffer solution, pH 9, and 2 drops 0.01% Victoria Blue B indicator. Titrate with 2×10^{-4} M tetradecyldimethylbenzylammonium chloride, with thorough shaking after each addition. The colour change is from blue to red in the organic layer.
5. Carry out a reagent blank.

There is no obvious reason why other quaternary surfactants should not be used as titrants.

The method must be calibrated against a pure sample of the nonionic to be determined, but for alkylphenol ethoxylates with between 8 and 40 EO units, 1 mg surfactant requires approximately 5.1 ml titrant, regardless of chain length. The authors do not say why they chose to use such small quantities or such low concentrations.

Tsubouchi *et al.* [4] titrated the potassium complexes of alkylphenol, fatty alcohol and sorbitan monopalmitate ethoxylates with sodium tetrakis(4-fluorophenyl)borate (TFPB) as follows.

1. Place 10 ml of a solution containing 0.3–3.0 mg nonionic in a titration vessel.
2. Add 1–2 drops 0.04% Victoria Blue B indicator, 5 ml 6 M potassium hydroxide and 5 ml 1,2-dichloroethane (caution: toxic vapour).
3. Titrate with 5×10^{-4} M TFPB, shaking thoroughly after each addition. The colour change is from blue to red in the organic layer.

The concentration of titrant chosen was found to be the optimum for clarity of end-point. Anionics do not interfere except by causing emulsification problems if their concentration in the titrated solution exceeds 0.001 M. Sodium, ammonium, aluminium, chloride, nitrate and sulphate do not interfere. Cationic surfactants must be absent. The volume of titrant needed per milligram of nonionic is almost constant at around 3.6 ml. The titrant must be standardised against the nonionic to be determined.

O'Connell [5] modified this method because of the very high cost of the titrant. He also advocated doing the titration in a centrifuge tube so that centrifuging could be used to speed separation. His method is as follows.

1. Prepare the 5×10^{-4} M titrant by vigorously boiling 0.025 g potassium tetrakis(4-chlorophenyl)borate (TCPB) in water, with continuous stirring, cooling and diluting to 100 ml.
2. Place a sample containing 0.3–3.0 mg nonionic in a 50 ml stoppered centrifuge tube.
3. Add exactly 50 μl 0.04% Victoria Blue B indicator, 5 ml 6 M potassium hydroxide and 5 ml 1,2-dichloroethane (caution: toxic vapour).
4. Titrate with 5×10^{-4} M TCPB, shaking thoroughly after each addition. Near the end-point, centrifuge if necessary to obtain an absolutely clear organic layer. The colour change is from blue to red, and the end-point is taken to be the intermediate purple colour.

This variation was successfully used to determine alcohol ethoxylates with 5–15 EO units. Interferences are as for the original version above. The titrant must be standardised against the nonionic to be determined.

6.2.3 Determination by potentiometric titration

A number of authors have described the potentiometric titration with sodium tetraphenylborate (TPB), $NaB(C_6H_5)_4$, of the pseudocationic complexes formed by ethylene oxide condensates with divalent metals whose ionic radius exceeds 100 picometres ($1\,pm = 1 \times 10^{-12}\,m$). The method described here is that of Vytřas *et al.* [6]. It is applicable to ethoxylates containing 4–450 EO units. The electrode is by far the simplest of the several types proposed. The ratio of TPB to divalent metal in the complex formed is constant and has the value 2:1, but the ratio of TPB or divalent metal to nonionic varies considerably with the chain length of the nonionic, and the amount of nonionic titrated by 1 ml of titrant must be determined by experiment for each substance analysed. The end-point is fair to good.

Interferences from other surfactants were not studied, but cationics and amphoterics in acid solution would be expected to interfere quantitatively. Inorganic cations that have insoluble tetraphenylborates would also be expected to interfere, perhaps quantitatively. Anionics would not be expected to interfere.

Reagent. The reagent is 0.01 M sodium tetraphenylborate, pH adjusted to 9 by adding sodium hydroxide. This can be standardised by potentiometric titration against thallium (I) nitrate if desired, but must in any case be calibrated against a pure sample of the nonionic.

Electrodes

Sensing electrode. Aluminium wire coated with a membrane by dipping repeatedly into a solution of 0.09 g PVC and 0.2 ml 2,4-dinitrophenyl octyl ether in 3 ml tetrahydrofuran (caution: flammable; toxic vapour) and allowing the solvent to evaporate between treatments.

Reference electrode. Calomel or silver–silver chloride.

Procedure

1. Prepare a solution of the sample containing 50–100 mg nonionic in 100 ml. Place 20 ml of this solution in a beaker.
2. Add 20 ml 0.1 M (approximately 2%) barium chloride solution.
3. Insert the electrodes and titrate with 0.01 M NaTPB with continuous stirring. The titration is best done with an autotitrator, but can be done manually.
4. Calibrate the titrant against a pure sample of the nonionic, which must be exactly the same as the material being determined, using the procedure exactly as described.

$$\%\ \text{nonionic} = V \times E_{ni} \times \frac{100}{W}$$

where V = volume of 0.01 M NaTPB, E_{ni} = weight in grams of nonionic titrated by 1 ml NaTPB (step 4), and W = weight of sample in titration vessel (= $W' \times 20/V$, where W' = weight taken and V = dilution volume).

Other authors have used larger samples and higher concentrations of TPB, up to 0.1 M, and other types of electrode.

6.2.4 Determination by direct potentiometry

Membrane electrodes based on the TPB salts of barium–nonionic complexes have been used successfully to determine fatty alcohol-7EO condensates in detergent powders by direct potentiometry [7]. The electrodes were similar in construction to the first of the three types described in section 3.6.2. Two types of membrane were used, incorporating nonionics with EO chains ranging from four to 30 units. The most successful was prepared as follows.

1. Mix 100 ml of a 0.01 M aqueous suspension of a nonylphenol-4EO adduct (Antarox CO-430) with 10 ml 0.1 M barium chloride.
2. Add 150 ml 0.01 M NaTPB, with continuous stirring.
3. Whilst still stirring, slowly add 95% ethanol (up to 200 ml, or 75% by volume) until a flocculent white precipitate is formed.
4. Filter the precipitate, wash with 50% ethanol and dry under vacuum at 35°C.
5. Dissolve 0.1 g of the dried precipitate, 0.3 g *o*-nitrophenyl phenyl ether and 0.17 g PVC powder in 6 ml tetrahydrofuran (caution: flammable; toxic vapour).
6. Place a glass ring of internal diameter 30–35 mm with flat ground ends on a glass plate and pour the solution into it. Place a wad of filter paper and a weight of at least 500 g on top of the ring. Alternatively use a Petri dish of the same diameter. Leave for 48 h.
7. Cut discs of the membrane of suitable size for the electrode with a cork borer.
8. Construct an electrode like the first one illustrated in section 3.6.2. The filling solution is 0.1 M barium chloride saturated with silver chloride.

The originators [8] used an ordinary silver–silver chloride electrode fixed in the upper part of a glass sheath. A PVC tube was securely fitted over the lower end, and the membrane was glued on to the end of this, using a solution of PVC in tetrahydrofuran. The electrodes were immersed in water for at least 15 h before use.

The determination of 7EO adducts in both model systems and commercial detergent powders was done by both the method of standard additions and its converse, i.e. by adding small increments of a solution of the analate to a larger volume of a standard solution of nonionic. The results obtained by the two methods were in good agreement. The standard additions method (section 3.7.1) was carried out as follows.

1. Prepare a solution of the test sample containing about 0.015% of the nonionic.
2. Pipette 25 ml into a 100 ml beaker, insert surfactant-sensitive and calomel electrodes and commence stirring.
3. Record the potential when it becomes constant.
4. Add small volumes of a 1% solution of the nonionic, recording the potential when it becomes constant after each addition.
5. Determine the slope of the plot of potential versus log concentration of nonionic and calculate the concentration of nonionic in the product.

The 'small volumes' at step 4 were not specified, but 1 ml would be satisfactory. Once the method is established it would be sufficient to add a single increment of 2–5 ml. The time taken for the potential reading to become constant after each addition of nonionic was about 15 min.

The method is somewhat time-consuming, but it is straightforward, and results on model systems indicate that it should be equally applicable to nonionics of other chain lengths.

6.2.5 Determination by the cobaltothiocyanate colorimetric method

Numerous versions of this colorimetric procedure have been published, using various compositions for the reagent, shaking times and extracting solvents. The method given here is essentially that described by Milwidsky [9]. Only adducts with six or more ethylene oxide units yield the coloured complex, and the colour intensity depends on the length and distribution of the EO chain. Many materials with nominally less than six EO units can be analysed, however, because their EO distribution includes sufficient molecules with six or more EO units. For very long EO chains the calibration curve is linear only at low concentrations, because such compounds have limited solubility in dichloromethane.

The method can be used for raw materials and for products containing anionic surfactants, but not cationics or amphoterics. It is also valid for polypropoxylates.

Colour reagent. Dissolve 30 g cobalt (II) nitrate, $Co(NO_3)_2 . 6H_2O$, 143 g ammonium chloride and 256 g potassium thiocyanate in water, dilute to 1 l and mix. Alternatively dissolve 30 g cobalt (II) nitrate, 200 g potassium chloride and 200 g ammonium thiocyanate in water, dilute to 1 l and mix.

Calibration curve

1. Dissolve 1.000 g of pure nonionic surfactant of the type to be determined in water, ensuring complete dissolution, transfer to a 250 ml volumetric flask, dilute to volume and mix.
2. Titrate 50 ml of this solution to methyl red with 0.1 M hydrochloric acid. Note the volume used and discard the solution.
3. Place a second 50 ml portion in a 100 ml stoppered cylinder, add the volume of acid used by the first portion, dilute to 100 ml and mix. This solution contains 2.000 mg/ml of the nonionic.
4. Measure 20.0 ml dichloromethane (caution: toxic vapour) into each of five 100 ml separating funnels. Add 20.0 ml colour reagent to each.
5. Add 2, 4, 6, 8 and 10 ml of the 2 mg/ml nonionic solution (step 1), plus 18, 16, 14, 12 and 10 ml water respectively.
6. Stopper the funnels and shake each for exactly 1 min. Allow the layers to separate.
7. Place 10 ml dry propan-2-ol in each of five 25 ml volumetric flasks. Run off about 1 ml of each of the dichloromethane extracts, then run the extract into one of the flasks exactly to the mark. Mix.
8. Measure the absorbance in 1 cm cells at 640 nm against dichloromethane as reference.
9. Plot absorbance against milligrams of nonionic, or calculate the equation to the regression line.

Analysis of sample

1. Take a 10 g sample, dissolve in water, transfer to a 500 ml volumetric flask, dilute to volume and mix.
2. Titrate 75 ml of this solution to methyl red with 1 M or 0.1 M hydrochloric acid. Note the volume used and discard the solution.
3. Place a second 75 ml portion in a 100 ml stoppered cylinder, add the volume of acid used by the first portion, dilute to 100 ml and mix.
4. Into a 100 ml separating funnel put 20 ml dichloromethane, 20 ml cobaltothiocyanate reagent and 20 ml sample solution (step 3). Stopper the funnel and shake for exactly 1 min.
5. Allow the layers to separate. Place 10 ml dry propan-2-ol in a dry 25 ml volumetric flask. Run off about 1 ml of the dichloromethane layer, then run the extract into the flask exactly to the mark and mix.
6. Read the absorbance in a 1 cm cell at 640 nm against dichloromethane as reference.
7. Read the milligrams of surfactant from the calibration curve, or calculate it from the regression equation.

$$\%\ \text{nonionic (NI)} = \frac{W_2}{1000} \times \frac{100}{20} \times \frac{500}{75} \times \frac{100}{W_1} = \frac{3.33W_2}{W_1}$$

where W_2 = milligrams nonionic (analysis, step 3), and W_1 = weight of sample in grams (analysis, step 1).

This method can be used in the presence of anionic surfactants and hydrotropes, but anionics may depress the colour intensity.

This difficulty can be overcome as follows. Take several 10 ml portions of the 2 mg/ml standard solution of the nonionic and add the anionic in incremental amounts, taking colour readings until further increments of anionic do not cause any further depression of the reading. Note the lowest concentration of anionic at which constancy is attained. Prepare a calibration curve and carry out all analyses in solutions containing the concentration so determined. When the analysis is done as described, hydrotropes do not interfere.

Cationics and amphoterics also yield a blue colour with the reagent and must be absent.

The colour reaction can also be used as a spot test for both polyethoxylates and polypropoxylates. The reagent is diluted tenfold, 5 ml of a 1% solution of the product is added and the mixture is shaken. A blue colour or precipitate indicates the presence of polyethoxylate or polypropoxylate. Polyethylene glycols do not react.

6.2.6 Determination of total nonionics and polyethylene glycols

Polyethoxylates may contain polyethylene glycols as well as ethoxylated alcohols, alkylphenols, etc. Both types of component are determined by the Weibull method [10], which has been adopted as an international standard [11], reference to which is advised. The following is an abbreviated version. The method can be used for fatty acid ethoxylates, but the ethyl acetate extract will include any free fatty acid present in the sample.

Procedure

1. Weigh about 5 g of the sample and dissolve it in 100 ml ethyl acetate (caution: flammable) at 35°C in a separating funnel.
2. Extract with three 100 ml portions of 30% sodium chloride solution at 35°C.
3. Wash each extract twice with 100 ml ethyl acetate, also at 35°C.
4. Evaporate the ethyl acetate to dryness in a tared beaker and dry to constant weight.
5. Combine the sodium chloride extracts and extract with three 100 ml portions of chloroform (caution: toxic vapour).
6. Wash each chloroform extract with 100 ml 30% sodium chloride solution.

7. Evaporate the chloroform to dryness in a tared beaker and dry to constant weight.

$$\% \text{ nonionic active matter} = 100\, W_2/W_1$$

$$\% \text{ polyethylene glycols} = 100\, W_3/W_1$$

where W_2 = weight of dried residue (step 4), W_3 = weight of dried residue (step 7), and W_1 = weight of sample.

ISO 2268 suggests that the extraction at 35°C can most easily be done by keeping the solvents and reagents in a large cupboard or small room maintained at that temperature, and doing the analysis in the same enclosure. The extraction must be done at this temperature to ensure the extraction into the ethyl acetate layer of higher-molecular-weight nonionics. The chloroform extraction can be done at normal room temperature.

6.2.7 Volumetric determination of polyethylene glycols

Ivanov *et al.* [12] have reported a volumetric method for polyethylene glycols. Their procedure is as follows. The indicating electrode is a membrane electrode in which the filling solution is 0.004 M cetylpyridinium chloride. The reference electrode is not specified; calomel or silver–silver chloride would be satisfactory.

1. Dissolve 1 g of the nonionic in 20 ml ethyl acetate.
2. Extract with two 20 ml portions of sodium chloride (cf. Weibull [10]).
3. Dilute the sodium chloride extracts to a convenient volume and take a portion containing 10–30 mg polyethylene glycols.
4. Add 10 ml 0.1 M barium chloride and titrate potentiometrically with 0.02 M phosphotungstic acid.

They also used 0.02 M NaTPB for the same purpose.

Phosphotungstic acid is known as a precipitating titrant for cationics. Its use here is novel.

6.2.8 Determination of oxyethylene groups

This method is a modification of that originally described by Siggia [13] and is similar to the method given in ISO 2270 [14], reference to which is advised. It is applicable to ethylene oxide adducts of fatty alcohols, saturated fatty acids and alkylphenols. It is not applicable to compounds containing oxypropylene groups, oxygen or halogen on adjacent carbon atoms, compounds containing sulphur or nitrogen, aldehydes, acetals or sterols and their derivatives. Notwithstanding this caution about nitrogen compounds, Longman [15], who gives a detailed discussion of this and other methods, states that the method can be used for ethoxylated alkanolamides. The method can be applied to the ethoxylates obtained by acid hydrolysis of alkylether sulphates and sulphated alkylphenol ethoxylates and subsequent extraction with chloroform or by deionisation. The chemical process is hydrolysis of the oxyethylene groups by nascent hydriodic acid, thermal decomposition of the ethylene di-iodide formed and titration of the iodine with thiosulphate.

$$RO{-}(CH_2{-}CH_2O)_nH + 2nHI \rightarrow nICH_2{-}CH_2I + ROH + nH_2O$$

$$ICH_2{-}CH_2I \rightarrow C_2H_4 + I_2$$

$$I_2 + S_2O_3^{2-} \rightarrow 2I^- + S_4O_6^{2-}$$

1. Weigh a sample containing not more than 100 mg ethylene oxide into a boiling rod (a glass rod having a belled end into which samples can be weighed).
2. Place the boiling rod in a 50 ml round-bottomed flask having two ground glass necks (see Figure 6.1). This is best done by inverting the flask over the rod, then turning it upright and allowing the rod to slide gently in, with the cavity downwards.
3. Add 3 g potassium iodide. Attach a reflux condenser to the vertical neck and a gas inlet tube to the angled neck.
4. Flush the flask with nitrogen or carbon dioxide at a flow rate greater than 0.5 ml/sec for at least 20 min, then place the flask in an oil bath at 165°C and reduce the gas flow to 0.1–0.5 ml/sec.
5. Add via the condenser 5 ml phosphoric acid of refractive index *c.* 1.70.
6. Fill a bubble trap of 7–10 ml capacity with 10% potassium iodide solution and attach it to the top of the condenser.
7. Allow the reaction to continue for 30 min, then cool to below 80°C.
8. Pour the potassium iodide solution from the bubble trap down the condenser, rinse the bubbler with the same solution and pour this down the condenser, and remove and rinse the gas inlet tube.
9. Transfer the contents of the flask quantitatively to a 250 ml conical flask. If insoluble matter is present add a few ml of methanol to dissolve it.
10. Titrate with 0.1 M sodium thiosulphate, adding a little starch indicator near the end-point.
11. Carry out a blank determination in exactly the same way but without the sample.

$$\%\ C_2H_4O = (V_s - V_b) \times C \times 2.2/W$$

where V_s = volume of sodium thiosulphate used by sample, V_b = volume of sodium thiosulphate used by blank, C = exact

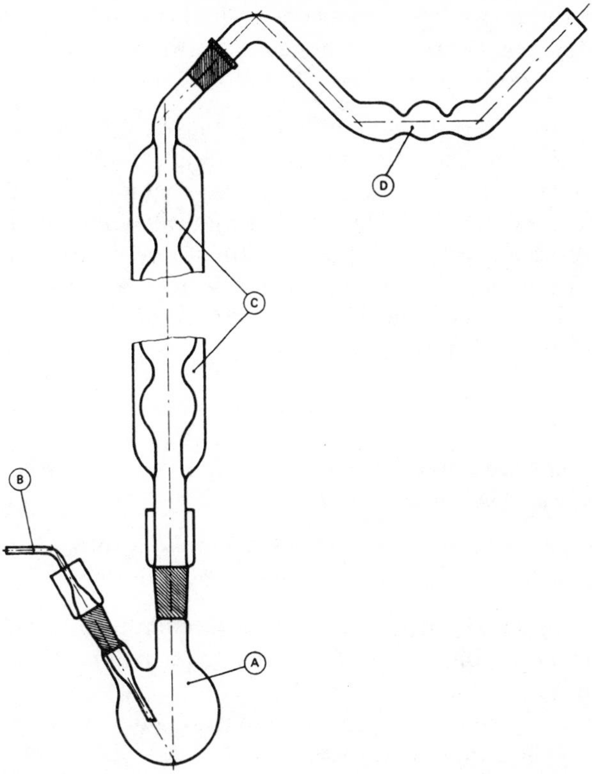

Figure 6.1 Apparatus for determination of oxyethylene groups. A: 50 ml flask; B. Gas inlet tube; C: Reflux condenser; D: Bubbler, capacity 7–10 ml (reproduced by kind permission of the International Organization for Standardization, Geneva).

concentration of the sodium thiosulphate in mol/l, and W = weight of sample.

A quite different approach is that of Tsuji and Konishi [16], who used the mixed anhydride of toluenesulphonic and acetic acids to prepare alcohol ethoxylates for gas chromatography. The reagent is prepared as follows.

1. Add 80 g acetic anhydride dropwise to 120 g *p*-toluenesulphonic acid in a 300 ml round-bottomed flask.
2. Heat under reflux at 120°C for 1 h and allow to cool.
3. Store in a brown bottle and discard after 1 month.

The ethoxylate is treated as follows.

1. Mix 0.1 g of the ethoxylate with 2 ml mixed anhydride in a 20 ml flask.
2. Heat under reflux at 120°C for 2 h.

3. Cool, neutralise with 50% aqueous sodium hydroxide, extract with 20 ml diethyl ether (caution: flammable), wash with several portions of water and evaporate to dryness.

The fatty alcohol is converted to the acetate and both ethylene and propylene oxides are converted to their diacetates, apparently quantitatively. The mixture is analysed by gas chromatography, which separates the glycol diacetates sufficiently for quantitative determination. If the method were scaled up by a factor of 10 or 20, it would be easy to separate the reaction products and analyse them by alkaline hydrolysis to determine, for example, the molecular weight of the alcohol and the alkylene oxide chain length.

6.2.9 Fatty acid ethoxylates: determination of polyethylene glycols, free fatty acid and mono- and diester

These determinations are readily carried out using procedures already described.

1. Extract the nonionics with ethyl acetate (section 6.2.6), but make the solution in 30% sodium chloride alkaline to phenolphthalein with 10 M sodium hydroxide before extraction.
2. Acidify the aqueous layer to bromophenol blue with concentrated hydrochloric acid and extract the free fatty acid with petroleum ether. Evaporate to dryness (caution: flammable vapour), dry, cool and weigh.
3. Extract the polyethylene glycols with chloroform (section 6.2.6). Evaporate to dryness (caution: toxic vapour), dry, cool and weigh.
4. Evaporate the ethyl acetate extract to dryness, dry for 1 h in a vacuum oven at 50°C, cool and weigh. Dissolve the residue in methanol, dilute to 250 ml in a volumetric flask with the same solvent and mix.
5. Use 100 ml of the solution from step 4 to determine the saponification value (section 3.4.1), using 0.25 M alkali, to determine total esters.
6. Pipette a second 100 ml portion into a round-bottomed flask. Evaporate to dryness and dry for 1 h at 50°C in a vacuum oven. Determine the hydroxyl value (section 3.4.3), using 0.25 M alkali, to determine the monoester.
7. If the molecular weight of the fatty acid is unknown, extract the fatty acid from the saponified solution from step 5, evaporate to dryness, dry and weigh.

The free fatty acid content was measured at step 2, the polyethylene glycols content at step 3 and the total ester content at step 4.

Assuming that the molecular weight of the monoester is known, the molecular weight of the diester is found as follows:

$$M_{de} = (800W' - 2V_h CM_{me})/C(V_s - V_h)$$

where M_{de} = molecular weight of diester, W' = weight of esters (step 4), M_{me} = molecular weight of monoester, V_s = net volume of alkali used in determination of saponification value (step 5), V_h = net volume of alkali used in determination of hydroxyl value (step 6), and C = exact concentration of alkali in mol/l.

The 'net volume of alkali' means the difference between the sample and blank titrations.

The percentage of each ester can then be calculated. The basis of this calculation is that the monoester has one terminal hydroxyl group and one saponifiable ester group. The net volume of alkali used to saponify the monoester is therefore equal to the net volume used in the determination of hydroxyl value.

This method of calculation is valid for samples made by either synthetic route.

The sample can be pretreated in other ways. For example, it can be passed through a strongly basic anion exchanger hydroxide to remove the free fatty acid, which can subsequently be recovered by elution with aqueous acetic acid. The acetic acid can be removed by evaporation and the fatty acid examined as required. Or it can be passed through a strongly acidic cation exchanger in the acid form so that the free fatty acid can be titrated without interference from alkaline catalysts.

The free fatty acid can be directly determined by two-phase titration with benzethonium chloride, in which case the presence of basic catalyst is irrelevant. Bares and Zajic [17] used this approach, with bromocresol green as indicator, and taking as the end-point the final disappearance of colour from the aqueous layer. A blank test was necessary. They showed that quantitative titration only occurred when the pH was at least 11.3. The bromophenol blue method would serve equally well.

6.3 Block copolymers of ethylene and propylene oxides

Hei and Janisch [18] reported that titration with sodium tetrakis(4-fluorophenyl)borate according to the method of Tsubouchi *et al.* [4] worked satisfactorily for block copolymers.

The application of the mixed anhydride method of Tsuji and Konishi [16] to both propoxylates and block copolymers was described in section 6.2.8.

6.4 Esters of polyhydroxy compounds and ethoxylated derivatives

6.4.1 Synthesis and composition

Sorbitan mono-, sesqui-, di- and triesters are made by heating sorbitol with a fatty acid. The sorbitol dehydrates to give chiefly 1,4-sorbitan, which esterifies mostly on the primary hydroxyl group in the case of monoesters, and on that group and any other hydroxyl groups in the case of other esters. Ethoxylation occurs on any hydroxyl group, and as a result of transesterification it can even occur on already esterified hydroxyl groups. Products made by esterifying ethoxylated sorbitans are different from those made by ethoxylating sorbitan esters, having the ester groups chiefly on the terminal groups of the ethylene oxide chains. Commercial products may also contain unreacted fatty acid, sorbitan or ethoxylate and polyethylene glycols.

Sucrose esters are made by transesterification of sucrose in solution with methyl ester or in aqueous dispersion with a glyceride oil or fat. Esterification can take place on any hydroxyl group, but in the monoesters it is mainly on the primary hydroxyl of the glucose residue; isomers with the ester group on either of the primary hydroxyl groups of the fructose residue are also formed. Esterification on any hydroxyl group is possible, however, and commercial products may be quite complex mixtures including several isomeric mono- and diesters. They may also contain unreacted sugar plus residual methyl ester or mono- and diglycerides and free glycerol.

Comprehensive analysis of these materials is obviously quite complicated, and only methods for general characterisation are considered here.

6.4.2 Analysis

The following are useful analytical parameters.

Free fatty acid: standard procedure, section 3.4.1.
Saponification value: sorbitan esters—standard procedure, section 3.4.2; sucrose esters—allow to stand overnight in cold ethanolic 1.0 M alkali [19]. Value found includes glycerides or unreacted methyl ester.
Total fatty acid: saponify, evaporate off ethanol, dissolve residue in water, acidify, extract (see below).
Total sorbitan: see below.
Hydroxyl value: standard procedure, section 3.4.3.
Ethylene oxide content: standard procedure, section 6.2.8.
Polyethylene glycols content: standard procedure, section 6.2.6.

Provided that the saponification value or total fatty acid content of the raw material is known, the saponification value or total fatty acid content

Table 6.1 Total fatty acid contents of pure sorbitan esters

Monolaurate	57.8%
Monostearate	66.0%
Dilaurate	75.8%
Distearate	81.6%
Trilaurate	84.5%
Tristearate	88.6%

of a product affords a route for its determination in products, provided that no other ingredient is present that undergoes the same reaction.

The saponification values of sorbitan esters do not by themselves characterise the esters very well. The total fatty acid content, however, is more useful, even when allowance is made for the fact that each nominal type contains a proportion of its neighbours, e.g. monoesters contain some diester. The total fatty acid contents of some pure compounds are shown in Table 6.1. The figures are for the esters with single homologue acids. The values for oleates are almost identical with those for stearates.

Determination of total fatty acid followed by determination of the molecular weight and iodine value of the fatty acid serves to characterise sorbitan esters quite well.

The following procedure is also useful for characterising sorbitan esters.

Determination of sorbitan in sorbitan esters. This method depends on the fact that saponification of sorbitan esters hydrolyses not only the ester groups but also the ether linkage in the sorbitan itself, to produce sorbitol. The latter is a hexahydric alcohol, and can be oxidised by periodate to give four molecules of formic acid and two of formaldehyde, either of which can be measured. The method cannot be used for ethoxylates.

1. Take a sample containing the equivalent of about 2 g sorbitan. Add 40 ml ethanolic 0.5 M sodium or potassium hydroxide to a monoester, 75 ml to a diester or 100 ml to a triester, attach a reflux condenser and boil under reflux for 1 h. Cool and titrate with ethanolic 0.5 M hydrochloric acid. Calculate the saponification value.
2. Evaporate to dryness on a steam bath.
3. Dissolve the residue in water, transfer to a separating funnel, acidify with aqueous hydrochloric acid and extract the fatty acid with petroleum ether. Keep the petroleum ether extracts for further examination if required.
4. Transfer the aqueous layer to a 200 ml volumetric flask, dilute to volume with water and mix.

5. Pipette 10 ml into a stoppered conical flask. Neutralise to phenol red with 1 M sodium hydroxide, then add 1 ml 1 M hydrochloric acid.
6. Boil for 2 min, cool and carefully neutralise with 0.1 M sodium hydroxide.
7. Add 20 ml 5% sodium metaperiodate solution, swirl to mix, stopper and place in a dark cupboard for 30 min.
8. Add 5 ml ethane-1,2-diol (ethylene glycol), swirl to mix, stopper and replace in the dark cupboard for 20 min.
9. Titrate with 0.1 M sodium hydroxide.

$$\% \text{ sorbitan} = V \times \frac{C}{1000} \times 41 \times \frac{100}{W} \times \frac{200}{10}$$

where all symbols have their usual meanings.

6.5 Alkylpolyglycosides

The most important analytical parameter of alkylpolyglycosides is the degree of substitution. This is easily determined by liquid chromatography, which gives the polymer distribution, or ^{1}H NMR spectroscopy, which gives the mean number of glycoside units per molecule.

These materials are decomposed by boiling under reflux with acid, but subsequent chemical analysis is reported not to be straightforward. There would appear to be no difficulty in extracting the fatty alcohol with petroleum ether or diethyl ether, evaporating and weighing, but attempts to determine the glucose by periodate oxidation have been unsuccessful.

6.6 Alkanolamides and ethoxylated alkanolamides

6.6.1 Composition

Alkanolamides are usually ethanolamides, but isopropanolamides are sometimes found. Coconut fatty acids are the commonest hydrophobes. Mono- and diethanolamine are alcohols as well as amines, and synthesis by heating with fatty acids or by transesterification can both produce esters as well as amides. Glycerides are sometimes used for the synthesis instead of methyl esters, and the product then contains glycerol and possibly mono- and diglycerides as well as unreacted triglyceride. Depending on the manufacturing method, monoethanolamides may contain, in addition to the monoethanolamide itself, the following:

Free fatty acid or fatty acid methyl ester or glycerides and glycerol;
Free monoethanolamine;

Monoethanolamine fatty ester ('ester amine', $RCOOC_2H_4NH_2$)
Monoethanolamide ester ('amide ester', $RCONHC_2H_4OOCR$)

Diethanolamides may contain, in addition to the diethanolamide itself, the following:

Free fatty acid or fatty acid methyl ester or glycerides and glycerol;
Free diethanolamine;
Di(*N*-hydroxyethyl)piperazine ('DHEP', from condensation of two molecules of diethanolamine);
Monoester amine, $RCOOC_2H_4NHC_2H_4OH$, and possibly diester amine, $(RCOOC_2H_4)_2NH$;
Amide monoester, $RCON(C_2H_4OH)C_2H_4OOCR$, and possibly amide diester, $RCON(C_2H_4OOCR)_2$

Diethanolamides made by the Kritchevsky process (heating acid and amine together) contain a large amount (up to 30%) of free diethanolamine.

Schemes, e.g. [20], have been constructed for the determination of most of these, but for raw material analysis or the characterisation of nonionic fractions obtained by deionisation it is often sufficient to determine total nitrogen by the Kjeldahl method (section 2.5.2), or total fatty acid and total ethanolamine (section 6.6.2).

Of the ethoxylates, only the monoethanolamide derivatives are commercially significant. Their ethylene oxide content is determined as described in section 6.2.8. Otherwise their analytical chemistry is similar to that of the ethanolamides from which they are derived. The analyses of mono- and diethanolamines are identical in most respects.

The methods that follow permit the determination of most of the possible constituents in analyses of varying degrees of complexity. They are a small selection of the many methods and variations possible, but should be sufficient for almost all purposes. There is obviously a certain degree of overlap, and analysts must choose the most appropriate procedures for their needs.

6.6.2 Determination of total fatty acid and total ethanolamine

In the experience of the present author, diethanolamides but not monoethanolamides are saponified by boiling with 0.5 M ethanolic alkali, although this is contrary to the general belief. Both mono- and diethanolamides are saponified by boiling with 1 M alkali in ethylene glycol, boiling point 197°C [21], but results obtained by this method are not very accurate, and the usual hydrolysis procedure is prolonged boiling with strong acid.

1. Weigh a 5 g sample in a round-bottomed flask. Add 25 ml 6 M hydrochloric acid, attach a reflux condenser and boil under reflux for 8 h.
2. Cool and transfer quantitatively to a 250 ml separating funnel, using water and petroleum ether (caution: flammable) alternately to effect the transfer.
3. Make the total volumes of aqueous solution and petroleum ether each up to 100 ml and extract. Extract with two more 100 ml portions of petroleum ether, collecting the aqueous layer in a 500 ml conical flask. Combine the petroleum ether extracts and wash with 50 ml water, collecting the washings in the same flask. Keep both extracts.
4. Transfer the petroleum ether to a tared beaker and evaporate to dryness on a steam bath (caution: flammable vapour), dry, cool and weigh. If desired, dissolve the fatty acid in neutral ethanol and titrate to phenolphthalein with ethanolic 1.0 M sodium hydroxide.
5. Neutralise the aqueous extract from step 3 to phenolphthalein with 6 M sodium hydroxide and add about 1 ml excess.
6. Titrate with 1.0 M hydrochloric acid either first to phenolphthalein and then to bromophenol blue, or potentiometrically through two end-points at high and low pH. The acid consumed between the end-points measures the ethanolamine.

If desired the aqueous solution from step 6 may be used to determine glycerol as described in section 5.5.4.

6.6.3 Determination of total amide and total amine

The amides present are the ethanolamide itself and the amide ester(s). The amines are the ester amine(s), the mono- or diethanolamine, and possibly DHEP. Amines only are titrated with perchloric acid in glacial acetic acid. In acetic anhydride the same titrant titrates amides as well [23, 24].

A suitable titrant is made as follows: dissolve 28–29 g of 70–72% perchloric acid in 1 l of glacial acetic acid (caution: corrosive, irritant vapour), add 46–47 ml acetic anhydride (caution: corrosive, irritant vapour), dilute to 2 l and mix. The acetic anhydride combines with the 28–30% water in the perchloric acid, so guaranteeing an anyhdrous reagent. Standardise against pure potassium hydrogen phthalate.

1. Amines: dissolve a 5 g sample in glacial acetic acid (caution: irritant vapour) and titrate potentiometrically with 0.1 M perchloric acid in glacial acetic acid. Or use 1% methyl violet in glacial acetic acid as indicator. The colour change is from violet to a light blue–green.
2. Amides plus amines: dissolve a 0.6–0.7 g sample in acetic anhydride

(caution: irritant vapour) and titrate potentiometrically with 0.1 M perchloric acid in glacial acetic acid. Or use methyl violet as in step 1.

The procedure gives total amines and total amides in mol per 100 g.

6.6.4 Determination of free ethanolamine and ester amine

The free ethanolamine is particularly important in Kritchevsky materials. The ethanolamine (and DHEP if present) is extracted into an aqueous solution, leaving any ester amine in the organic layer [22]. If free fatty acid is present it is extracted as soap, but it does not interfere. This is a simple and direct approach for use when the more comprehensive analysis described in the next two sections is not required.

1. Weigh a 5 g sample in a 250 ml separating funnel. Add 100 ml diethyl ether and 100 ml 10% sodium chloride. Shake and allow to separate. Re-extract the aqueous layer with 50 ml ether and collect the aqueous layer in a 400 ml beaker. Wash the ether layers in turn with 50 ml of the salt solution, collecting the washings in the same beaker.
2. Make the aqueous solution just alkaline to phenolphthalein and titrate potentiometrically with 0.5 M hydrochloric acid through two end-points at high and low pH, or first to phenolphthalein and then to bromophenol blue.
3. Evaporate the ether to dryness (caution: flammable vapour). Dissolve the residue in glacial acetic acid and titrate with 0.1 M perchloric acid (see section 6.6.4).

The titration at step 2 determines the free ethanolamine and the titration at step 3 the ester amine.

The ester amine is cationic in acid solution and may be determined directly as follows.

1. Dissolve a 1 g sample in 25 ml chloroform in a vessel suitable for two-phase titration.
2. Add 20 ml water and acid mixed indicator and titrate with 0.004 M sodium dodecyl sulphate.

6.6.5 Determination of free ethanolamine, ester amine and free fatty acid or glycerol

The methods described in this section and the next give a more or less complete analysis of the sample. The extractions in steps 1 and 2 can be done in the reverse order, in which case the first aqueous extract will contain the free ethanolamine plus the free fatty acid as soap, separation and determination of which is straightforward. The ester amine is determined by two-phase titration with sodium dodecyl sulphate.

1. Place a 5 g sample in a 250 ml separating funnel. Add 50.0 ml 0.1 M hydrochloric acid and 100 ml diethyl ether (caution: flammable) and shake thoroughly, cautiously releasing the excess pressure from time to time. Allow the layers to separate and run the lower layer into a second separating funnel. Extract again with 50 ml diethyl ether and run the aqueous layer into a 250 ml volumetric flask.
2. Wash the diethyl ether extracts in turn with 50 ml water and run the water into the same flask. Keep the ether layers.
3. Titrate the hydrochloric acid extract to bromophenol blue with 0.1 M sodium hydroxide. The amount of 0.1 M acid consumed measures the total amines.
4. Dilute to volume and mix.
5. Pipette 50 ml into a suitable titration vessel, add acid mixed indicator and chloroform and titrate with 0.004 M sodium dodecyl sulphate. This measures the ester amine.
6. If glycerol is present, pipette 100 ml of the solution from step 4 into a 500 ml conical flask and determine glycerol as described in section 5.5.4.
7. If free fatty acid is present, extract the ether layers from step 2 in turn with 50 ml 0.1 M sodium hydroxide. Run the aqueous layer into a third separating funnel. Wash the ether layers in turn with 50 ml water and run this into the third separating funnel. Keep the ether layers.
8. Acidify the sodium hydroxide extract and extract the fatty acid with petroleum ether. Evaporate to dryness, dry, cool and weigh. If desired, dissolve in neutral ethanol and titrate to phenolphthalein with ethanolic 0.1 M sodium hydroxide.

6.6.6 Determination of ethanolamide, amide ester and methyl ester or glycerides

1. Combine the ether extracts from step 2 or step 7 of the previous procedure in a tared round-bottomed flask. If methyl ester is present, evaporate almost to dryness on a steam bath and go to step 2. If not, evaporate completely to dryness and go to step 4.
2. Complete the drying under a jet of air without heating. Add 5 ml acetone, evaporate almost to dryness and complete the drying under a jet of air without heating. Repeat the acetone treatment.
3. Dry in vacuum oven at room temperature, weighing at half-hourly intervals, until two weighings differ by not more than 2 mg.
4. Dry for 1 h in a vacuum oven at 105°C, cool and weigh. The loss in weight is the methyl ester.
5. Add 50 ml 6 M hydrochloric acid and boil under reflux for 8 h. Complete the determination of total fatty acid and total ethanolamine as described in section 6.6.2.

6. If glycerides were present, determine glycerol in the aqueous layer from step 6 by the procedure described in section 5.5.4.

Calculate the molar ratio of fatty acid to ethanolamine, making an estimated allowance for fatty acid from any glycerides present. The excess fatty acid over a ratio of 1.00 is due to the amide ester. In the case of diethanolamides calculate as monoester, although diester may be present.

Alternatively, the two amide species can be separated by chromatography and determined by weighing or titration with perchloric acid [20]. Use the residue from step 4 of the previous procedure, or start again as follows.

1. Dissolve a 1 g sample in methanol.
2. Pass the solution through a column of 25 ml mixed-bed ion-exchange resin in methanol.
3. Evaporate to dryness on a steam bath (caution: flammable vapour). Cool and dissolve the residue in 3 ml 9:1 chloroform–butanol-1-ol.

The solution contains only the ethanolamide, the amide ester and the methyl ester or glycerides. The separation and determination are performed as follows [20].

1. Heat sufficient 0.05–0.2 mm silica gel overnight at 110°C and cool in a vacuum desiccator. Prepare a 25 cm × 2.6 cm column in a 9:1 mixture of chloroform and butan-1-ol.
2. Use the solution prepared as described above, or dissolve the residue from step 4 of the previous procedure in 3 ml 9:1 chloroform–butan-1-ol. Pass the solution through the column at 0.5–1.0 ml/min.
3. Wash the column with 250 ml of the same solvent and then with 600 ml methanol. The chloroform–butan-1-ol contains the amide ester and the methanol the ethanolamide.
4. Evaporate both eluates to dryness, dry, cool and weigh. If the methanol contains silica gel, dissolve the residue in dry methanol and centrifuge, then evaporate, dry and weigh.
5. If desired, dissolve the residues in acetic anhydride (caution: irritant vapour) and titrate potentiometrically with 0.1 M perchloric acid in glacial acetic acid.

References

1. Porter, M. R. *Handbook of Surfactants*, Blackie, Glasgow, 1991, p. 118.
2. Porter, M. R. *Handbook of Surfactants*, Blackie, Glasgow, 1991, p. 155.
3. Tsubouchi, M. and Tanaka, Y. *Talanta*, **31**(8) (1984), 633–634.
4. Tsubouchi, M., Yamasaki, N. and Yanagisawa, K. *Anal. Chem.*, **57** (1985), 783–784.
5. O'Connell, A. W. *Anal. Chem.*, **58** (1986), 669–670.
6. Vytřas, K., Dvorakova, V. and Zeman, I. *Analyst*, **114** (1989), 1435–1441.

7. Jones, D. L., Moody, G. J., Thomas, J. D. R. and Birch, B. J. *Analyst*, **106** (1981), 974–984.
8. Moody, G. J. and Thomas, J. D. R. *Nonionic Surfactants: Chemical Analysis*, Surfactant Science Series, vol. 19, Marcel Dekker, 1987, pp. 117–136.
9. Milwidsky, B. M. *Soap, Cosmet. Chem. Spec.*, **47**(12) (1971), 66.
10. Weibull, B. *Proc. IIIrd International Congress on Surface Active Substances*, Cologne, 1960, vol. C, pp. 121–124.
11. ISO 2268: 1972: Surface active agents (non-ionic)—Determination of polyethylene glycols and non-ionic active matter (adducts)—Weibull method. International Organization for Standardization, Geneva.
12. Ivanov, V. N., Pravshin, Y. S. and Bavykina, N. I. *Zh. Anal. Khim.*, **43**(7) (1988), 1313–1317.
13. Siggia, S. *et al. Anal. Chem.*, **30** (1958), 115.
14. ISO 2270: 1989: Non-ionic surface active agents—Polyethoxylated derivatives—Iodometric determination of oxyethylene groups. International Organization for Standardization, Geneva.
15. Longman, G. F. *The Analysis of Detergents and Detergent Products*, John Wiley & Sons, 1975, pp. 285–286.
16. Tsuji, K. and Konishi, K. *Analyst*, **99** (1974), 54.
17. Bares, M. and Zajic, J. *Tenside*, **11** (1974), 251–254.
18. Hei, R. D. and Janisch, N. M. *Tenside*, **26** (1989), 288–290.
19. Tsuda, T. and Nakashini, H. *JAOAC*, **66** (1983), 1050–1052.
20. Mutter, M., van Galen, G. W. and Hendrikse, P. W. *Tenside*, **5** (1968), 33–36.
21. Olsen, S. *Die Chemie*, **56** (1943), 202
22. Kroll, H. and Nadeau, H. *JAOCS*, **4** (1957), 323–326.
23. Wimer, J. *Anal. Chem.*, **30** (1958), 77.
24. Krusche, G. *Tenside*, **10** (1973), 182–185.

7 Analysis of cationics and amphoterics

7.1 Introduction

As in previous chapters, this chapter deals with the analysis of cationics and amphoterics either alone, as raw materials or as fractions isolated by ion exchange or otherwise, or in formulated products. Fractions isolated by ion exchange are likely to contain other materials, analogously with anionics. Amines, ethoxylated amines and amine oxides are included in this chapter because they are bases and capable of a cationic function. They are retained as cations by ion-exchange columns and do not appear in the nonionic fraction of separated mixtures, they can be titrated with acids and, in acid solution, with sodium dodecyl sulphate, provided the ethylene oxide chains of ethoxylates are not too long.

7.2 Quaternary ammonium salts

There are many methods for the determination of quaternary ammonium salts (quats), of which a small, reasonably representative, selection is given here. With two exceptions, the methods fall into four broad categories: two-phase or potentiometric titration with an anionic surfactant or with sodium tetraphenylborate (NaTPB). Gravimetric methods were favoured at one time, but have largely gone out of fashion because they have mostly been found to be not very quantitative, and they take longer than volumetric methods. Gravimetric determination with NaTPB is quantitative for most quats, but it is more tedious and no more accurate or precise than titration. Many cations, both organic and inorganic, have insoluble or sparingly soluble TPB salts, and interfere in titrations of quats.

7.2.1 Two-phase titration with sodium dodecyl sulphate

The most generally applicable procedures are those described in the two ISO methods [1,2]. The first of these is for high-molecular-weight quats, which may be broadly defined as those containing two hydrocarbon chains. These materials are not very soluble in water, and this procedure is more reliable for them than the simpler version used for low-molecular-weight quats. The analysis is performed as follows. It is virtually identical with ISO 2271 for anionics.

1. Dissolve a sample containing 2–4 millimol of active in 20 ml propan-2-ol.
2. Add 50 ml water and stir. Transfer quantitatively to a 1 l volumetric flask, dilute to volume with water and mix.
3. Pipette 10 ml 0.004 M sodium dodecyl sulphate solution into a stoppered flask or cylinder or the mechanical apparatus shown in Figure 3.2. Add 10 ml water, 15 ml chloroform and 10 ml acid mixed indicator and titrate with the solution prepared at step 2, with thorough shaking or stirring after each addition. The chloroform is initially pink, becoming colourless or greyish at the end-point. A blue colour indicates that the end-point has been overshot.

The titration is done the other way round for low-molecular-weight quats, i.e. those with only one alkyl chain, and whose molecular weights are 200–500. It is the exact converse of ISO 2271.

1. Dissolve a sample containing 2–3 millimol of active in water, transfer quantitatively to a 1 l volumetric flask, dilute to volume and mix.
2. Pipette 25 ml of this solution into a suitable titration vessel. Add 25 ml chloroform, 25 ml water and 10 ml acid mixed indicator and titrate with 0.004 M sodium dodecyl sulphate, shaking thoroughly after each addition. The chloroform is initially blue, becoming colourless or greyish at the end-point. A pink colour indicates that the end-point has been overshot.

A variant of the last method [3] is to prepare the solution as described, take 25 ml, add 1 ml 0.004% methylene blue and 25 ml chloroform and titrate with the same titrant, shaking after each addition, until the upper layer is clear and colourless. The relation between weight of quat and volume of titrant is stoichiometric for many if not all quats.

7.2.2 Potentiometric titration with sodium dodecyl sulphate

Quats may be titrated potentiometrically with sodium dodecyl sulphate. The procedure is straightforward, being the converse of that described for anionics in section 5.2.2, and end-points are mostly sharp. Quat solutions and titrant may be up to 0.04 M, solubility permitting. Any of the electrode types described in section 3.6.2 is suitable, as are many others, with calomel or silver–silver chloride as reference.

7.2.3 Two-phase titration with sodium tetraphenylborate

The best known two-phase method is that of Cross [4]. By titrating at three different pHs, it permits a degree of classification of the analyte. It requires two buffer solutions and two anionic indicators.

Buffer solution, pH 3.0: mix equal volumes of 0.5 M citric acid and 0.2 M disodium hydrogen phosphate.

Buffer solution, pH 10.0: mix 100 ml 0.2 M disodium hydrogen phosphate with 6 ml 0.25 M trisodium phosphate.

Indicators: 0.15% methyl orange and 0.2% ethanolic bromophenol blue.

1. Weigh a sample containing about 1.0 g active. Dissolve in water, transfer quantitatively to a 500 ml volumetric flask, dilute to volume and mix. Alcohol may be added to increase the solubility.
2. Pipette 50 ml into a stoppered titration vessel and titrate by the appropriate method (see Note 1).
3. Titration at pH 3.0: add 5 ml pH 3.0 buffer, two drops methyl orange and 30 ml chloroform. Titrate with 0.01 M NaTPB, with thorough shaking after each addition. At the end-point the yellow colour disappears from the chloroform layer and the water layer becomes pink.
4. Titration at pH 10.0: add 5 ml pH 10.0 buffer, six drops bromophenol blue and 30 ml chloroform. Titrate as in step 3. At the end-point the blue colour disappears from the chloroform layer and the water layer becomes purple (cf. [5]).
5. Titration at pH 13.0: add 50 ml 1 M sodium hydroxide, six drops bromophenol blue and 30 ml chloroform. The titration and end-point are as in step 4.

   Note 1. At pH 3.0, all quats and species with weakly basic nitrogen are titrated. The latter category includes amines, and although Cross did not study them, ethoxylated amines and amine oxides.

   At pH 10.0, all quats are titrated, but compounds with weakly basic nitrogen are not.

   At pH 13.0, most quats are titrated, but those with the quaternary nitrogen in an aromatic ring (pyridinium, quinolinium, isoquinolinium and biguanidinium salts) are not. These materials cause a small positive error at pH 13.0, which is overcome by heating the alkaline solution. This opens the ring and destroys the quaternary nitrogen function:

   $$C_5H_5NR + H_2O \rightarrow CHO{-}CH{=}CH{-}CH_2{-}CHO + RNH_2$$

   This sometimes causes discoloration, which can be removed by extracting with petroleum ether.

The NaTPB is standardised gravimetrically against pure potassium chloride.

The behaviour of imidazolinium quats in this method is uncertain, because many of their structures are uncertain. Those having an imidazoline ring structure are very alkali-labile and are irreversibly converted at high pH to tertiary bases:

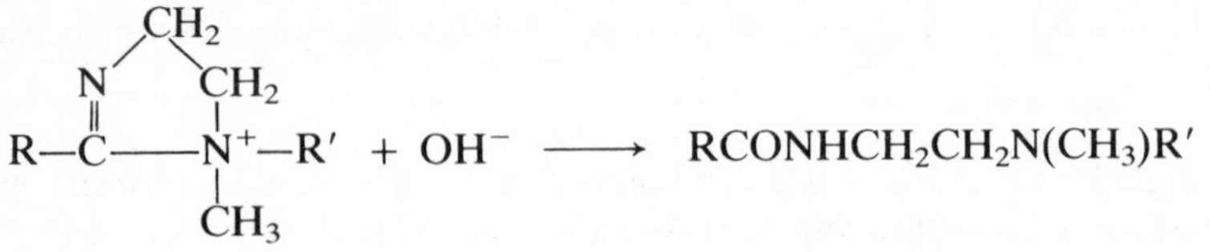

The ring opens at an appreciable rate at pH 10 and very rapidly at pH 13, and such compounds can be titrated only at pH 3. Imidazoline-derived quats having an amide structure with no ring, which may be the majority, would be expected to be titratable at all pHs, unless some other part of the structure is alkali-labile.

A variant, said to be more objective, is the method of Tsubouchi *et al.* [6], in which the colour remains in the organic layer throughout. The colour change is very sharp. The authors do not say why they used such dilute solutions or such small volumes. With dilute solutions of surfactants, especially quats, losses of solute through adsorption on to the glassware can cause serious errors. With small volumes, small absolute errors are large in relative terms. However, the procedure as reported is as follows.

Phosphate buffer solution, pH 6.0: insert pH electrodes into a stirred 0.3 M solution of disodium hydrogen phosphate and add 5 M sulphuric acid until the pH is 6.0.

Indicator: potassium salt of tetrabromophenophthalein ethyl ester, 0.1% in ethanol.

1. Pipette up to 10 ml of a 5×10^{-5} M solution of the quat into a stoppered bottle.
2. Add 5 ml phosphate buffer, pH 6.0, one drop indicator and 0.5–1.5 ml 1,2-dichloroethane, and dilute to about 20 ml with water.
3. Titrate with 5×10^{-5} M NaTPB, shaking after each addition. The colour remains in the organic layer throughout, and changes sharply from sky blue to pale yellow at the end-point.

7.2.4 Single-phase titration with sodium tetraphenylborate

The following method is due to Metcalfe *et al.* [7]. The authors do not say why they chose to use such an odd concentration of titrant, although it is of no importance.

1. Weigh in a 150 ml beaker a sample containing 2–3 millimol of active.

2. Dissolve in 40 ml water. If the sample is not very soluble, dissolve it in the minimum of propan-2-ol and then add 40 ml water. Complete dissolution is not essential.
3. Add 0.5 g powdered sucrose and a few drops of 0.2% ethanolic 2,7-dichlorofluorescein and titrate with 0.067 M NaTPB in 0.01 M sodium hydroxide. The solution is initially pink. Near the end-point it becomes deep pink, and at the end-point it changes sharply to yellow.

The purpose of the sucrose is to assist coagulation of the precipitate.

7.2.5 Potentiometric titration with sodium tetraphenylborate

There are many published variations on this principle, only a few of which are mentioned here. The first is a method developed by Akzo Chemie [8]. It is suitable for all types of quat except diquaternaries and amphoteric quats (betaines). The titrant is 0.05 M NaTPB containing 15 ml 0.1 M sodium hydroxide per litre and standardised against pure benzethonium chloride dried at 105°C for 3 h. The one molecule of water of hydration is lost during this drying. The other reagent required is glacial acetic acid containing 3 g/l benzethonium chloride.

The indicator electrode is a platinum sheet electrode. The reference electrode may be silver–silver chloride or a calomel electrode filled with 2 M sodium chloride instead of potassium chloride.

1. Condition the indicator electrode by doing a blank titration, but ignore the result.
2. Weigh into a 150 ml beaker a sample containing about 0.5 millimol of active. Dissolve in 15.00 ml glacial acetic acid containing benzethonium chloride, heating if necessary.
3. Dilute to about 100 ml with boiling water. Insert the electrodes and titrate potentiometrically, with continuous stirring, with 0.05 M NaTPB to just beyond the point of inflection.
4. Carry out a blank titration on 15.00 ml of the glacial acetic acid containing benzethonium chloride.

The use of glacial acetic acid containing benzethonium chloride is to obtain a linear plot between net volume of titrant and weight of active.

There are two other methods originating from Akzo Chemie [9,10], differing from the above method only in detail. They use a platinum ring indicating electrode as an alternative to the platinum sheet, and a platinum wire reference electrode inserted into the plastic tube that delivers the titrant. The burette tip must dip below the surface of the solution in the titration vessel. An alternative reference electrode is silver–silver chloride, mercury (I) sulphate or calomel, in a double junction arrangement, e.g. in a beaker of 2 M sodium chloride connected to the titration vessel via a salt bridge. The sample may contain up to about 2 millimol of active.

Other methods differ chiefly in their choice of electrodes. Christopoulos *et al.* [11] used a liquid membrane electrode containing tetrapentylammonium TPB and a silver–silver chloride electrode to titrate quats and other organic cations in pharmaceuticals. Pinzauti and La Porta [12] used a 30 cm × 1 mm silver wire electrode wound on a glass rod and a mercury (I) sulphate electrode for a similar purpose, using very low concentrations. The silver electrode must be scrupulously clean; the authors recommend dipping it in 6 M nitric acid containing a little sodium nitrite, rinsing, dipping in stirred 2 M potassium cyanide for 2 min and rinsing again. They obtained excellent titration curves with 0.0005 or 0.001 M NaTPB. If this method is used the analyst must remember that cyanides are extremely poisonous and extremely fast-acting, and extreme care must be taken to avoid ingesting the solution or vapour.

There are many other published variants.

Any of the electrodes described in section 3.6.2 may also be used, as may the Orion 93-42 or any other commercially available surfactant electrode.

None of the electrodes mentioned, nor any of the much larger number described in the literature, has any obvious general advantage over the others.

An electrode system that does seem to offer the advantage of added sharpness of the end-point was reported by Shoukry *et al.* [13,14]. He and his co-workers made membrane electrodes selective for TPB and hexadecylpyridinium ions. The first membrane contained hexadecylpyridinium TPB and the second hexadecylpyridinium phosphotungstate, in a PVC-dioctyl phthalate matrix. The electrodes were used together, with either one functioning as indicating electrode and the other as 'reference' electrode. When 50 ml 0.001 M hexadecylpyridinium bromide was titrated with 0.01 M NaTPB, the titration curves showed either a sharp peak or a sharp trough at the end-point. It was not necessary to add particularly small increments of titrant near the end-point, which could be located very accurately by plotting intersecting straight lines through three or four points on either side. The system would be well worth trying for other quats, although results were reported only for hexadecylpyridinium bromide.

The phosphotungstate electrode was also successfully used with a calomel reference electrode to determine the quat by the method of standard additions [15]. Response was instantaneous and Nernstian over the concentration range 6.3×10^{-6} to 3.1×10^{-3} M, and the actual potential was only slightly affected by pH over the range 2.5–8.0.

7.2.6 Determination of free amine and amine hydrochloride

Free amine and amine hydrochloride respond to all the preceding titration procedures when they are carried out in acid solution. They are easily determined by the following procedure [16], and subsequently corrected for.

Free amine

1. Dissolve a 10 g sample in 100 ml propan-2-ol, heating if necessary.
2. Add 1 ml 0.2% bromophenol blue in propan-2-ol (see Note 1).
3. Titrate with 0.2 M hydrochloric acid in propan-2-ol to a bright yellow colour.
   Note 1. If the solution is yellow at this point, the presence of free hydrochloric acid is indicated. In this case take a fresh sample, dissolve in propan-2-ol, add a measured excess of 0.1 M sodium hydroxide in propan-2-ol and titrate to phenolphthalein with 0.2 M hydrochloric acid in the same solvent. Then determine amine hydrochloride by titrating to bromophenol blue as described.

Amine hydrochloride

1. Dissolve a 10 g sample in 100 ml propan-2-ol, heating if necessary.
2. Add 1 ml 1% phenolphthalein in ethanol (see Note 2).
3. Purge with a stream of nitrogen for 5 min and continue to purge during the titration.
4. Titrate with 0.1 M sodium hydroxide in propan-2-ol.
5. Carry out a blank experiment.
   Note 2. If the solution becomes pink at this point, the presence of free caustic alkali is indicated. In this case neutralise to phenolphthalein with 0.1 M hydrochloric acid in propan-2-ol and determine free amine as described above.

It is not possible to have free hydrochloric acid and free amine together, nor free caustic alkali and amine hydrochloride.

7.3 Carboxybetaines

The most common commercial betaines are of two main types, having the structures

$$\begin{array}{c}R'\\|\\RN^+CH_2COO^-\\|\\R'\end{array}\quad\text{and}\quad RCONH(CH_2)_3\begin{array}{c}R'\\|\\N^+\\|\\R'\end{array}CH_2COO^-$$

in which R′ is usually methyl but sometimes 2-hydroxyethyl. Many other structures are possible, but these are the most important, and the analytical chemistry of others is mostly similar.

7.3.1 Titration with sodium dodecyl sulphate

Both types of betaine can be titrated in a two-phase system by the method of ISO 2871-2, but the concentration of hydrogen ion in the solution must be at least 0.1 M. Even then the end-point is not very well defined.

Rosen *et al.* [17] developed a variant of this procedure in which a small amount of ethanol was added to the solution. The method had to be calibrated for each individual type of betaine, by titrating pure material and adjusting the volume of ethanol until a result of 100% was obtained. Using 0.001 M sodium dodecanesulphonate as titrant they successfully titrated the two main types, 2-pyridinium fatty acids of different chain lengths, *N*-dodecyl-*N*-benzyl-*N*-methylglycine and nine commercial betaines.

Potentiometric titration is much more satisfactory, using any of the electrodes described in section 3.6.2. Sample solutions and titrant can be up to 0.04 M. The solution must be at least 0.1 M in hydrogen ion.

7.3.2 Titration with sodium tetraphenylborate

Provided the solution is at least 0.1 M in hydrogen ion, betaines can be titrated potentiometrically with NaTPB. Vytřas *et al.* [18] successfully titrated cocodimethylbetaine, cocoamidopropylbetaine and an imidazoline derivative with NaTPB using the coated aluminium wire described in section 3.6.3. The potential changes at the end-point were not very large, and the method would probably be useful mainly for fairly pure samples such as raw materials rather than for products.

It is to be expected that other types of surfactant-sensitive electrodes can be used with equal or greater success, and it would not be surprising if the two-phase methods, e.g. that of Cross (section 7.1.3) could also be used, although the author is unaware of any published evidence.

Buschmann [22] has reported an unusual titration method for betaines, which is more fully discussed in section 7.4.

7.3.3 Potentiometric acid–base titration

All amphoterics, including betaines, can be titrated between end-points at high and low pH. Unfortunately the raw materials contain other weak acids and bases that also titrate. The following method [19, 20] avoids this difficulty.

1. Weigh a sample containing about 1 millimol of betaine in a 250 ml beaker. Add 15 ml 0.1 M hydrochloric acid in propan-2-ol and dissolve.
2. Add 150 ml methyl isobutyl ketone and purge with a stream of nitrogen for 5 min. Maintain purging during the titration.
3. Insert pH electrodes and titrate potentiometrically with 0.1 M ethanolic potassium hydroxide through two points of inflection.

The first end-point measures the excess hydrochloric acid. The additional titrant to the second end-point measures the betaine. A third point of inflection, corresponding with the titration of glycolic acid, monochloroacetic acid and free amine, is found if the titration is continued further. These extraneous substances, being titrated over a different part of the curve, do not interfere, and the authors [19] confirmed by experiment that this was so.

7.3.4 Acid hydrolysis of amidopropylbetaines

An alternative approach for amidobetaines is to acid-hydrolyse the amido group and extract and weigh the fatty acid.

1. Weigh a sample containing about 2 g active in a round-bottomed flask.
2. Add 50 ml 6 M hydrochloric acid, attach a reflux condenser and boil for 8 h.
3. Cool and transfer to a separating funnel, rinsing the flask several times with petroleum ether.
4. Extract the fatty acid with petroleum ether, evaporate to dryness, dry at 105°C for 1 h, cool and weigh.

7.3.5 Determination of free amine

The same authors described a method [19, 21] for measuring the free amine, presumed to be tertiary, in betaines.

1. Weigh a 10 g sample in a 250 ml beaker. Dissolve in 150 ml 50% aqueous propan-2-ol. Purge with a stream of nitrogen for 5 min and maintain purging during the titration.
2. Add 10.00 ml 0.02 M tri-*n*-butylamine in propan-2-ol.
3. Neutralise to thymol blue with 0.1 M sodium hydroxide and add 1 ml excess.
4. Insert pH electrodes and titrate with 0.1 M hydrochloric acid through two points of inflection.
5. Carry out a blank experiment in exactly the same way, but omitting the sample.

The tri-*n*-butylamine ensures that there is a recognisable interval between the points of inflection when the amine content is very low. The sodium hydroxide ensures that the amine is entirely in the free base form and provides an initial point of inflection from which to measure the titration. Presence of 1.0% of either monochloroacetic acid or glycolic acid did not interfere with the determination of 0.6% free amine.

It would presumably be possible to combine the last two methods to determine monochloroacetic acid and glycolic acid together.

7.4 Amphoterics with weakly basic nitrogen

There are four possible approaches to the determination of WW type amphoterics. They are:

A. Titration with sodium dodecyl sulphate in acid solution. The solution must be at least 0.1 M in hydrogen ion, but otherwise the procedure is straightforward. Potentiometric titration is preferred.
B. Titration with benzethonium chloride in alkaline solution. The solution must be at least 0.1 M in hydroxide ion. Potentiometric titration is preferred, and the bromophenol blue method is probably better than the mixed indicator procedure. The present author's limited experience suggests that WW amphoterics are titrated rather more successfully as cationics than as anionics.

In two-phase titration using either A or B, it may be more effective to add a measured excess of titrant, shake very thoroughly, allow the layers to separate, run off the chloroform, add fresh chloroform and back-titrate the excess.

C. Titration with NaTPB. The solution must be at least 0.1 M in hydrogen ion. Potentiometric titration is preferred, but two-phase titration should be satisfactory. It may be more effective to add a measured excess of NaTPB, filter or centrifuge (or run off the chloroform layer) and determine the excess by precipitation with potassium, filtration and weighing.
D. Acid–base titration. All amphoterics can be titrated through two end-points, either potentiometrically or with indicators, in ethanol or propan-2-ol. This is not very useful, because other weak acids and bases, which are almost always present, also titrate. The method described in section 7.2.3 may be generally applicable. The following is an alternative.

Equal amounts of the sample are passed through a strongly acidic cation exchanger in the acid form and a strongly basic anion exchanger hydroxide. The amphoteric is retained by both columns. All anions emerge from the cation-exchange column as strong or weak acids. All cations emerge from the anion-exchange column as hydroxides or free weak bases. The effluents and a third equal amount of the sample are titrated potentiometrically with acid or alkali, as appropriate. Total weak acids, total weak bases and amphoteric content can then be calculated.

1. Prepare a column containing 10 milliequivalents of a strongly acidic cation-exchange resin in the acid form, in 80% ethanol. Wash with 100 ml 80% ethanol.
2. Prepare a column containing 10 milliequivalents of a strongly basic anion-exchange resin hydroxide, in 80% ethanol. Wash with 100 ml 80% ethanol.
3. Weigh a sample containing 12–15 millimol (24–30 milliequivalents) active. Dissolve in ethanol and transfer to a 500 ml volumetric flask containing 100 ml water, using ethanol to effect the transfer. Dilute to volume with ethanol and mix.
4. Pass 50 ml of the sample solution through each of the ion-exchange columns. Rinse in with 80% ethanol and wash each column with 100 ml of the same solvent. Do not exceed a flow rate of 2 ml/min/cm^2. Collect the effluents in 400 ml beakers.
5. Titrate the effluent from the cation-exchange column potentiometrically with ethanolic 0.1 M sodium hydroxide through two points of inflection at pHs below and above 7.
6. Titrate the effluent from the anion-exchange column potentiometrically with ethanolic 0.1 M hydrochloric acid, through two points of inflection at pHs above and below 7.
7. Pipette 50 ml of the original solution (step 3) into a 150 ml beaker. Either acidify to bromophenol blue with a few millilitres of 1 M hydrochloric acid and titrate potentiometrically with ethanolic 0.1 M sodium hydroxide as in step 5, or make alkaline to phenolphthalein with a few millilitres of 1 M sodium hydroxide and titrate potentiometrically with ethanolic 0.1 M hydrochloric acid as in step 6.

The alkali consumed between the end-points at step 5 measures the weak acids.

The acid consumed between the end-points at step 6 measures the amines.

The acid or alkali consumed between the end-points at step 7 measures the sum of weak acids, amines and amphoteric. The amphoteric consumes two mol of acid or alkali, assuming it has one acid and one basic group.

The points of inflection may be rather flat. If so, the end-points may be accurately found by doing the titration with an autotitrator programmed to draw the first derivative curve.

7.5 Sulphobetaines

SW sulphobetaines are more like anionics than cationics, and can be titrated with benzethonium chloride in alkaline solution, by either two-phase or potentiometric procedures. They cannot be titrated with NaTPB.

True sulphobetaines, with quaternary nitrogen, cannot be titrated with either sodium dodecyl sulphate or benzethonium chloride. They do not interact with ion exchangers, and so appear in the nonionic fraction of mixtures. They cannot be titrated with NaTPB by any of the procedures described earlier, but Buschmann [22] has reported successful titration of sulphobetaines by a technique which he calls 'electrochemically indicated amphimetry'.

According to Buschmann, in very acid solution the ionisation of the sulphonic acid group is sufficiently suppressed for sulphobetaines to function as cationics, forming TPB salts which can be extracted by 1,2-dichloroethane. If the TPB salt is titrated in such a two-phase system with a quat, the sulphobetaine is displaced and returns to the aqueous phase. The indicator is the dodecyl sulphate of the iron (II)–1,10-phenanthroline complex, which is initially in the organic layer. The titration is done by the technique of AC voltammetry. The current rises during the titration. At the end-point the quat reacts with the dodecyl sulphate, releasing the iron complex, which passes into the aqueous layer, causing the current to fall.

The paper [22] leaves some questions unanswered. The author states that sulphobetaines behave like cationics only at acid concentrations exceeding 1.5 M, but the procedure described involves sulphuric acid at less than 0.05 M. It is not clear why the colour transfer is not a satisfactory indication of the end-point; the iron (II)–1,10-phenanthroline complex is deep red. Although the author claims to have successfully titrated 3-(dodecyldimethylammonium)propyl sulphonate by this procedure, he does not cite any supporting data. Nevertheless, this is an interesting development and the method is worth investigation. The AC voltammetry apparatus described is no longer available, and an alternative indicator mechanism is required, preferably visual.

The same paper [22] also describes a similar technique by which the author claims to have successfully titrated *N*-lauroylsarcosine, *N*-lauroylmethyltaurine, carboxybetaines, a WW amphoteric and amine oxides. In these cases the extracted TPB complex was titrated with 0.01 M dodecyl hydrogen sulphate, using the same indicator and technique as before. Titration curves (current versus volume) for a lauryliminopropionic acid, an amine oxide and *N*-lauroylmethyltaurine show well-defined peaks at the end-point. Acyl sarcosines and methyltaurines are apparently protonated on the amide nitrogen atom under acid conditions, which is unexpected. Again, an alternative indicator mechanism is needed.

7.6 Amines

This section includes methods for determining the total, primary, secondary and tertiary amine contents of fatty amines which are nominally primary, secondary or tertiary or mixtures. For determination, another approach is to titrate with sodium dodecyl sulphate. Either two-phase or potentiometric titration may be used. The solution must be distinctly acid, pH 3 or lower. Two-phase or potentiometric titration with NaTPB is also applicable. Procedures for all of these are given in section 7.1. The difficulty with these approaches is in deciding what value to use for the molecular weight.

7.6.1 Determination of molecular weight and total, primary, secondary and tertiary amines

The analysis has three parts. In the first the total amine content is determined by acid titration. In the second the secondary and tertiary amines are determined together after converting the primary amine to a Schiff's base, which is not titrated. In the third, only the tertiary amine is determined, after converting the primary and secondary amines to thiourea derivatives.

Primary amines are converted to a Schiff's base by reaction with salicylaldehyde:

$$HOC_6H_4CHO + H_2NR \rightarrow HOC_6H_4CH{=}NR + H_2O$$

Primary and secondary amines are converted to thiourea derivatives by reaction with phenyl isothiocyanate:

$$C_6H_5N{=}C{=}S + H_2NR \rightarrow C_6H_5NH\overset{\overset{\displaystyle S}{\|}}{C}NHR$$

Secondary amines give a similar product, except that the end-group is $—NR_2$.

The analytical procedure [23] is as follows. The propan-2-ol must contain not more than 0.05% water. The phenyl isothiocyanate must have a base content equivalent to not more than 0.006 ml of 0.2 M hydrochloric acid per millilitre. The hydrochloric acid is in propan-2-ol. The solvent referred to is 150 ml ethanediol (ethylene glycol) containing not more than 0.1% water, 250 ml *n*-heptane and 400 ml propan-2-ol.

Total amine

1. Weigh an appropriate weight of sample (see below) in a 250 ml beaker.

2. Add 60 ml propan-2-ol and a few glass beads and boil for 1 min (caution: flammable vapour).
3. Add 80 ml solvent.
4. Insert pH electrodes and titrate potentiometrically with 0.2 M hydrochloric acid to just beyond the point of inflection.

Secondary and tertiary amine

1. As for total amine, steps 1 and 2.
2. Add 5 ml salicylaldehyde, mix and allow to stand for 5 min at 60–70°C.
3. Add 80 ml solvent.
4. As for total amine, step 4.

Tertiary amine

1. As for total amine, steps 1 and 2.
2. Add 5 ml phenyl isothiocyanate, mix and allow to stand for 15 min at 60–70°C.
3. Add 80 ml solvent.
4. As for total amine, step 4.

Guideline sample weights are as follows.

For total amine, use 0.6–0.7 g for primary amines (group A) and tertiary amines with two methyl groups (group B), 1.1 g for secondary and tertiary amines with two C_{12}–C_{14} fatty chains (group C) and 1.5 g for secondary and tertiary amines with two C_{16}–C_{18} fatty chains (group D).

For secondary and tertiary amine, use 2.5–3.0 g for group A, 0.6–0.7 g for group B, 1.1 g for group C and 1.5 g for group D.

For tertiary amine, use 3.0–4.0 g for group A, 0.6–0.7 g for group B, 1.1 g for group C and 1.5–2.0 g for group D.

Use weights at the lower end of the ranges given for lower-molecular-weight amines and at the higher end for higher-molecular-weight amines.

The calculations below give concentrations in the sample in meq/g. To find the weight percentage, multiply by the molecular weight and divide by 10.

$$\text{Apparent molecular (equivalent) weight} = 1000W_1/V_1C$$

$$\text{Total amine} = V_1C/W_1$$

$$\text{Primary amine} = C(V_1/W_1 - V_2/W_2)$$

$$\text{Secondary amine} = C(V_2/W_2 - V_3/W_3)$$

$$\text{Tertiary amine} = V_3C/W_3$$

where V_1 = volume of 0.2 M acid used by total amine, W_1 = weight of sample taken for total amine determination, V_2 = volume of 0.2 M acid used by secondary and tertiary amine, W_2 = weight of sample taken for secondary and tertiary amine determination, V_3 = volume of 0.2 M acid used by tertiary amine, W_3 = weight of sample taken for tertiary amine determination, and C = exact concentration of the acid in mol/l.

The same analysis can be done by visual titration [24], using 1.0 ml bromophenol blue in propan-2-ol, increasing the volume of prop-an-2-ol to 150 ml and titrating to a clear yellow colour.

7.6.2 Tertiary amines: determination of primary plus secondary amines

This [25] is an alternative method for determining the sum of primary and secondary amines, without distinguishing between them. The primary and secondary amines are converted to alkyl dithiocarbamic acids by reaction with carbon disulphide:

$$RNH_2 + CS_2 \rightarrow RNHCSSH; \; R_2NH + CS_2 \rightarrow R_2NCSSH$$

These acids are titrated potentiometrically in a largely nonaqueous medium.

The propan-2-ol must not contain any primary alcohols. The solvent used at step 1 is a mixture of 800 ml tetrahydrofuran (caution: flammable), 100 ml propan-2-ol and 100 ml water. In this medium the above reaction is very rapid.

Most samples show more than one point of inflection. To avoid errors arising from this source, two determinations are done on different amounts of sample, and the difference between them is used to calculate the result. A blank titration is necessary for each determination. This means doing four titrations for each determination, which makes the method rather cumbersome. The analyst must decide in individual cases, on the basis of experience, whether the additional accuracy gained is sufficient to justify the three titrations of steps 4 and 5.

Because of the extremely offensive smell and high toxicity of carbon disulphide, it is strongly recommended that the analysis be done in a well-ventilated fume cupboard. It must in any case be done in a titration vessel that can be closed to the atmosphere, to exclude moisture. A simple way of doing this is to fit a wide-necked titration flask with a rubber bung through which pass the glass electrode and a salt bridge, a flexible capillary to deliver the titrant and a glass tube closed by a smaller rubber bung, through which the carbon disulphide is added. However, an all-glass apparatus is preferable.

Before the analysis the glass electrode must be conditioned by immersion in 0.1 M hydrochloric acid for 30 min. The electrode should be positioned so that it is at least 1.5 cm away from the burette tip in the titration vessel.

The reference electrode is a calomel electrode dipping into 1.5 M lithium chloride in 50% aqueous methanol, and connected to the test solution via the salt bridge.

It is very desirable, though not absolutely essential, to do the titration automatically, with the titrator programmed to measure millivolts and to plot the first derivative curve, i.e. dE/dV against V, where E is the potential and V the volume of titrant.

1. Weigh a 5 g sample into the titration vessel. Add 100 ml solvent and dissolve. Assemble the titration vessel. Stir the solution magnetically throughout the titration.
2. Add 5 ml carbon disulphide (caution: flammable; toxic and obnoxious vapour) from a dispenser, e.g. a glass syringe. Close the titration vessel.
3. Wait until the indicated potential or pH is constant, then titrate potentiometrically with 0.1 M sodium hydroxide. Plot the first derivative curve.
4. Repeat with a 2.5 g sample.
5. Carry out blank determinations on a 5 g and a 2.5 g sample, following the same procedure but omitting the carbon disulphide.

The volume of titrant used at step 3 must not exceed 4 ml.

$$\text{Primary + secondary amine (meq/g)} = C[(V_1 - V_2)/(W_1 - W_2) - (V_3 - V_4)/(W_3 - W_4)]$$

where V_1 = volume of 0.1 M alkali used at step 3, V_2 = volume of 0.1 M alkali used at step 4, W_1 = weight of c. 5 g sample titrated at step 3, W_2 = weight of c. 2.5 g sample titrated at step 4, V_3 = volume of alkali used by larger sample at step 5, V_4 = volume of alkali used by smaller sample at step 5, W_3 = weight of larger sample titrated at step 5, W_4 = weight of smaller sample titrated at step 5, and C = exact concentration of alkali in mol/l.

The volumes are found from the first derivative plots by selecting the tallest or most clearly distinguished corresponding peaks in the four plots.

The result may be expressed as a percentage of secondary amine by multiplying the mol/kg result by the molecular weight of the secondary amine and dividing by 10.

7.7 Fatty diamines: determination

Fatty diamines have the general structure $RNHR'NH_2$, where R is a fatty chain with 12–18 carbon atoms and R′ is usually C_3H_6. The preferred method of determination [26] is to follow the indicator methods for determination of total amine and secondary and tertiary amine given in section 7.2.1, using about 0.5 g sample for total amine and about 1.0 g for secondary and tertiary amine. The volume of propan-2-ol to be used in each analysis is 50 ml.

The total amine method determines both the primary and the secondary amine groups, together with any tertiary amine that might be present as an impurity. The secondary and tertiary amine method determines what it says. The difference is the primary amine, which is taken as a measure of the diamine. The same results can alternatively be used to calculate the equivalent weight, which is half the molecular weight.

$$\% \text{ diamine} = (V_1/W_1 - V_2/W_2) \times C \times M/10$$

$$\text{Equivalent weight} = 1000W_1/V_1C$$

where V_1 and W_1 refer to the total amine determination and V_2 and W_2 to the secondary and tertiary amine determination.

One would expect diamines to be titratable with both sodium dodecyl sulphate and NaTPB in acid solution, requiring two molecules per molecule, but the present author has no experience of such titrations and is not aware of any reports of them in the literature.

7.8 Ethoxylated amines and diamines

7.8.1 Determination of total amine or equivalent weight

Ethoxylated amines with short ethylene oxide chains can be titrated with sodium dodecyl sulphate and NaTPB in acid solution, but the titrations become increasingly unsatisfactory as the chain length increases. The same is no doubt true of ethoxylated diamines. Commercial ethoxylated amines have typically 12–60 EO units, and the diamines have up to 25.

The recommended method is the indicator method for total amine given in section 7.2.1. Use a sample of 4–5 meq, dissolved in 150 ml propan-2-ol. The procedure is as described. The method used by Akzo [27] uses rather small sample weights and obtains rather small titrations; this is a scaled-up adaptation.

7.8.2 Determination of primary plus secondary amine

The ethoxylates of both mono- and diamines consist principally of tertiary amines, most of the primary nitrogen atoms adding on two ethylene oxide chains and the secondary nitrogen atoms of the diamines adding on one. Any surviving primary amine is the starting material, RNH_2, and any secondary amine is either unreacted diamine or primary amine that has added on only one EO chain. Such chains are always short, mostly only one EO unit. The level of primary and secondary amines is low.

This method [28] is a simpler version of the potentiometric method for primary and secondary amines given in section 7.2.2. The acid is added at step 4 to ensure the presence of an observable point of inflection. Observe appropriate precautions for carbon disulphide.

1. Weigh 5 g sample into the titration vessel.
2. Dissolve in 50 ml propan-2-ol.
3. Add 5 ml carbon disulphide, stir and wait 1 min.
4. Add 3.00 ml 0.2 M hydrochloric acid in propan-2-ol.
5. Titrate potentiometrically with 0.1 M sodium hydroxide to just beyond the point of inflection.

$$\text{Primary + secondary amine (eq/kg)} = (V_1 C_1 - 3C_2)/W$$

where V_1 = volume of 0.1 M alkali used, C_1 = exact concentration of alkali in mol/l, C_2 = exact concentration of hydrochloric acid in mol/l, and W = weight of sample.

7.8.3 Determination of ethylene oxide chain length

ISO 2270 states that the modified Siggia method cannot be used for nitrogenous compounds. However, Longman [29] maintains that it can, but that the —N—C— bond of the first ethylene oxide unit is not broken, and the result calculated in the standard way therefore gives a result one unit too low for each chain, i.e. two units per molecule, assuming a negligible level of primary amine adding on only one chain.

7.8.4 Determination of polyethylene glycol

Polyethylene glycols and total ethoxylate active may be determined by the Weibull method (section 6.2.6).

An alternative method, which is quicker and easier to carry out, is to remove the ethoxylated amines on a strongly acidic cation-exchange resin. The polyethylene glycols pass through and are weighed.

1. Prepare an ion-exchange column containing about 40 meq of a strongly acidic cation-exchange resin in the acid form, in methanol.
2. Dissolve 4–5 g of the sample in 50 ml methanol.
3. Pass the solution through the column at not more than 2 ml/min/cm^2, collecting the effluent in a tared beaker.
4. Wash the column with three bed-volumes of methanol, collecting the effluent in the same beaker.
5. Evaporate the methanol to dryness on a steam bath (caution: flammable vapour), dry the residue at 105°C for 1 h, cool and weigh.

ISO 6384 [30] describes a similar method.

7.9 Amine oxides

The determination of amine oxides is straightforward: they can be titrated in acid solution with sodium dodecyl sulphate or NaTPB, either potentiometrically or in a two-phase system, for which purposes the solution must be at least 0.1 M in hydrogen ion, and they can be titrated with acid. However, for quality-control purposes it is necessary to determine the free tertiary amine content. This can be done by potentiometric titration, either in a solvent which permits discrimination on the basis of base strength, or before and after removal of the amine oxide. Methods are described for both of these.

In the first method [31] the sample is titrated with hydrochloric acid in 50% propan-2-ol. In water or propan-2-ol alone only one point of inflection is observed, but in the mixture the tertiary amine is the stronger base and is titrated first, well separated from the amine oxide.

Solvent: 50% aqueous propan-2-ol, carbon dioxide free. Mix the solvents and either heat to boiling and cool (caution: flammable vapour) or purge with a stream of nitrogen for half an hour.

'Spike' solution: dissolve 0.4 g sodium hydroxide and 2 g tri-*n*-butylamine in 500 ml solvent.

1. Weigh a sample containing about 1 millimol amine oxide in a 250 ml beaker. Dissolve in 100 ml solvent.
2. Add 10.00 ml 'spike' solution.
3. Titrate potentiometrically with aqueous 0.1 M hydrochloric acid.
4. Repeat on 10 ml 'spike' solution in 100 ml solvent.

The titration curve shows three points of inflection. The first represents the titration of the sodium hydroxide, the second the titration of the tertiary amine and the third the titration of the amine oxide.

The addition of the 'spike' serves to ensure firstly a marker inflection from which to measure, and secondly an appreciable interval between the points of inflection for the sodium hydroxide and the tertiary amine. The calculation is as follows. In each case the volume is the total volume of titrant added to reach that point.

Sample titration

First end-point at V_1 ml = sodium hydroxide in sample + sodium hydroxide in spike

Second end-point at V_2 ml = sodium hydroxide + tertiary amine in sample + tertiary amine in spike

Third end-point at V_3 ml = sodium hydroxide + tertiary amine + amine oxide

Blank titration

First end-point at V_4 ml = sodium hydroxide in spike

Second end-point at V_5 ml = sodium hydroxide + tertiary amine in spike.

Tertiary amine in sample is measured by $(V_2 - V_1) - (V_5 - V_4)$. Amine oxide in sample is measured by $(V_3 - V_2) - (V_4 - V_1)$.

An alternative procedure [32] uses a different approach, making use of the Polonovski reaction, which converts the amine oxide into an *N*,*N*-disubstituted acetamide:

$$RN(CH_3)_2O + (CH_3CO)_2O \rightarrow RN(CH_3)COCH_3 + HCHO + CH_3COOH$$

1. Weigh a sample containing about 1.5 millimol amine oxide in a 250 ml beaker. Add 50 ml glacial acetic acid and dissolve.
2. Titrate potentiometrically with 0.1 M perchloric acid in glacial acetic acid to just beyond the point of inflection.

This measures the total bases in the sample.

3. Weigh about 5 g sample in a 100 ml conical flask. Add 25 ml acetic anhydride and dissolve.
4. Add some glass beads, attach a reflux condenser and boil on a hot-plate for 10 min.
5. Cool and add 50 ml glacial acetic acid.
6. Titrate potentiometrically with 0.1 M perchloric acid in glacial acetic acid to just beyond the point of inflection.

This measures the tertiary amine.

References

1. ISO 2871-1: 1988: Surface active agents—Detergents—Determination of cation-active matter content—Part 1: High-molecular-mass cation-active matter. International Organization for Standardization, Geneva.
2. ISO 2871-2: 1990: Surface active agents—Detergents—Determination of cation-active matter content—Part 2: Cation-active matter of low molecular mass (between 200 and 500). International Organization for Standardization, Geneva.
3. Cullum, D. C. Proc. IIIrd International Congress on Surface Active Substances, Cologne, 1960, Vol. C, pp. 42–50.
4. Cross, J. T. *Analyst*, **90** (1965), 315–324.
5. Patel, D. M. and Anderson, R. A. *Drug Standards*, **26** (1958) 189.
6. Tsubouchi, M., Mitsushio, H. and Yamasaki, N. *Anal. Chem.*, **53** (1981), 1957–1959.
7. Metcalfe, L. D., Martin, R. J. and Schmitz, A. A. *JAOCS*, **43** (1966), 355–357.
8. Akzo Chemie Method VE/2.007: Fatty quaternary ammonium salts: Determination of activity. Akzo Chemie Research Center, Deventer, Netherlands.
9. Donkerbroek, J. J. and Wang, C. N. *Proc. 2nd World Surfactants Congress*, Paris, 1988.
10. Wang, C. N., Metcalfe, L. D., Donkerbroek, J. J. and Cosijn, A. H. M. *JAOCS*, **66** (1989), 1831–1833.

11. Christopoulos, T. K., Diamandis, E. P. and Hadjiioannou, T. P. *Anal. Chim. Acta*, **143** (1982), 143–151.
12. Pinzauti, S. and La Porta, E. *Analyst*, **102** (1977), 938–942.
13. Shoukry, A. F. *Analyst*, **113** (1988), 1305–1308.
14. Shoukry, A. F., Badawy, S. S. and Issa, Y. M. *Anal. Chem.*, **59** (1987), 1078.
15. Shoukry, A. F., Badawy, S. S. and Farghali, R. A. *Anal. Chem.*, **60** (1988), 2399–2402.
16. Akzo Chemie Method VV/2.002: Fatty quaternary ammonium salts: Determination of amine hydrochloride and free amine.
17. Rosen, M. J., Zhao, F. and Murphy, D. S. *JAOCS*, **64** (1987), 439–441.
18. Vytřas, K., Kalous, J. and Symersky, J. *Anal. Chim. Acta*, **177** (1985), 219–223.
19. Plantinga, J. M., Donkerbroek, J. J. and Mulder, R. J. *JAOCS*, **70** (1993), 97–99.
20. Akzo Chemie Method VE/2.009: Alkyldimethylbetaines: Determination of betaine.
21. Akzo Chemie Method VE/2.010: Alkyldimethylbetaines: Determination of free amine.
22. Buschmann, N. *Tenside*, **28** (1991), 329–332.
23. Akzo Chemie Method VE/1.005: Fatty amines: Determination of primary, secondary and tertiary amine.
24. Akzo Chemie Method VV/1.001: Fatty amines: Determination of primary, secondary and tertiary amine.
25. Akzo Chemie Method VE/2.005: Fatty tertiary amines: Determination of primary and secondary amine.
26. Akzo Chemie Method VV/4.001: Fatty diamines: Determination of diamine.
27. Akzo Chemie Method VV3.001: Ethoxylated fatty amines/diamines: Determination of equivalent mass.
28. Akzo Chemie Method VE/3.001: Ethoxylated fatty amines/diamines: Determination of primary + secondary amine.
29. Longman, G. F. *The Analysis of Detergents and Detergent Products*, John Wiley, 1975, pp. 285–286.
30. ISO 6384: 1981: Surface active agents—Technical ethoxylated fatty amines—Methods of analysis. International Organization for Standardization, Geneva.
31. Donkerbroek, J. J. and Wang, C. N. *Proc. 2nd World Surfactants Congress*, Paris, 1988.
32. Akzo Chemie Method VE/2.008: Fatty amine oxides: Determination of amine oxide and of free tertiary base.

8 Analysis of mixtures without separation

8.1 Introduction

Many mixtures of surfactants can be analysed only after prior separation into the main classes, or by methods that are intrinsically separative, notably the various chromatographic techniques. Even in those cases where chromatography is not the only possible approach, it is often the most convenient and cost-effective.

Nevertheless, there are many other mixtures of surfactants that can be analysed by classical methods without prior separation. Mixtures of surfactants of one type, e.g. the anionic, cationic and nonionic fractions recovered from an ion-exchange column, can often be so treated, the most notable exception being mixtures of different kinds of sulphonated hydrocarbons, which require high-tech methods. Mixtures of nonionics and of cationics are, however, not often encountered in practice, but a fair proportion of mixtures of surfactants of different classes can be analysed without recourse to high technology, and this is of considerable practical use.

It is not feasible, nor would it serve any useful purpose, to deal with every possible combination. This chapter gives a broad selection of examples illustrating the general approach. The intention is to encourage the analyst to think along the right lines rather than to cater for every contingency. It is assumed throughout that the surfactants present in the mixtures have already been identified by some form of spectroscopy.

Because all the techniques required have already been covered in earlier chapters, only outline approaches are given here. Because of the enormous range of possible surfactant structures and combinations of components, it is inevitable that the generalised guidelines given here will sometimes fail, or succeed for only part of the analysis.

Quite arbitrarily, solvent extraction of weak acids or bases as a first step in the analysis is considered for the purposes of this chapter to be prior separation. Solvent extraction of hydrolysis products during the analysis is, by definition, not prior separation.

In the majority of cases the results must first be calculated in mol/100 g or in meq/g, and subsequently converted to a weight percentage by multiplying by the appropriate molecular weight in the first case and by one-tenth of the molecular weight in the second.

8.2 Mixtures of anionics

8.2.1 Introduction

All sulphate and sulphonate anionics except hydrocarbon sulphonates can be acid-hydrolysed. In every case but one, namely α-sulphonated fatty esters, the hydrolysis destroys their surface activity and renders them incapable of titration with benzethonium chloride. In that exceptional case, the titration per molecule is doubled by the hydrolysis. All esters are decomposed by alkaline hydrolysis or saponification, but in the case of α-sulphonated fatty esters this is not a useful measurement because of the slow hydrolysis of the sulphonate group. Also, alkanolamides, particularly dialkanolamides, are hydrolysed slowly by the normal saponification process, and although this is not a useful method of analysis for them, the possibility of generating measurable amounts of soap and alkanolamine cannot be ignored when alkali-hydrolysing anionic esters.

The solvent-extractable hydrolysis products are often fatty acids, but may be fatty alcohols, ethoxylated fatty alcohols, ethoxylated alkylphenols or fatty amines (sulphosuccinamates).

The non-extractable hydrolysis products are characteristic of the surfactant type, and include hydrogen ions, sulphate ions, glycerol, alkanolamines, sulphonated amino acids and sulphosuccinic acid.

Among carboxylates only sarcosinates are susceptible to hydrolysis.

Anions of strong acids undergo titration with benzethonium chloride at all pHs. Anions of weak acids do so only in alkaline solution.

In all cases the first action is to determine total activity by titration with benzethonium chloride (BEC), and this is to be taken as mandatory in each of the following cases. After acid hydrolysis a second BEC titration to determine the sulphonated hydrocarbon is probably all that is required in the majority of cases, but the following are alternatives. In the case of α-sulphonated fatty esters, the second BEC titration measures the total α-sulphonated fatty acid plus the sulphonated hydrocarbon (see below).

8.2.2 Sulphonated hydrocarbons with any other anionic

General. Acid-hydrolyse by the appropriate method and determine the surviving titratable surfactant (sulphonated hydrocarbon) or one of the hydrolysis products.

With alkyl sulphate, alkylether sulphate or alkylphenyl ether sulphate. Acid hydrolysis (section 2.4.2). Then any of the following.

Determination of liberated acidity (section 5.3.1)
Extraction of fatty alcohol with petroleum ether (section 5.3.2), or of

ethoxylated alcohol or alkylphenol by deionisation (section 5.3.5)
Determination of sulphate ion (section 5.3.4)

With monoalkanolamide or ethoxylated monoalkanolamide sulphate. Acid hydrolysis (section 2.4.4). Then any of the following.

Extraction of fatty acid (section 5.4.2)
Determination of alkanolamine or ethoxylated alkanolamine (section 5.4.2)

With mono- or diglyceride sulphate. Acid hydrolysis (section 2.4.3), then as for alkyl sulphates, etc., above, or

Determination of glycerol (section 5.5.4)

Alkaline hydrolysis (section 3.4.1) saponifies the ester groups and may give all the information required, but does not liberate glycerol or a sulphate ion. The fatty acid can be extracted if desired.

With any ester except α-sulphonated fatty ester. Either acid or alkaline hydrolysis. Then any of the following.

Extraction of fatty acid (acyl isethionates) or of fatty alcohol (sulphosuccinates)
Determination of sulphosuccinic acid (sulphosuccinates)

With α-sulphonated fatty ester. Acid hydrolysis (section 5.11.4). Then BEC titration in alkaline solution. The increase over the result in acid solution measures the sum of α-sulphonated fatty acid already present and the hydrolysed α-sulphonated fatty ester.

With sulphonated amide. Acid hydrolysis (section 5.13.1). Then any of the following.

Extraction of fatty acid (acyl taurates and methyltaurates)
Determination of taurine or methyltaurine (section 5.13.1)
Extraction of fatty amine (sulphosuccinamates, section 5.13.2)
Determination of sulphosuccinic acid (section 5.13.2)

8.2.3 Combinations of species other than sulphonated hydrocarbons

The presence of sulphonated hydrocarbons does not really make much difference. In the following examples it is assumed that a sulphonated hydrocarbon is present.

General. Acid hydrolysis by the procedure appropriate to the most resistant species present, then a second BEC titration plus the determination of at least one hydrolysis product.

Any sulphate with any sulphonated ester except α-sulphonated fatty ester. Determine one of the following to measure the sulphate:

Fatty alcohol (unless the ester is a sulphosuccinate)
Ethoxylated alcohol or alkylphenol
Liberated hydrogen ion
Liberated sulphate ion
Fatty acid (from sulphated alkanolamide)

Determine one of the following to measure the ester

Fatty acid
Fatty alcohol (if the ester is a sulphosuccinate)
Sulphosuccinic acid

or determine saponification value and repeat BEC titration to obtain two measurements of the ester and, separately, acid-hydrolyse and repeat BEC titration to measure the sulphate.

With a glyceride sulphate and any other sulphate together, this is the simplest approach.

Any sulphate with α-sulphonated fatty ester. BEC titration in alkaline solution after acid hydrolysis gives twice the α-sulphonated ester (approximately—see above) plus any sulphonated hydrocarbon, less the sulphate. Determine the sulphate separately by measuring any one of its decomposition products.

Any sulphate with any amide. Determine any one hydrolysis product from each. As the two species have no hydrolysis product in common, it does not matter which; choose the most convenient.

Alpha-sulphonated fatty ester with any other ester or amide. BEC titration in alkaline solution after acid hydrolysis gives twice the α-sulphonated ester (approximately—see above) plus any sulphonated hydrocarbon, less the other ester or amide. Determine whichever is appropriate:

Fatty acid (acyl isethionate or amide)
Fatty alcohol (sulphosuccinate)
Fatty amine (sulphosuccinamate)
Sulphosuccinic acid
Taurine or methyltaurine

The results then give total active (A), other ester or amide (B) and twice the α-sulphonated fatty ester plus any sulphonated hydrocarbon (C). $(C - A + B)$ then gives (twice) the α-sulphonated ester, and any sulphonate hydrocarbon is found by difference.

8.2.4 Any three species together, with or without sulphonated hydrocarbons

The possibilities are very numerous, and as it is unlikely that such mixtures will be encountered in practice, only two examples are given to illustrate the approach. Readers will be able to work out others, if needed, for themselves.

Alkyl ether sulphate, glyceride sulphate, sulphosuccinamate

BEC titration gives total active
BEC titration after alkaline hydrolysis gives glyceride sulphate
Sulphate ion or liberated acidity after acid hydrolysis gives sum of ether sulphate and glyceride sulphate and hence ether sulphate (and hence sulphosuccinamate)
Total nitrogen (section 2.5.2) confirms sulphosuccinamate
BEC titration after acid hydrolysis gives sulphonated hydrocarbon, if present

Alternative analyses after acid hydrolysis that might be considered include:

Glycerol determination gives glyceride sulphate
Extraction of ethoxylated alcohol (by deionisation) gives ether sulphate
Fatty acid by petroleum ether extraction and alkali titration (lower ethoxylates also extracted) gives glyceride sulphate
Extraction of fatty amine and acid titration (lower ethoxylates also extracted) gives sulphosuccinamate
Sulphosuccinic acid gives sulphosuccinamate, but fatty acid and fatty amine must be removed first.

There is no need to determine more than one hydrolysis product for each species.

Acyl isethionate, α-sulphonated fatty ester, acyl methyltaurate. BEC titration gives total active. After acid hydrolysis:

BEC titration in alkaline solution gives the sulphonated hydrocarbon plus twice the α-sulphonated fatty ester plus soap from the isethionate and acyl methyltaurate.
BEC titration in acid solution gives the sulphonated hydrocarbon plus the α-sulphonated fatty ester.

Extraction of fatty acid gives the acyl isethionate plus the acyl methyltaurate.
After removal of fatty acid, potentiometric titration of the aqueous layer gives the methyltaurine.

Total nitrogen on the original mixture gives the acyl methyltaurate.

8.2.5 *Any sulphate or sulphonate with any carboxylate*

BEC titration in alkaline solution measures both species. BEC titration in acid solution measures only the sulphate or sulphonate. Potentiometric acid-base titration in alcoholic solution measures only the carboxylate.

The presence of a carboxylate in any of the mixtures discussed earlier is therefore not a real complication.

Any carboxylate can be determined in any mixture of anionics by extraction of the acid form from acidified 50% ethanol, using petroleum ether for fatty acids and acyl sarcosines and chloroform for alkylether carboxylic acids.

Determination of total nitrogen is an alternative for mixtures of a sarcosinate with any sulphate or sulphate species except amides.

8.3 Mixtures of nonionics

Sulphobetaines are considered to be nonionic in the present context, because they appear in the nonionic fraction in ion-exchange separations. Mixtures of nonionics are not very often found. To keep this section short, sulphobetaines are considered separately.

8.3.1 *Sulphobetaines with any other nonionic*

With alkanolamides or their ethoxylates, determine the alkanolamide by acid hydrolysis and extraction of the fatty acid. Deionise the aqueous solution to isolate and determine the sulphobetaine.

With any non-nitrogenous nonionic or mixture, determine the total nitrogen content by Kjeldahl (section 2.5.2).

8.3.2 *Other nonionics*

The remaining types considered here are these:

Ethoxylated alcohols and alkylphenols
Sorbitan esters and their ethoxylates
Sucrose esters and their ethoxylates

Alkylpolyglycosides
Alkanolamides and their ethoxylates.

Useful analytical approaches are the following.

All ethoxylates: titration with NaTPB (section 6.2.2), or the cobaltothiocyanate method (section 6.2.5), or determination of total oxyethylene groups (section 6.2.8).
All esters: determination of saponification value (section 3.4.1), extraction of fatty acid. Sorbitan esters: determination of sorbitol (section 6.4.2).
Alkylpolyglycosides: acid hydrolysis and extraction of fatty alcohol.
Alkanolamides and their ethoxylates: acid hydrolysis, extraction of fatty acid, titration of alkanolamine or ethoxylated alkanolamine.

Total nitrogen by Kjeldahl (section 2.5.2).
Titration with perchloric acid in acetic anhydride (section 6.6.3) to determine total amine plus total amide.

Selection of suitable procedures permits the analysis of many mixtures of two or even three components, using the same general principles as in section 8.2. When both fatty alcohol and fatty acid are to be extracted, extract the alcohol first from an alkaline solution, then acidify and extract the acid. If ethoxylates are present as well, the ones with shorter ethylene oxide chains are wholly or partially extracted with the fatty alcohol.

If a sorbitan ester and a sorbitan ester ethoxylate are present as separate components, they can be determined provided the degrees of ethoxylation and esterification are known.

Mixtures of an alkanolamide ethoxylate with any other ethoxylate or with unethoxylated alkanolamide can be analysed provided the degree of ethoxylation of the alkanolamide ethoxylate is known.

8.4 Mixtures of cationics and amphoterics

8.4.1 Mixtures containing any two quaternary ammonium salts

The quats considered here fall into five groups:

1. Mono- and dialkylammonium salts
2. Benzalkonium salts
3. Alkylpyridinium salts (and others with nitrogen in an aromatic ring)
4. Imidazolinium quats
5. Quats whose hydrocarbon chains contain acid- or alkali-labile groups such as ester or amide.

Mono- and dialkylammonium quats cannot be determined in mixtures with each other without separation; chromatography is the best approach. The following guidelines will be sufficient for most other cases. Imidazolinium quats are likely to give problems of interpretation, because they are often mixtures of two or more species.

Total active is determined by two-phase or potentiometric titration with sodium dodecyl sulphate (SDS), or with NaTPB at pH 3. This is the mandatory first step in all cases.

Two-phase and potentiometric NaTPB titration at different pHs (section 7.1) will deal with some mixtures. Imidazolinium quats containing a ring structure are alkali-labile and titrate quantitatively only at pH 3 (but may give a variable and partial response at pH 10), alkylpyridinium and other aromatic nitrogen quats titrate at pH 3 and 10 and all others at pH 3, 10 and 13. The response of imidazoline-derived quats not containing a ring structure is uncertain; they probably titrate at all pHs, but this would need to be confirmed for individual cases.

For the sake of brevity, the groups identified above are referred to by their group numbers in the following guidelines.

Mixtures of groups 1 and 2. De-quaternisation of the benzalkonium salt by treatment with iodide, extraction of benzyl iodide and tertiary amine, titration with iodate [1]. The tertiary amine can be titrated with acid if useful.

Group 1 or group 2 with group 3. NaTPB titration at pH 3 and 13. All groups titrate at pH 3, only group 1 or 2 at pH 13.

Group 1 or group 2 with group 4. SDS or NaTPB in alkaline solution (0.1 M alkali) if the imidazoline contains a ring structure. SDS in alkaline solution after acid hydrolysis (all cases). Either of these determines group 1 or 2. Extraction and determination of fatty acid after acid hydrolysis determines group 4.

Group 1 or group 2 with group 5. Acid hydrolyse and extract fatty acid to determine group 5. SDS in acid solution on hydrolysed solution to determine group 1 or 2.

Group 3 with group 4 or 5. Acid hydrolyse. Determine group 3 by SDS titration, group 4 or 5 by extraction of fatty acid.

Group 4 with group 5. This combination could be difficult because of the uncertain composition of group 4 quats. Try NaTPB at pH 10 to determine group 5 alone. Otherwise, if the group 5 compound is an ester, acid hydrolyse, extract the fatty acids and determine amine in the hydrolysed solution by acid–base titration to determine group 4.

If the group 5 compound has more than one chain containing an ester or amide group, acid hydrolyse, extract the fatty acids, dissolve in neutral ethanol and titrate with ethanolic alkali. The molar ratio of fatty acid to total active permits calculation of the mole ratio of the two actives.

Other cases cannot be analysed without separation.

8.4.2 Quaternary ammonium salts with weakly basic cationics

Several approaches are possible. Complications arise if the quat contains appreciable levels of weak base as impurity.

1. Titrate with SDS in acid and alkaline solution, or with NaTPB at pH 3 and 10 or 13. The weak base titrates in the acid but not in the alkaline medium.
2. Dissolve in neutral ethanol or titrate with ethanolic hydrochloric acid, either potentiometrically or to bromophenol blue, to determine the weak base.
3. Extract the weak base with diethyl ether (caution: flammable) from alkaline 33% ethanol. To minimise the risk of decomposing labile quats, make the solution just alkaline to phenolphthalein. If desired, titrate the quat in the aqueous layer with SDS, and weigh the extracted weak base and titrate it with acid.

8.4.3 Amphoterics

Mixtures of amphoterics are not likely to be encountered, but a basic approach is outlined which will succeed in many cases. Acid–base titration is unlikely to be useful because of the presence of other weak acids and bases. The procedure therefore relies on titration with SDS and/or benzethonium chloride (BEC) and will not always be completely successful. In all cases, if two-phase titration fails, potentiometric titration is still likely to succeed. Acid solutions must be at least 0.1 M in H^+ and alkaline solutions at least 0.1 M in OH^-. It may be possible to replace SDS with NaTPB in at least some cases. Sulphobetaines cannot be determined in the presence of other amphoterics without separation.

WW with WS. Titrate with SDS in acid solution to determine total activity and with BEC in alkaline solution to determine the WW amphoteric.

WW with SW. This is the converse of the previous case. Titrate with BEC in alkaline solution to determine total active and with SDS in acid solution to determine the WW amphoteric.

WS with SW. Titrate with SDS in acid solution to determine the WS and with BEC in alkaline solution to determine the SW.

8.4.4 Quaternary ammonium salts with amphoterics

The possible procedures involving titration with SDS or BEC all depend on pH control, and are therefore not applicable to mixtures containing alkali-labile quats. It is better to assume that such mixtures will necessitate separation, and the guidelines given below assume that alkali-labile quats are absent, and that only one quat and one amphoteric are present, although the astute analyst will be able to work out schemes for some mixtures containing more than one surfactant of each class.

As before, potentiometric titration may sometimes be more successful than two-phase, and acid and alkaline solutions must be at least 0.1 M in H^+ or OH^-. Because of the variety of structures possible in amphoterics, it is advisable to confirm the practicability of these suggested procedures in individual cases. It is assumed that all amphoterics contain only one acidic and one basic group. Similar procedures may be applicable in other cases, but would need to be investigated by experiment.

Titration with SDS and BEC. Titration in acid and alkaline solution gives the kinds of measurement shown in Table 8.1.

Table 8.1 Titration of mixtures of quats and amphoterics

	SDS acid	SDS alkaline	BEC alkaline
Quat (Q) with WW amphoteric (A)	Q + A	(Q − A)**	(A − Q)*
Quat (Q) with WS amphoteric (A)	Q + A	Q	—
Quat (Q) with SW amphoteric (A)	Q	(Q − A)**	(A − Q)*

*If the amphoteric is present in molar excess over the quat, the excess is measured. If it is not, no titration is observed.
**If the quat is present in molar excess over the amphoteric, the excess is measured. If it is not, no titration is observed.

Titration with NaTPB. Mixtures of quats and WW or WS amphoterics can be analysed by titration in acid solution and at pH 10.

Potentiometric acid–base titration. With reservations about the effects of other weak acids and bases, amphoterics can be determined in any mixture with quats by potentiometric acid–base titration in ethanol or propan-2-ol. WW amphoterics consume two mols of acid or base per mol. WS and SW amphoterics consume one mol per mol. Mixtures can be analysed by determining the amphoteric in this way and the quat by SDS or NaTPB titration in alkaline solution.

8.4.5 Weakly basic cationics with amphoterics

This short section is included for the sake of completeness, since it is not likely that this combination will be encountered.

Mixtures of weakly basic cationics with WW or SW amphoterics can be analysed by titration with SDS in acid solution and BEC in alkaline solution. The measurements obtained are shown in Table 8.2.

Table 8.2 Titration of mixtures of weakly basic cationics and amphoterics

	SDS acid	BEC alkaline
Weak base (C) with WW amphoteric (A)	C + A	A
Weak base (C) with SW amphoteric (A)	C	A

Mixtures of weak bases with WS amphoterics (carboxybetaines) can be analysed by the method described in section 7.2.4 [2]. The betaine is determined by potentiometric titration of an initially acid solution in methyl isobutyl ketone with alkali, and the weak base by potentiometric titration of an initially alkaline solution in 50% aqueous propan-2-ol with acid. This procedure distinguishes between tertiary amines and amine oxides (section 7.8).

8.5 Mixtures of anionics and cationics

In the analysis of unknowns it is usually necessary or desirable to isolate the surfactants for complete characterisation. With mixtures of known composition, however, it is more often than not possible to determine surfactants without prior separation. Nonionics are not considered, because their isolation by deionisation is simple and because they rarely if ever interfere in the determination of ionic species.

8.5.1 Anionics with quaternary ammonium salts

Quat tetraphenylborates are more stable than their salts with anionics at high pHs, but the converse is true at intermediate pHs. As a consequence, quats can be determined by NaTPB titration even in the presence of anionics. This is the basis of a pair of methods [3, 4] for the analysis of mixtures of anionics and quats.

Anionic present in excess

Buffer solution, pH 9. Mix equal volumes of 0.3 M sodium dihydrogen phosphate and 0.05 M sodium borate.

Indicator. Victoria blue B, 0.01% in ethanol.

Standard quat solution. 5×10^{-5} or 1×10^{-4} M benzethonium chloride. Benzalkonium chloride of chain length 14 or 8–18 carbon atoms may also be used.

NaTPB solution. 5×10^{-5} or 1×10^{-4} M.

1. Place 10 ml of a solution containing up to 1.2×10^{-4} mol/l of the anionic in a stoppered titration vessel. Add 5 ml buffer solution, 1–2 drops indicator and 3 ml 1,2-dichloroethane.
2. Titrate with standard quat solution, shaking vigorously after each addition. The colour change is from blue to red in the organic layer, the aqueous layer remaining colourless throughout.
3. Place a second 10 ml portion of the same test solution in a stoppered titration vessel. Add 5 ml 6 M sodium hydroxide, 1–2 drops indicator and 3 ml 1,2-dichloroethane.
4. Titrate with standard NaTPB solution, shaking vigorously after each addition. The end-point is as in step 2.

The titration at step 2 measures the excess anionic. The titration at step 4 measures the quat. The total concentration of anionic is the sum of these two molar concentrations.

Quat present in excess

Buffer solutions, pH 6.0 and 12.5. Prepare a solution containing 0.3 mol/l sodium dihydrogen phosphate and 0.05 mol/l sodium borate. Divide into two. To one portion add 5 M sulphuric acid dropwise until the pH is 6.0 and to the other add 10 M sodium hydroxide until the pH is 12.5.

Indicator. Potassium tetrabromophenolphthalein ethyl ester, 0.03% in ethanol.

Titrant. Prepare and standardise a 0.02 M solution of NaTPB. Dilute accurately to 5×10^{-5} M with 5×10^{-5} M sodium hydroxide.

1. Place 5 ml of a solution containing up to 10^{-4} mol/l quat in a stoppered titration vessel. Add 5 ml pH 12.5 buffer, one drop indicator and 3 ml 1,2-dichloroethane.

2. Titrate with standard NaTPB, shaking vigorously after each addition. The colour change is from blue to colourless in the organic layer, the aqueous layer remaining colourless throughout.
3. Place 5 ml of the same solution in a stoppered titration vessel. Add 5 ml pH 6.0 buffer, one drop indicator and 3 ml 1,2-dichloroethane.
4. Titrate with standard NaTPB, shaking vigorously after each addition. The colour change is from blue to pale yellow in the organic layer, the aqueous layer remaining colourless throughout.

The titration at step 2 measures the total quat. The titration at step 4 measures the excess quat. The concentration of anionic is the difference between these two molar concentrations.

Neither procedure is applicable to alkali-labile quats. It could be an advantage to work with higher concentrations, but the authors may have chosen to work at high dilution to avoid emulsification problems.

8.5.2 Anionics with weakly basic cationics or amphoterics

Some of these pairs occur fairly often; others are somewhat improbable. There are two basic approaches, BEC titration and acid–base titration.

BEC titration. Most mixtures of anionics with weakly basic cationics or amphoterics can be analysed by BEC titration in acid and alkaline solution. Two-phase titration is more satisfactory than potentiometric. 'Acid' and 'alkaline' have the same meanings as before. The measurements produced are shown in Table 8.3.

Table 8.3 Titration of mixtures of anionics with weakly basic cationics or amphoterics

	Acid	Alkaline
Anionic (A) with weakly basic cationic (C)	A − C	A
Anionic (A) with WW amphoteric	A − WW	A + WW
Anionic (A) with WS amphoteric	A − WS	A
Anionic (A) with SW amphoteric	A	A + SW

This is more satisfactory for weakly basic cationics than for amphoterics, although the results with WW and WS amphoterics are usually considered reasonable for quality-control purposes.

Acid–base titration. This is good for weakly basic cationics, and can be satisfactory for amphoterics provided that other weak acids and bases are not present at too high a concentration, or if the titration can be done in such a way as to nullify their effects (cf. [2] for carboxybetaines).

8.5.3 Carboxylates with cationics

Such mixtures are not often found, but are easily analysed.

Carboxylates with quats. SDS titration in acid solution; BEC titration in alkaline solution; potentiometric acid–base titration in ethanol or propan-2-ol.

Carboxylates with weakly basic cationics. As for quats. Results of acid–base titration must be interpreted bearing in mind the degree of protonation or deprotonation in the product; acid titration measures free bases plus carboxylate anions, titration with alkali measures free acids plus protonated amine cations. If only weak acids and bases are present, acid titration measures the bases and alkali titration the acids.

Carboxylates with amphoterics. For WW and WS amphoterics, SDS titration in acid solution gives the amphoteric in both cases. In alkaline solution, BEC titration gives carboxylate plus WW and carboxylate alone, respectively.

That approach is useless for SW amphoterics; BEC titration measures the total active and SDS titration in acid solution measures nothing. Acid–base titration measures both the acid group of the carboxylate and the basic group of the amphoteric. If the sample is dried and dissolved in glacial acetic acid, the amine function of the SW amphoteric can be titrated with perchloric acid in glacial acetic acid, but it is likely that other weak bases will be present. Extraction of the carboxylic acid is the simplest way.

8.6 Mixtures with more than one component in each class

The possibilities are too numerous to consider in detail. The general approach is similar to that for mixtures of two or more anionics, namely:

1. Consider the information to be obtained from SDS/BEC or NaTPB titrations at high and low pH.
2. Consider what other functional groups are present which might offer a route to any of the components, e.g. hydroxyl groups (hydroxyl value), free acidic or basic groups that can be titrated, ester groups (saponification value) or polyalkenoxy chains (colorimetry, NaTPB titration of barium complexes), etc.
3. Consider what labile groups are present which would yield extractable hydrolysis products, e.g. amide groups yielding fatty acid, or whose decomposition would prevent their determination by some other procedure, e.g. SDS/BEC titration after acid hydrolysis of sulphates, esters or amides.

Not all mixtures are amenable to this approach, or are perhaps amenable in principle but would require an impracticably complicated scheme of analysis. But with the right strategy, and preferably with some cross-checking by using alternative routes or a mass balance, a high proportion of multicomponent mixtures can be analysed by the techniques described in the preceding chapters.

References

1. United States Pharmacopoeia, 22nd Revision.
2. Plantinga, J. M., Donkerbroek, J. J. and Mulder, R. J. *JAOCS*, **70** (1993), 97–99.
3. Tsubouchi, M. and Yamamoto, Y. *Anal. Chem.*, **55** (1983), 583–584.
4. Tsubouchi, M. and Mallory, J. H. *Analyst*, **108** (1983), 636–639.

9 Chromatographic analysis of surfactants

J. G. LAWRENCE

9.1 Introduction

Chromatography is a separation process, or rather a family of related but different separation processes which have developed from the work of Tswett in the first decade of this century. The various types of chromatography have names descriptive of their modes of separation.

Gas chromatography (GC) involves separation in the gaseous phase. It is applied to volatile molecules or molecules which can be made volatile at temperatures up to 400°C.

High performance liquid chromatography (HPLC) involves separation in the liquid phase in a column. It is applied to non-volatile molecules, ionic molecules and thermally labile molecules. It should be noted that both GC and HPLC can be applied to many molecules.

Thin layer chromatography (TLC) involves separation with a liquid moving phase and a thin (open) layer of the stationary separation phase. Its advantages are that many samples and standards can be run simultaneously and that all the sample, apart from very volatile molecules, can be visualised on the surface of the layer.

Ion chromatography (IC) involves separation of ionic species in the liquid phase. It is of particular importance for small cations and anions.

Gel permeation chromatography (GPC) involves molecular size separation in the liquid phase by permeation through a porous packing. It leads to information on molecular weight distribution of the polymer.

The main features of all chromatographic separations are that molecules of different structures are separated from one another, and that molecules within a structural family can be separated from one another on the basis of carbon chain length, or, for molecules of the same chain length, on the basis of positional isomers or chain branching. How these features apply to surfactant analysis will now be discussed in more detail.

Why then do we need chromatography for surfactant analysis? Many procedures for analysis of surfactants give average values for the property determined. As examples, consider (i) two-phase titration for active level, which uses a mean molecular weight for the surfactant molecule to calculate an average active level; or (ii) determination of the degree of ethoxylation of an alcohol or alkyl phenol ethoxylate by proton nuclear magnetic resonance, which gives an average ethoxylation figure per molecule with no information on the ethoxamer distribution. Chromatography can generally supply the mean molecular weight to use in the active level calculation and the detailed ethoxamer distribution of ethoxylates. Often, for understanding of surfactants and their properties, a knowledge of the distribution of molecular structures/weights is required. As has been observed above, chromatographic techniques are excellent for supplying information about molecular structure—both chain length distribution and structural information.

A second area of surfactant analysis to which chromatographic analysis can contribute is in the determination of the level of surfactant in a product, though this is not as common. A third area of importance is in determining low levels of impurities or contaminants resulting from the manufacturing process. Such analyses are becoming more and more common as legislative and consumer pressures for cleaner, safer, more environmentally friendly raw materials and products are increasing. Contaminant analysis, together with the other areas, will be dealt with later in the chapter.

What factors do we then have to consider when creating a chromatographic separation? There is a wide range of parameters over which we have control, many of which are specific to individual techniques. One of the decisions to be made is the mode of chromatography to be used—what is most appropriate for the surfactant molecule? Alternative modes of chromatography may be selected depending on the information required, and the time and instrumentation available. For example, for a simple comparative screening of an alcohol ethoxylate, the low-cost technique of TLC might be used; to obtain more detailed information on the hydrophobe (alkyl chain) distribution, GLC might be used; and to obtain information on the ethoxamer distribution, HPLC might be used. Both GLC and HPLC are significantly more costly to install in a laboratory than is TLC, the relative costs depending on the final complexity of the instrumentation (e.g. for GLC whether autosamplers and selective detectors are included, for HPLC whether a series of interconnected modules or an integrated system is used). A second decision to be made is what has to be done to the surfactant molecule to make it suitable for chromatographic separation. A simple example of making a surfactant molecule amenable to chromatography is soap, which exists as a range of sodium salts of fatty acids of chain lengths generally in the range C_{10} to C_{18}. Information on the chain-length distribution of the soap is readily obtained by protonation of the soap to its acids and derivatisation of the acids to their methyl esters, which are readily separated and quantified by gas liquid chromatography.

In the following sections, the different modes of chromatography already introduced will be described in greater detail, and examples of their application to many types of surfactant molecules presented.

9.2 Gas chromatography

9.2.1 General description

In gas chromatography, the mixture to be separated is vaporised and swept into a column containing a solid adsorbent or a high-boiling liquid. All or part of the mixture will condense at the inlet of the column which is normally at a much lower temperature than that at which the vaporisation took place. Those compounds which do not condense will be carried through the column in a steam of inert gas (commonly helium or oxygen-free nitrogen) with which they undergo no reaction. Within the column they may interact with the adsorbent or stationary liquid to different extents, which will cause them to be retained for varying times in the column, thus giving separation at the column outlet. By increasing the column temperature, further components of the mixture will vaporise and undergo their specific interactions in the column in turn. At the column outlet, the mass or concentration of each component in the inert carrier gas is measured and recorded. A gas chromatograph consists of a source of supply of an inert carrier gas, a heated injector to transfer the sample into the column, a separation column in a heated oven, a detector and a means of recording the detector signal. There are two types of gas chromatography; (i) gas solid chromatography (GSC) where separation takes place through a combination of temperature effects and interaction with a solid adsorbent; and (ii) gas liquid chromatography (GLC) where the separation takes place through a combination of temperature effects and partition into a stationary liquid phase. Of the two, GLC is by far the most common.

9.2.2 Resolution

There are three main contributions to resolution in gas chromatography. These are: (i) the column length; (ii) the selectivity of the column contents for the materials of interest; and (iii) the relative capacity (a measure of the retention of the material in the column) of the column for the materials of interest. Of these factors, column length and selectivity can be controlled readily and must be considered when developing a separation. As a general rule, non-polar components such as hydrocarbons are separated by boiling point on non-polar stationary phases; more polar materials such as fatty acid esters are separated on more polar columns by a combination of temperature and stationary phase interaction.

9.2.3 The column

Gas chromatographic columns come in two main types known as packed and capillary. Generally, packed columns are of glass (rigid) or metal (non-rigid, stainless steel). They vary in external diameter from 1/8 to 1/4 inch; in internal diameter from 2 to 4.6 mm; and in length from 1 m upwards, the most common length being 2 m. Glass columns are normally dedicated to a single gas chromatograph having the correct geometry of injection and detection ports, whereas metal columns can be transferred between instruments by suitable adjustment of their geometry. Capillary columns are long, typically 20–60 m or more, with internal diameters of approximately 0.2–0.3 mm, and are normally fabricated from quartz of low metal-ion content and coated with a polyimide (or similar polymer) film to give them great flexibility. Commercial glass and stainless steel capillaries are available but are less commonly used. Megabore columns are large-diameter capillary (commonly 0.53 mm) columns and are used in lengths of 10–25 m. Open tubular capillary columns have a film of non-volatile liquid (e.g. polydimethylsiloxane) coated or bonded to their inner surface. When selecting which column type to use, the packed column can take a high loading of the material to be separated and comes in a wider variety of stationary liquids or adsorbents than the capillary column. The capillary column gives better separation with narrower peaks eluted from the column (due to more favourable dispersion processes in the column). Sample injection onto capillary columns is more complex. Megabore columns have advantages in more efficient separation than packed columns and greater sample loading capacity than capillary columns. The most common type of column used in the development of new separations is the capillary column.

9.2.4 The solid support

For the packed column, an inert diatomaceous earth (e.g. Kieselguhr) having some porosity and carefully sized into narrow particle diameter distribution (e.g. 80–100 or 100–120 BS mesh) is coated with a thin film of the liquid stationary phase. Depending on the porosity of the support used, different maximum amounts of stationary phase can be accommodated. These supports are known by trade names such as Chromosorb, with qualifiers such as W, P and R which refer to different surface areas and porosities. Examples of such supports with their stationary phase loading will be given in the Applications section (9.2.12). Coating of the support is carried out by dissolving the stationary phase in a suitable solvent, adding the support, and removing the solvent slowly while gently agitating the solid–liquid slurry to ensure uniform deposition. The coated phase should finally be sized to remove any agglomerates. Although diatomaceous earths are inert, in chromatographic terms they have adsorptive capacity and it is common to treat the support with dimethylchlorosilane or a similar silylating reagent before coating to eliminate any adsorption sites in the packed column. Such treated supports are identified as DMCS (dimethylchlorosilane treated).

9.2.5 The stationary phase

The stationary phase is critical in GLC as it has so much effect on the separation. Over the forty or so years since GC was first introduced, hundreds of liquids have been used as stationary phases. Currently, most laboratories use only a very small number (less than ten) of phases together with a few for particular separations (e.g. inert gases). The reduction in stationary phase numbers was triggered by the work of Kovats, Rohrschneider and McReynolds who did much to categorise the wide range of phases in terms of their interactions with a range of solvents of different polarities (benzene, butanol, 2-pentanone, nitropropane, pyridine). Their compilations demonstrated that in reality only a few phases are of sufficiently different properties to be needed in developing separations. A short list of necessary phases would include methyl silicones (OV1/SE30) for non-polar boiling point separations, phenylmethyl silicones (OV7/OV17), trifluoropropylmethyl silicones (QF1), cyanopropylphenyl silicones (OV225), polyethylene glycols (Carbowax 20M) and polyesters (polyethylene glycol adipate/diethylene glycol succinate). These are in order of increasing polarity. The labels OV1, SE30, etc. are supplier's codes for the stationary phases. OV1 and SE30 are methyl silicones from different suppliers. OV7 and OV17 are mixed methylphenyl silicones having different proportions of phenyl substitution. The reader is referred to the brochures of the various suppliers for more detailed information. This comment applies equally to the applications section (9.2.12) in which reference is made to equivalent phases. Silicone stationary phases are the most common phases used in capillary columns where a thin film of phase is coated on or bonded to the wall of the capillary.

9.2.6 Injection

The amounts of sample analysed by gas chromatography are small. For packed columns where the sample is vaporised in the injection port or injected directly onto the column, the sample volume injected is of the order of 1 microlitre. For a 1% solution of the sample in a suitable solvent this corresponds to a mass of about 10 microgram sample. The normal means of sample introduction is through a septum using a micro-syringe. The septum must be replaced at regular intervals to prevent leaks and any chance of contamination. For capillary columns, there is a wider range of injection options. As capillary columns are 0.3 mm and less, on-column injection requires specialised equipment and techniques. The more common types of capillary injection are 'split' and 'splitless'. In 'split' injection the sample is volatilised, the major portion is vented and a small amount passes into the column. Split ratios of 100:1 where only about 1% of the sample injected passes onto the column are common. In 'splitless' injection the sample solution is injected into a relatively cool injection port and substantially all the solvent and sample are transferred to the column. The preferred injection technique for quantitative analysis is 'on-column' since this applies all the sample as a sharp band on the column and leads to efficient chromatography. Cool on-column injection, where the injection port and column are cool allowing the solvent to volatilise before the sample, is becoming recognised as the best practice. Automatic injectors capable of on-column injection to capillary columns and of any other injection technique are commercially available and are recommended for improved precision in quantitative analysis. Sample valves as used in HPLC are used in GC for gas samples only.

9.2.7 Column temperature

There are two types of column oven control in gas chromatographic analysis: (i) isothermal where the column temperature is constant throughout the analysis; and (ii) temperature programmed where it is increased. The programme can consist of an initial isothermal stage, an increase at a controlled rate to a higher temperature, and a further isothermal stage. This sequence may be repeated to yet higher temperatures at different programmed rates. The temperature programme selected will depend on the range of molecules in the sample.

9.2.8 Detectors

There is a wide range of detectors for gas chromatography. Their important operating parameters are their absolute detection limit for the molecule of interest and their linearity of response to increasing sample amounts. The most common, almost universal, detector is the flame ionisation detector (FID). This consists of a jet of hydrogen burning in a sheath of air. The column carrier gas is mixed with the hydrogen. A potential of greater than 240 V is applied between the jet and a collector electrode. When carbon-containing compounds from the column pass through the flame, an ionisation phenomenon occurs and a current passes between the electrodes. This current is detected, amplified and recorded

Table 9.1 Characteristics of detectors for gas chromatography

Detector	Selective to	Minimum detection limit	Linear dynamic range
Flame ionisation	Almost all carbon compounds	5 pg C/sec	10E7
Thermal conductivity	Universal	400 pg/ml carrier	10E6
Electron capture	Halogens/electrophores	0.1 pg Cl/sec (variable)	10E4
Alkali flame	Nitrogen and phosphorus compounds	0.4 pg N/sec 0.2 pg P/sec	10E4
Flame photometric	Sulphur and phosphorus compounds	20 pg S/sec 1 pg P/sec	10E3
Photoionisation	Compounds ionised by UV	2 pg C/sec	10E7
FT infrared	Monitors molecular vibrations	1 ng/sec	10E3
Mass spectrometer	Universal or selectable	Down to 10 pg/sec	10E5

using a potentiometric recorder or an integration device. Other detectors of particular application to surfactant analysis are the flame photometric detector (FPD) and the mass spectrometer which offers both high sensitivity and selectivity. Table 9.1 compares a wide range of detectors for selectivity, detection limit and sensitivity.

Of the detectors in Table 9.1, many can be more sensitive for certain molecules than the FID. Electron capture can be useful for sensitive detection of derivatised molecules but has a limited dynamic range and is not a good general detector for quantitative analysis.

9.2.9 Qualitative and quantitative analysis

Gas chromatography by itself does not provide unambiguous component identification. However, if a component does not elute from a column at the point expected from prior experience, it is almost certain that the component is not present in the mixture. If a peak is obtained at the correct time, it does not confirm identity, only makes it possible. If the sample is then analysed on a second stationary phase and it is found that the component of interest also has the correct elution time for the second phase, then confidence in the identification increases. Coupling the gas chromatograph to a mass spectrometer and/or a Fourier transform infrared spectrometer is the optimum means for unambiguous component identification.

Gas chromatography is an excellent quantitative technique as the peak areas (or heights for uniform-shaped peaks) are a measure of the amount of material present. There are numerous ways of carrying out quantitative analysis, of which the best are the inclusion in the sample of internal standard to which the components of interest are related and the use of external calibration curves of the pure material when autoinjectors with reproducible injection volumes are used. If analysing multicomponent mixtures, it may be necessary to correct for varying detector responses to different components.

9.2.10 Derivatisation

Derivatisation is carried out to make molecules more volatile, more easily separated, more symmetrical in peak shape, more easy to detect (possibly at lower levels), and less likely to undergo any thermal degradation. There is a vast literature on derivatisation. The most common derivatives for surfactant analysis are esterification of fatty acids and silylation of hydroxyl compounds.

9.2.11 Head-space analysis

The previous discussion has dealt with liquid injection. There are other more specialised injection techniques which are important for low-level and contaminant analysis such as adsorption of the sample from the vapour phase onto a solid substrate followed by thermal desorption of the adsorbed volatile onto a gas chromatographic column for analysis and head-space analysis. In head-space analysis the sample is allowed to come to thermal equilibrium at a controlled temperature in a sealed vial. The gaseous phase in the vial is sampled and analysed. This technique has two major advantages: (i) only the volatiles in the sample are transferred to the column, thus reducing contamination; and (ii) the components of interest are usually at relatively high concentration in the vapour as opposed to in the sample (which may be in any physical form, solid, liquid or paste). Quantification is complicated and is best done using standard additions where this is possible.

9.2.12 Applications

9.2.12.1 Anionics. The many approaches in the literature to linear alkylbenzene sulphonate (LAS) analysis include the following, which demonstrate a number of sample preparation and detection options.

—Desulphonation of the LAS (with phosphoric acid) to the corresponding linear alkyl benzene (LAB) followed by separation on a fused silica capillary column (15 m × 0.32 mm i.d.) of DB-1 temperature programmed from 100°C to 170°C at 5°C per min with helium as carrier gas and FID

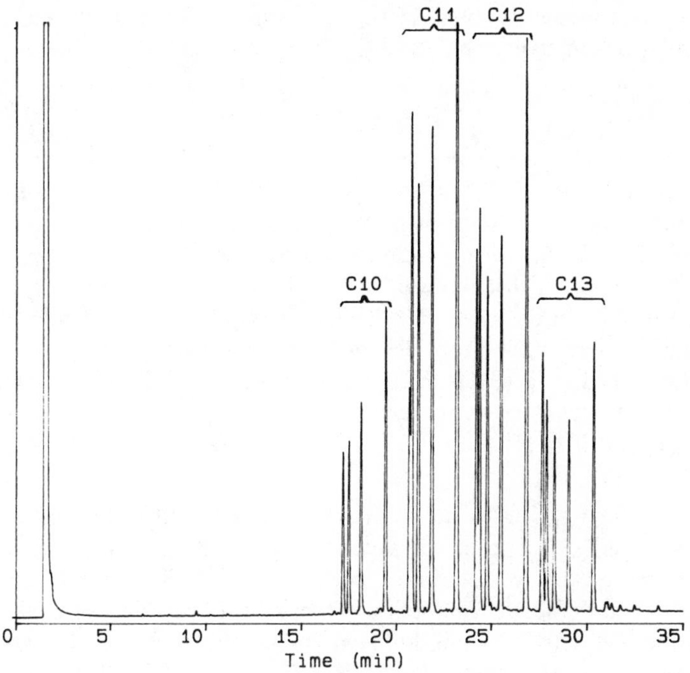

Figure 9.1 Separation of linear alkyl benzenes (column 25 m × 0.2 mm i.d., HP101, programmed from 120°C to 240°C at 3°C/min, injector/detector temperature 270°C, helium carrier, split injection, flame ionisation detector).

[1]. This gives detailed chain length and positional isomer information (see Figure 9.1).

—LAB or desulphonated LAS is prefactionated using argentation TLC, in which the TLC plate is coated with silver nitrate to modify the separation process, and analysed on a 30 m × 0.25 mm i.d. column of DB-5 (see comment in section 9.2.5) programmed from 35°C (10 min hold) to 290°C at 4°C per min with helium as carrier gas and GC/electron impact mass spectrometry (EIMS) detection [2].

—LAS and dialkyltetralin sulphonates are converted to their sulphonyl chlorides by reaction with phosphorus pentachloride and then to their trifluoroethyl derivatives by reaction with trifluoroethanol. These derivatives are separated on a 15 m × 0.25 mm i.d. (0.25 micron film) of DB-5 temperature programmed from 125°C (1 min) to 230°C at 5°C per min with a final hold of 5 min. Helium is the carrier gas and detection is by negative-ion chemical ionisation mass spectrometry [3].

—LAS is converted to its dibutyl sulphonamide derivative by ion-pair extraction and amidation. The derivatives are separated on a packed glass column (1 m × 3 mm i.d.) of 2% SE-30 on Uniport HP (100–120 mesh) (a solid support) isothermally at 250°C with nitrogen at 50 ml/min as carrier and flame photometric detection [4].

Alcohol sulphates can be readily determined by acid hydrolysis, recovery of the parent alcohols, and separation either as is or after conversion to their trimethylsilyl ether derivatives. Most manufacturers demonstrate such separations in their applications literature. Typical conditions are a 10 m × 0.53 mm i.d. column of methylsilicone programmed from 70°C to 240°C at 5°C per min with helium as carrier and FID.

The hydrophobe distribution of alkylethoxylated sulphates can be obtained by reaction with 30% HBr in glacial acetic acid at 90°C overnight to give alkyl bromides, followed by separation on a column (6 ft × 1/4 in o.d.) of 10% OV-17 on Chromosorb W with temperature programming from 100°C to 250°C at 8°C per minute with helium as carrier gas and FID [5].

9.2.12.2 Nonionics. Alcohol ethoxylates and alkylphenol ethoxylates were separated on a fused silica column (30 m × 0.25 mm (0.25 micron film)) of SE-54 using helium as carrier gas and EI or CI (chemical ionisation) (methane) MS detection. Temperature programming was from 70°C (1 min) to 300°C (10 min hold) at 3°C per minute. Ethoxylates up to 6 EO units could be detected [6].

Alternatively, ethoxylated alcohols were separated on a 50 cm × 0.125 inch column of 10% SE-30 on Chromosorb W DMCS (80–100 mesh) programmed from 50°C to 300°C at 6°C per min with nitrogen as carrier gas and FID. A detailed study of structure/retention time/response factors has been presented [7].

Separation of alkylphenol ethoxylates and alcohol ethoxylates on an aluminium clad fused silica column (10 m × 0.53 mm i.d.) of OV-1 with helium as carrier gas, FID and temperature programming to 325°C is presented in [8]. Separations are compared with those from SFC.

The hydrophobe distribution can be obtained as described in [5]. Silylation of alcohol ethoxylates to their silyl ether derivatives, when combined with temperature programmed GLC in a non-polar methylsilicone column, gives an extremely complex pattern of peaks.

9.2.12.3 Cationics. Cationic surfactants are non-volatile. The contribution which GC can make to their analysis is in determination of their alkyl substitution. This is achieved by a degradation reaction, examples of which are given in the following paragraphs.

Alkyltrimethylammonium and dialkyldimethylammonium cationics are converted by Hoffmann degradation to their alk-1-enes by heating on a water bath for 30 min with potassium *t*-butoxide in benzene-DMSO (4:1). After extraction and clean-up, the alkenes are separated on a glass column (2 m × 3 mm i.d.) of 5% SE-30 on Chromosorb W AW-DMCS

(80–100 mesh) temperature programmed from 160°C to 270°C at 6°C per min with nitrogen as carrier gas and FID. For quantification, isothermal operation in the given temperature range is used [9].

Alkyl chain distribution can also be obtained by thermal decomposition in the chromatograph injection port [10]. Separation is on a 1 m × 2 mm i.d. column packed with 8% Carbowax 20 M (KOH treated) on acid-washed Chromosorb W (80–100 mesh) programmed from 70°C to 210°C at 8°C per min with nitrogen as carrier gas and FID.

Alternatively, the quaternary cationic is mixed with silver oxide and the solution analysed on a 2 m × 0.125 inch column of 10% UC-W98 on acid-washed Chromosorb W DMCS (60–80 mesh) at 200°C with FID. Mechanisms for the dealkylation and elimination reactions that occur in the GC injection port are postulated [11].

9.2.12.4 Soap. Soap, the sodium salt of fatty acids with a chain length of C_{10} to C_{18}, is analysed by protonation of the salts to their acids, followed by derivatisation of the acids with boron trifluoride/methanol to give fatty acid methyl esters. A wide range of stationary phases and conditions has been used for the separation. One example is on a wall-coated open tubular column (25 m × 0.22 mm i.d.) of SE-30 programmed from 50°C (2 min) to 250°C at 30°C per min with a final temperature hold. Detection is by FID and carrier gas in nitrogen [12]. Alternatively, manufacturers of chromatography supplies give examples in their literature. An example of the separation of soap fatty acids is given in Figure 9.2.

9.2.12.5 Contaminants. Three common contaminants determined by GC are ethylene oxide, 1,4-dioxane and sultones.

Ethylene oxide in alcohol ethoxylates is determined by equilibrium head-space analysis. An aliquot of the vapour is analysed on a column (8 ft × 0.125 inch) of Chromosorb 102 (80–100 mesh) (a gas solid phase) programmed from 120°C (5 min) to 190°C (10 min) at 8°C per min, with helium as carrier gas and FID. A detection limit of 1 ppm is achieved [13].

Alternatively, a method is described for alkyl ether sulphates, which would also be applicable to alcohol ethoxylates, in which the equilibrium head-space is analysed on a 3 m × 1.8 mm i.d. column of 0.8% THEED/Carbopack C (80–100 mesh) [14]. A detection limit with FID of 0.1 ppm is claimed. For improved quantitation, both methods can be adapted to a method of standard additions.

1,4-dioxane in alkyl ether sulphates is determined by head-space analysis. The standard industrial method involves sample preparation using method of standard additions for quantification though there is some flexibility as to whether capillary or packed columns are used and as to the actual phase. One publication [15] describes the use of a totally deuterated 1,4-dioxane analogue with isotope dilution and MS detection

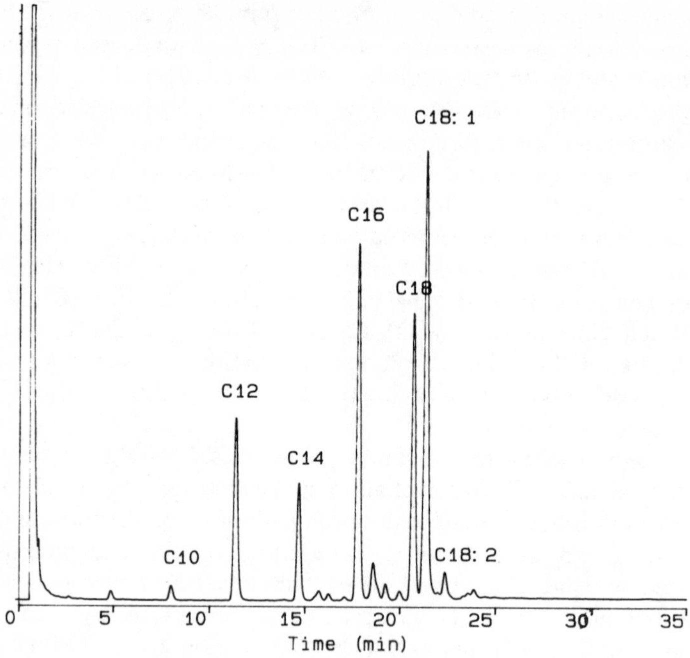

Figure 9.2 Separation of fatty acid methyl esters (column 2 m × 2 mm i.d. of 10% SP2330 on Chromosorb W AW (100–120 mesh) programmed from 100°C to 220°C at 5°C/min, helium carrier, flame ionisation detector).

to minimise matrix effects. The separation was carried out on a 60 m × 0.32 mm i.d. column of Supelcowax 10, temperature programmed from 50°C (2 min) to 100°C at 5°C per min.

1,3-sultones are determined in α-olefin sulphonates by extraction with diethyl ether, trapping from a silica column, and GC on a column (1 m × 3 mm) of 2% DEGS on Chromosorb W AW-DMCS (60–80 mesh) at 220°C with helium as carrier and flame photometric (sulphur mode) detection. Alternatively, detection limits down to 0.2 ng/g were obtained with negative chemical ionisation MS with methane as reagent gas [16].

9.3 High performance liquid chromatography

9.3.1 General introduction

Chromatography in open columns with the liquid flow driven by gravity was the first form of chromatography. Modern high performance liquid chromatography (HPLC) began in the late 1960s and has had a spectacular development. The HPLC chromatograph consists of a solvent reservoir (or a number of reservoirs), normally degassed, a high-pressure pump, a column, possibly thermostatted, a detector and a recorder or integration device. In addition, a solvent mixing device may be used before a single pump to give low-pressure mixing of solvents, or after a pair of pumps delivering different solvents to give high-pressure mixing of solvents. After sample injection, the components of the sample are separated by their relative interactions with the column packing material and with the mobile phase. The mobile phase composition may be altered in the course of an analysis (gradient elution, in some ways analogous to temperature programming in GC) to elute from the column those components which interact strongly with the stationary phase. There are two main types of HPLC: (i) normal phase; and (ii) reverse phase. In normal phase, the column packing is silica or another adsorbent or a substrate to which is bonded a polar layer (e.g. diol, amino) and the mobile phase is increased in polarity during gradient elution (e.g. from hydrocarbon to hydrocarbon/ether). In reverse phase mode, the column packing is commonly a long-chain hydrocarbon (C_{18} is common) bonded by a siloxane bridge to a rigid silica substrate. C_{18} phases are known as ODS (octadecyl silane) by most manufacturers and are so noted in the applications section (9.3.11). The mobile phase reduces in polarity during gradient elution (e.g. from water to acetonitrile/water). In HPLC there is clearly great scope for controlling separation through the mobile phase, and the use of more complex mobile phases containing buffers, ion-pairing reagents or surfactant micelles is common. Resolution in HPLC is governed by the same factors as in GC: column length (or efficiency), a retention or capacity factor, and a relative selectivity factor.

9.3.2 The column

Columns in HPLC are predominantly stainless steel 1/4 inch o.d., 4.6 mm i.d. and 10, 15 or 25 cm long. Longer columns are made by coupling the standard columns. Columns can be supplied as cartridges which seal into a standard casing for simple interchangeability. An alternative dimension of column which is growing in popularity is 1 mm (or narrower) of stainless steel or a polymer such as Teflon in lengths up to 1 m. Columns are commonly operated at ambient temperature with no temperature control. More reproducible long-term behaviour is obtained through the use of thermostatted ovens.

9.3.3 Column packings

Column packings for normal phase columns are mainly porous silica (spherical or irregular) particles of 5–10 micron average diameter and of tight particle size distribution. For reverse phase, bonded phase materials based on 5–10 micron particle size silica are used. The silica is reacted to give a siloxane bond to an alkyl chain. C_{18} is the most common but C_8, C_6 and shorter chains are also used. The objective of this bonding reaction is to enable separation to be carried out through partition with a stationary hydrocarbon phase. To ensure that the molecules to be separated only encounter the hydrocarbon phase, various additional bonding chemistries with short-chain methyl silicones are used to prevent occurrence of free silanol groups on the silica substrate surface. Among other types of packing, the pellicular packing in which a thin layer of reverse phase is bonded to a silica layer on the surface of a rigid non-porous inert support is worthy of mention. This type of packing is used to minimise the analysis time and speed up the kinetics of the separation processes. In section 9.3.11, various commercial packings are given. Names such as LiChrosorb, Hypersil, Nucleosil, Zorbax and so on are the manufacturer's silica to which bonding has been carried out.

9.3.4 Packing the column

In order to obtain a dense, uniform packing of the small-diameter particles, it is recommended that the packing is slurried in a liquid of similar density to the packing. This slurry is then rapidly pumped into the column to form a uniform bed. If specialised column packing equipment is not available, it is recommended that pre-packed commercial columns are purchased.

9.3.5 Pumps

The ideal HPLC pump delivers a constant, unvarying flow at column back pressures of up to 6000 psi. The two main types of pump are syringe pumps where a syringe is mechanically driven to deliver a smooth flow and reciprocating piston pumps where two pump heads are connected in parallel in such a way that a constant flow is delivered. For syringe pumps, gradient elution is always carried out with high-pressure mixing. Reciprocating piston pumps give the option of mixing at the high-pressure side of the pump, or with a suitable low-volume proportioning valve, on the low-pressure side. Low-pressure mixing gives the possibility of increasing the numbers of solvents in the gradient.

9.3.6 The mobile phase

Unlike GC, the mobile phase in HPLC can play a major role in the separation process. There are no major restrictions apart from those imposed by the stability of the stationary phase (or column itself) to solvent or pH and the requirements of mutual solubility of the components of the phase. It is important in HPLC to ensure that the mobile phase is thoroughly degassed before use.

9.3.7 Injection

Injection in HPLC in normal diameter columns uses an injection valve with a loop of known volume. The loop volume can vary from 10 (or less) to 100 (or more) microlitres. The loop is filled at atmospheric pressure using a micro-syringe of suitable volume before being moved into the high-pressure mobile phase flow. Most reproducible injection is achieved by overfilling as opposed to underfilling the loop with a set volume from the micro-syringe. Automatic injection systems are common. Such systems may incorporate a sample pretreatment stage such as derivatisation.

9.3.8 Detectors

There is a wide range of selective and sensitive detectors for HPLC. Ranking detectors for sensitivity is impractical as response is a function of molecular structure. The most common detectors will now be described.

UV photometric. Such detectors are applicable only to molecules with a UV chromophore. They can either use a filter to select a particular wavelength or be tunable to any wavelength. Radiation is absorbed at that wavelength to an extent determined by the extinction coefficient of the molecule and thus calibration is required for individual molecules. By using highly purified solvents, it is possible to extend their use down to 210 nm or below, where many molecules with no chromophore absorb weakly. An alternative approach is to incorporate a chromophore as part of the solvent and monitor a reduction in absorption (indirect photometric detection).

Diode array detector. The diode array detector records in about 10 msec the complete spectrum from 190 to 600 nm using an array of over 200 photodiodes. The advantage of this is that the complete spectrum at any point on a peak can be obtained and compared with that at any other point. This gives confidence (or not) as to peak homogeneity and also as to peak identity. However, it is less sensitive than the variable wavelength detector.

Refractive index. This is the closest approach to a universal HPLC detector. It determines differences in refractive index between the mobile phase and the mobile phase containing any other component. Its drawbacks are that it is not particularly sensitive and it cannot be used with gradient elution, which limits its range of application.

Evaporative light scattering detector. The mobile phase from the column is nebulised and passed through a beam of light. Any particles of sample not volatilised pass through the beam, and the resulting light scattering is detected. Volatile components are not detected and quantification requires attention as response curves may not be linear. However it is sensitive and can be used with gradients and non-UV active components.

Electrochemical detectors. These come in various configurations and are inherently sensitive to electroactive molecules.

Fluorescence detectors. These are very sensitive and selective and can be used both with molecules which have native fluorescence and with molecules which have been derivatised to contain a fluorophore. This second class is most common.

Mass spectrometry. Coupling of HPLC to MS has been carried out through a variety of interfaces to a variety of MS modes. The most common is electrospray which gives both qualitative and quantitative information.

9.3.9 Qualitative and quantitative analysis

The comments under 9.2.9 above for GC are applicable to HPLC. The lack of a universal detector means more care is required in detector selection to ensure that the components of interest can be determined.

9.3.10 Derivatisation

Derivatisation may be carried out before the analysis as in GC to obtain molecules in a form more amenable to separation or detection. In addition derivatisation may be carried out on the separated molecules after the column to enhance selectivity or sensitivity of detection.

9.3.11 Applications

9.3.11.1 Anionics. Linear alkyl benzene sulphonates (LAS) can be determined by HPLC by a number of procedures. The following examples demonstrate different separation and detection conditions.

—A column (250 × 4.6 mm i.d.) of Zorbax C18 (10 micron) with an ion-pair (0.1 M tetrabutylammonium hydrogen sulphate (pH 5))/water/acetonitrile gradient system and both diode array and fast atom bombardment MS was used to determine the hydrophobe distribution of the LAS [17].

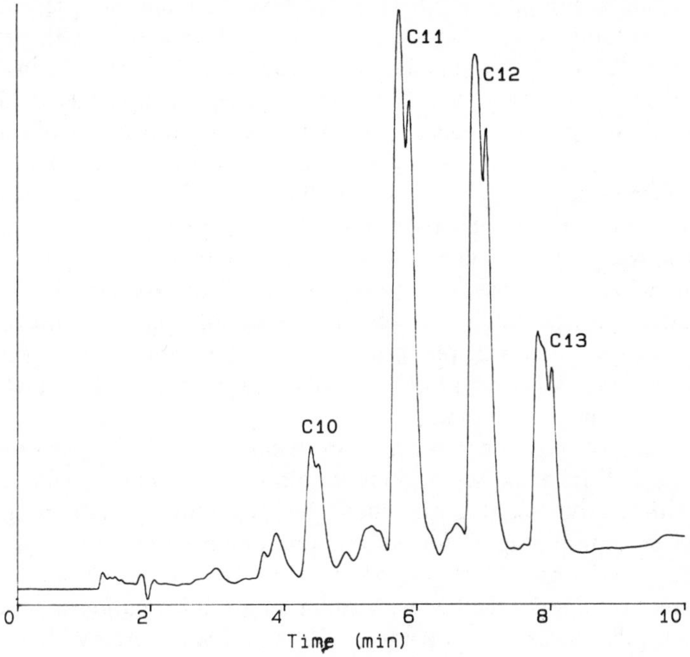

Figure 9.3 Separation of linear alkyl benzene sulphonate isomers (column 150 × 4.6 mm i.d. of 5 micron Hypersil SAS, gradient 35:65 acetonitrile: disodium hydrogen phosphate (aq.), to 75:25, UV detection at 254 nm).

—The hydrophobe distribution of LAS was determined on a column (300 × 3.9 mm) of microBondapak C18 (10 micron) using a linear gradient from 70/30 acetonitrile/0.15 M sodium perchlorate solution to 90/10 of the two solutions. Detection was at 230 nm. Peaks were eluted in order of increasing alkyl chain length [18].
—The positional isomer distribution is demonstrated in [19] using a column (250 × 4 mm i.d.) of Spherisorb ODS II (3 micron) with an iso-propanol/water/acetonitrile gradient with 0.02 M sodium perchlorate added and UV detection at 225 nm.

An example of the separation of LAS is given in Figure 9.3.

Alkyl ether sulphates can be analysed on a 2.5 cm × 2 mm i.d. column of C18 reverse phase material with a water/tetrahydrofuran gradient system [20]. In this example the detector was the evaporative light scattering detector, as a gradient system was being used with a molecule with no strong chromophore. Alternatively, to obtain more detailed distributions, the molecule could be desulphated and analysed as described for alcohol ethoxylates.

The separation of alkyl sulphates by carbon chain length is described in [21]. The column was 25 cm × 4.6 mm i.d. ODS material with gradient elution from 60 to 30% aqueous acetonitrile containing 0.01 M disodium hydrogen phosphate and 0.01 M sodium nitrate. Detection was at 242 nm. This is an example of inverse photometric where nitrate in the mobile phase absorbs a constant level of radiation apart from when the level of nitrate is reduced by the presence of the non-absorbing alkyl sulphate anion. It is the reduced absorbance which is monitored.

Alternatively, alkyl sulphates may be separated on a synthesised cross-linked amine–fluorocarbon polymer on silica column with 0.2 mM naphthalene disulphonate/35% acetonitrile mobile phase. Both indirect conductivity and indirect photometric detection can be demonstrated [22]. Conditions identical to those described in [22] for alkyl sulphates can be used for alkane sulphonates.

Separation of positional isomers of alkane monosulphonates is described in [23]. Baseline separations were obtained using Hypersil ODS I phase with acetonitrile/water gradient and *N*-methylpyridinium chloride as the visualisation reagent for indirect photometric detection.

The separation of α-olefin sulphonates into their hydroxy, alkene, and disulphonate isomers together with chain length information is described in [24]. A column (25 cm × 4.6 mm) of Zorbax TMS is used with a mobile phase of methanol/water (75:25, v/v) and refractive index detection. Equivalent columns and conditions were used in [25] with the major difference of an operating temperature of 55°C.

Sodium salts of alkyl (C_{10}–C_{14}) sulphosuccinates are separated on a 250 × 4.6 mm i.d. column of 10 micron Nucleosil C8 with aqueous 0.01 M tetrabutylammonium hydrogen sulphate/methanol (23:77) at pH 3 as mobile phase. Detection is by refractive index [26].

9.3.11.2 Nonionics. As alcohol ethoxylates have no strong UV chromophore, derivatisation to introduce a chromophore is an attractive option. The reaction to form phenyl isocyanate derivatives, which can then be detected by UV, is described in [27]. A micro-Bondapak C18 column was used for separation according to alkyl chain length and a micro-Bondapak amine column for separation according to degree of ethoxylation.

A similar analysis where both alkyl chain and ethoxamer separations are demonstrated, but without derivatisation, is described in [19]. The columns used were LiChrosorb RP8 for alkyl chain length and Hypersil aminosilica for ethoxamer distribution. Detection used absorption or indirect fluorescence.

The evaporative light scattering detector reduces the need for derivatisation. Its application to determination of degree of ethoxylation using a Zorbax amine column (250 × 4.6 mm i.d.) with a hexane/iso-propanol/

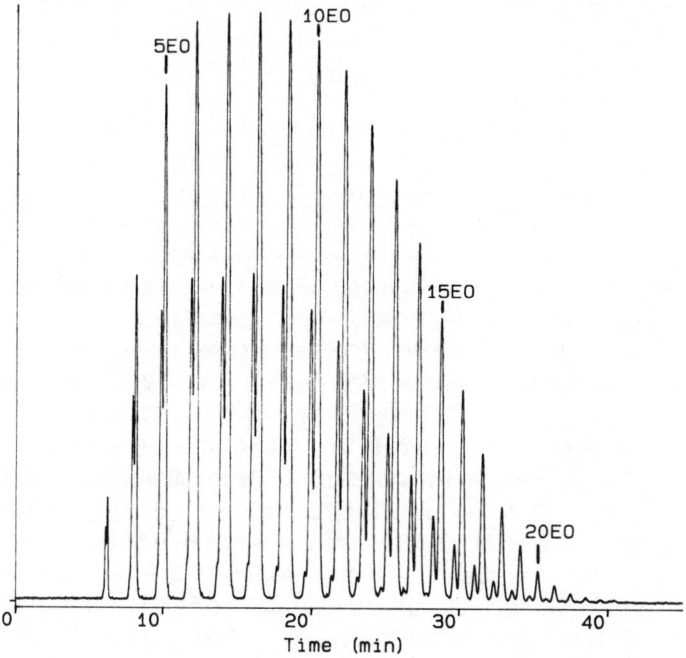

Figure 9.4 Separation of alcohol ethoxylate by ethoxamer (column 250 × 4.6 mm i.d. of 5 micron Nucleosil Si-50, gradient 99:1 ethyl acetate:water to 90:10 acetonitrile:water, evaporative light scattering detector).

water gradient is described in [20]. A typical chromatogram is shown in Figure 9.4.

Analysis of ethoxylated fatty acids using a column (250 × 4.6 mm i.d.) of Nucleosil DIOL with hexane/iso-propanol/water/acetic acid (105:95:10:1 by volume) is described in [28].

Alkylphenol ethoxylates can be readily detected by UV. Columns and separation conditions are similar to those for alcohol ethoxylates. Ethoxamer distribution can be determined on a LiChrosorb amine column (250 × 4.6 mm i.d.) with a hexane/iso-propanol to aqueous iso-propanol gradient system and UV detection at 277 nm [29].

An alternative separation using an equivalent column but a tetrahydrofuran/hexane to aqueous iso-propanol gradient and EI or CI mass spectrometric detection is described in [30].

9.3.11.3 Cationics. Cationic surfactants are more amenable to HPLC than to GC analysis as HPLC can analyse the intact molecule.

The separation of mono-, di- and trialkyl methylammonium quaternaries is described in [31]. The column used is a 250 × 4.6 mm 5 micron RSil

Polyphenol with guard column, with a mobile phase gradient from 90 hexane/10 THF/methanol to 10/90, both solvents containing 5 mM trifluoroacetic acid, and evaporative light scattering detection. Excellent separation is demonstrated.

Cationics of the structure alkylamidopropyl-*N*-(2,3-dihydroxy)-*NN*-dimethyl-ammonium chloride have been analysed on a column (150 × 4 mm i.d.) of micro-Bondapak CN with water/acetonitrile/THF (57:42:1 by volume) containing 0.1% trifluoroacetic acid as mobile phase and differential refractive index detection [32]. Quantification was by external standard and the method was applied to cosmetic products.

The separation of imidazoline-type cationics on a column (150 × 4.6 mm i.d.) of 3 micron Develosil ODS-3 with 0.1 M sodium perchlorate in methanol/acetonitrile/deionised water (60:60:5 by volume) as mobile phase and UV detection at 240 nm is described in [33]. Separation of different chain length alkyl substitution is demonstrated.

A final example of separation of dialkyldimethylammonium quaternaries on a column of 5 micron PLRP-S with a mobile phase of 5 mM methanesulphonic acid in 70% acetonitrile is given in [34]. This example uses post-column ion suppression (see section 9.5.1) and atmospheric pressure ionisation mass spectrometry for component identification. Also described in this paper is the analysis of alkyl sulphates and alkane sulphonates.

9.3.11.4 Betaines and soap. Betaines of the alkyldimethylammoniomethane carboxylate type were analysed on a column (250 × 4 mm i.d.) of 3 micron Develosil C8 at 30°C with a gradient from 80 to 20% of aqueous 45% acetonitrile to aqueous 90% acetonitrile, each containing 0.1 M sodium perchlorate [35]. Prior to analysis the betaines were derivatised with 4-bromo-7-methoxycoumarin to allow detection at 325 nm.

The fatty acids obtained from soap (section 9.2.12.4) can be analysed by HPLC without derivatisation. Use of a column (150 × 4 mm i.d.) of 5 micron Hitachi Gel 3056 at 50°C with methanol/5 mM tetrabutylammonium phosphate (3:1 by volume) at pH 7.5 with conductivity detection is described in [36].

9.3.11.5 Contaminants. Examples of contaminants in surfactants and surfactant products determined by HPLC are 1,4-dioxane, sultones and dialkyltetralins in LAB.

1,4-dioxane in alkylether sulphates was determined on a column (250 × 4.6 mm i.d.) of 5 micron LiChrospher C-8 with an aqueous acetonitrile gradient and UV detection at 200 nm. Details of quantification using a calibration curve are presented in [37].

1,3-sultones were extracted from α-olefin sulphonate and determined on a column (200 × 4.6 mm i.d.) of 5 micron CPS Hypersil with hexane/ethyl acetate (90:10, v/v) as mobile phase and differential refractive index detection [38]. Quantification was by external standard.

Dialkyl tetralins in LAB before sulphonation are determined on a column (250 × 4 mm i.d.) of 5 micron Lichrosorb Si60 with dry iso-octane as mobile phase and UV detection at 254 nm [39]. A reference standard is available to calibrate the method.

9.4 Thin layer chromatography

9.4.1 General description

Thin layer chromatography (TLC) was developed in the 1950s through the work of Stahl and Kirchner. In TLC, an adsorbent (silica or alumina) or reverse phase (often C_{18} alkyl bonded silica) layer (thickness about 250 micron) consisting of approximately 10 micron porous particles is coated on a rigid supporting plate (usually glass). The layer is activated by heating and samples are spotted about 1.5 cm from one edge of the plate. This edge is immersed in the developing solvent, which moves vertically up the plate by capillary action. The sample dissolves in the mobile phase and its components are carried up the plate at rates dependent on their relative strengths of interaction between the mobile phase and the layer. When the mobile phase has travelled to near the top of the plate, the plate is removed from the mobile phase; the solvent is then removed and the separated components can be visualised. Before visualisation, a further development with the same or a stronger mobile phase may be performed in the same direction to enhance the development. It should be noted that the development is carried out in an enclosed chamber and the plate may be presaturated with the developing solvent vapours. Though similar to HPLC in separation mechanisms, there are facets of the development which mean that it is not trivial to adapt a TLC separation to HPLC and vice versa. TLC can be a cheap simple technique, excellent for comparing large numbers of samples. Equally it gives scope for technical skill and understanding in creating separations.

9.4.2 HPTLC

HPTLC uses much finer (about 5 micron) particles and somewhat thinner layers (200 micron) than TLC. This leads to much less spreading on the plate. Smaller plates are used (10 × 10 cm as opposed to 20 × 20 cm) and this necessitates better sample application of smaller samples. This is best carried out with an automatic spotter rather than with a micro-syringe, and with a steady hand and eye. The qualitative and quantitative advances made possible by HPTLC and the instrumentation developed to utilise it have been substantial and have given TLC a new lease of life.

9.4.3 *Developing solvents*

The developing solvent has a critical role in TLC. It is often a complex multicomponent mixture of solvents of differing polarities, water and acid or base. Due to different interactions of the components of the mixture with the stationary layer, the composition of the mobile phase will not be constant at all points on its travel up the layer. This leads to additional opportunities in controlling separation.

9.4.4 *Visualisation*

On completion of development, the separated components of the mixture are visualised. If the components are coloured, they will be visible. If they are UV chromophores they can be visualised under light of the appropriate wavelength. Layers containing a green fluorescent indicator excited at 254 nm are often used. When exposed to radiation at 254 nm, UV chromophores absorbing at 254 nm reduce the background fluorescence and can be observed. Derivatisation reactions are common. The derivatising reagent is sprayed on the plate to induce the reaction, possibly with heating. General visualisation techniques, involving charring of the plate to give spots for all the components of a mixture, are used to ensure that no component is lost or overlooked, apart from very volatile components. This is an important facet of TLC—the knowledge that the whole mixture is visible on the plate. One further point is that, as the plate can analyse many samples simultaneously, it is possible to visualise different parts of the plate by different means to introduce additional selectivity.

9.4.5 *Quantitation*

Quantitation for individual components uses standards of various concentration run simultaneously with the sample. The concentration in the sample is obtained by visual inspection, provided the calibration range includes the sample. Instrumental techniques such as scanning densitometry improve quantitation.

9.4.6 *Additional information*

There are many ways of carrying out TLC not mentioned already in this section. A simple overview is contained in [40].

9.4.7 Applications

9.4.7.1 Anionics. The analysis of LAS using a Silica Gel G layer (a standard TLC silica) impregnated with 10% ammonium sulphate, with 2-methyl-4-pentanone/propyl alcohol/0.1 N acetic acid/acetonitrile (20:6:1.6:1, v/v/v/v) and visualisation by a spray of 5% phosphomolybdic acid in ethanol followed by charring by heating is described in [41]. The analysis of the parent LAB using layers of silica gel impregnated with 15% silver nitrate and double development with dichloromethane followed by recovery of the spots and analysis by GC/MS is described in [2].

The analysis of alkyl sulphates and alkane sulphonates using similar conditions to those given previously is described in [41]. Separation of alkyl sulphates and alkyl ether sulphates using a silica gel layer with acetone:tetrahydrofuran (9:1, v/v) and visualisation with Pinakryptol yellow is described in [42].

9.4.7.2 Nonionics. The determination of nonionic surfactants using a silanised silica gel GF254 layer with aqueous 80% methanol and the use of a scanning densitometer at 525 nm for detection is described in [43]. Analysis of alkylphenol ethoxylates using a Kieselgel F60 layer with chloroform:methanol as mobile phase and IR detection is described in [44].

9.4.7.3 Cationics. Methods for the analysis of cationics of imidazolinium and quaternary ammonium types are described in [45], [46] and [47].

9.5 Other techniques

9.5.1 Ion chromatography

The technique of eluent suppressed ion chromatography was introduced by Small, Stevens and Bauman in 1975 [48]. In ion-exchange chromatography, ionic species are separated on some form of ion-exchange resin using an ionic mobile phase. The conductivity of the mobile phase is high in comparison to that of the separated species. Ion chromatography as proposed by Small *et al.* [48] uses a suppressor column after the separation column which reduces the conductivity of the eluent and enables the sample ions to be detected. An example of this approach is the separation of anions on an anion-exchange column using sodium carbonate as eluent. By passing the column effluent through a cation-exchange column in the hydrogen form, the sodium is retained, the eluent is converted to weakly conducting carbonic acid and the anions are converted to their more conductive acid forms. There has been significant improvement in post-column ion suppression since its original conception but the basic principle remains unchanged. A further important advance was the use of 'pellicular' ion-exchange resins as column packings. These are thin films of ion-exchange resin bonded to a rigid non-permeable support. They do not swell or shrink with changing eluents, allow high flow rates at low column back pressures—which has the additional advantage of reducing equilibration times—and are ideally suited for fast separation of low levels of ions followed by sensitive conductivity detection. Although conductivity is the most common detection technique in ion chromatography, the column technology alone can be coupled with UV detection, electrochemical detection or post-column reaction.

Despite its potential application to charged surfactant analysis, there have been relatively few publications of ion chromatography applications. Dionex in their applications literature demonstrate several possible separations including alkyl sulphates; the separation of sodium isethionate from its alkyl isethionate ester on a Vydac 302 IC column with methanol:20 mM phthalic acid:water (3:5:12, 2.5) and conductivity detection is described in [49].

9.5.2 *Supercritical fluid chromatography*

Supercritical fluid chromatography (SFC) had its origins in the 1960s but only came to the fore as an analytical technique in the early 1980s when the possibility of carrying out SFC separations in capillary columns was demonstrated. As its name implies, SFC uses a supercritical fluid as the mobile phase in a chromatographic separation. The most common mobile phase is carbon dioxide. Columns may be either of GC- or HPLC-type phases and detectors can either be the universal FID from GC or UV from HPLC together with Fourier transform IR, MS and many others. Parameters which influence the separation include temperature and, more importantly, pressure. The applied pressure alters the density of the supercritical fluid and its solvating power for different molecules. Pressure (density) programming during SFC analysis is the equivalent of gradient elution or temperature programming. A further and extremely important feature of SFC is the incorporation of a modifier, a small polar organic molecule such as methanol, in the supercritical fluid to improve its solvating power for more polar molecules. SFC falls somewhere between GC and HPLC in its performance. It can be applied to molecules which cannot be analysed by GC, particularly those of higher molecular weight, or those which are thermally labile. It is more efficient and gives better separation than many HPLC applications. However, its use has been limited as there are few analyses where it gives unique advantages. The current move to SFC on packed micro-capillaries may lead to increased use of SFC in surfactant analysis.

9.5.2.1 Applications. Analysis of alkylbenzene sulphonates and alkyl sulphonates on a fused silica open tubular column (10 m × 0.53 or 0.25 mm i.d.) coated with 0.1 or 0.2 micron SE54 with carbon dioxide as mobile phase and FID is described in [8]. The anionics were derivatised before analysis. Among the many examples of analysis of nonionic surfactants are the following:

—Separation of ethoxylated alcohols on a 20 m × 0.1 mm i.d. column of poly(dimethylsiloxane) with density programmed carbon dioxide at 100°C as mobile phase and FID is described in [50]. Quantitative aspects including relative response factors are dealt with in detail.
—Ethoxylated alcohols were reacted with 50% HBr in glacial acetic acid to give their alkyl bromides. These were separated on a column (9 m × 50 micron) of SE-52XL at 140°C with density programmed carbon dioxide and FID. This analysis gives the alkyl chain distribution and total nonionic [51].
—An example of the coupling of the evaporative light scattering detector to SFC is given in [52]. Nonionic analysis without the need for derivatisation is discussed in [8].

9.5.3 Gel permeation chromatography

Gel permeation chromatography (GPC) involves the separation of molecules in solution based on their molecular size. The molecules to be separated, usually natural or synthetic polymers, are carried down a column of a porous gel with pores of varying sizes. Depending on the size of the polymer, it will pass into some of the pores. The more pores it passes into, the longer it will spend in the column. Thus the larger molecules are eluted first and the smaller last. GPC has little application in surfactant analysis apart from sample preparation where it is used to remove polymeric material from the components of interest prior to more detailed analysis.

9.5.4 Capillary zone electrophoresis

Capillary zone electrophoresis (CZE) is not strictly a chromatographic technique. However, CZE and its related techniques such as micellar electrokinetic chromatography give separations similar to chromatography but using totally different separation mechanisms. These are based on electrophoresis in narrow (50 micron and less) uncoated capillaries of about 1 m in length with an applied potential of 20 kV. CZE is complementary to chromatography and will grow in importance in surfactant analysis in the future.

9.6 Summary

The wide-ranging applications of chromatography to surfactant analysis have been illustrated. The family of techniques offers so many possibilities for any analysis that the practitioner is often spoilt for choice and must consider carefully what his objectives are—sensitivity of detection, qualitative identification, quantitative determination—and whether there are any time constraints. There is almost certain to be an existing chromatographic analysis which will lead to an acceptable solution.

This chapter represents the view of the author and is not necessarily the view of Unilever Research.

9.7 References

1. Osburn, Q. W. *JAOCS*, **63**(2) (1986), 257–263.
2. Eganhouse, R. P., Ruth, E. C. and Kaplan, I. R. *Anal. Chem.*, **55**(13) (1983), 2120–2126.
3. Trehy, M. L., Gledhill, W. E. and Orth, R. G. *Anal. Chem.*, **62**(23) (1990), 2581–2586.
4. Okazaki, T., Kataoka, H., Muroi, N. and Makita, M. *Bunseki Kaguka*, **38**(7) (1989), 312–315.
5. Neubecker, T. A. *Environ. Sci. Technol.*, **19**(12) (1985), 1232–1236.
6. Stephanou, E. *Chemosphere*, **13**(1) (1984), 43–51.
7. Czichocki, G., Gerhardt, E. W. and Blumberg, D. *Tenside Surf. Det.*, **25**(3) (1988), 169–173.
8. Sandra, P. and David, F. *J. High Resolut. Chromatog.*, **13**(6) (1990), 414–417.
9. Suzuki, S., Sakai, M., Ikeda, K. and Mori, K. *J. Chromatog.*, **362**(2) (1986), 227–234.
10. Cozzoli, O., Mariana, C., Faccetti, E., Fornasari, A., Latini, A., Sedea, L., Vignati, G. and Zappa, P. *Riv. Ital. Sostanze Grasse*, **62**(6) (1985), 307–316.
11. Ribaldo, E. J. *et al. J. Colloid Interface Sci.*, **97**(1) (1984), 115–119.
12. Korhonen, I. O. O. *Chromatographia*, **17**(2) (1983), 70–74.
13. Dahlgren, J. R. and Shingleton, C. R. *JAOAC*, **70**(5) (1987), 796–798.
14. Leskovsek, H., Grm, A. and Marsel, J. *Fresenius' Z. Anal. Chem.*, **341**(12) (1991), 720–722.
15. Rastogi, S. C. *Chromatographia*, **29**(9) (1990), 441–445.
16. Matsutani, S., Tsuchikane, H., Sugiyama, T. and Nagai, T. *Yukagaku*, **35**(2) (1986), 80–84.
17. Bear, G. R. *J. Chromatog.*, **371**(1) (1986), 387–402.
18. Matthijs, E. and De Henau, H. *Tenside Surf. Det.*, **24**(4) (1987), 193–199.
19. Marcomini, A. and Giger, W. *Anal. Chem.*, **59**(13) (1987), 1709–1715.
20. Bear, G. R. *J. Chromatog.*, **459**(1) (1988), 91–107.
21. Boiani, J. A. *Anal. Chem.*, **59**(21) (1987), 2583–2586.
22. Maki, S. A., Wangsa, J. and Danielson, N. D. *Anal. Chem.*, **64**(6) (1992), 583–589.
23. Liebscher, G., Eppert, G., Oberender, H., Berthold, H. and Hauthal, H. G. *Tenside Surf. Det.*, **26**(3) (1989), 195–197.
24. Johannessen, R. O., Dewitt, W. J., Smith, R. S. and Tuvell, M. E. *JAOCS*, **60**(4) (1983), 858–861.
25. Beranger, A. and Holt, T. *Tenside Surf. Det.*, **23**(5) (1986), 247–254.
26. Steinbrech, B., Neugebauer, D. and Zulauf, G. *Fresenius' Z. Anal. Chem.*, **324**(2) (1986), 154–157.
27. Schmitt, T. M., Allen, M. C., Brain, D. K., Guin, K. F., Lemmel, D. E. and Osburn, Q. W. *JAOCS*, **67**(2) (1990), 103–109.

28. Zeman, I., Silha, J. and Bares, M. *Tenside Surf. Det.*, **23**(4) (1986), 181–184.
29. Ahel, M. and Giger, W. *Anal. Chem.*, **57**(13) (1985), 2584–2590.
30. Levsen, K., Wagner-Redecker, W., Schaefer, K. H. and Dobberstein, P. *J. Chromatog.*, **323**(1) (1985), 135–141.
31. Wilkes, A. J., Walraven, G. and Talbot, J.-M. *JAOCS*, **69**(7) (1992), 609–613.
32. Caesar, R., Weightman, H. and Mintz, G. R. *J. Chromatog.*, **478**(1) (1989), 191–203.
33. Kawase, J., Takao, Y. and Tsuji, K. *J. Chromatog.*, **262**(1) (1983), 408–410.
34. Conboy, J. J., Henion, J. D., Martin, M. W. and Zweigenbaum, J. A. *Anal. Chem.*, **62**(8) (1990), 800–807.
35. Kondoh, Y. and Takano, S. *Anal. Sci.*, **2**(5) (1986), 467–471.
36. Tsuyama, Y., Uchida, Y. and Goto, T. *J. Chromatog.*, **596**(2) (1992), 181–184.
37. Scalia, S. *J. Pharm. Biomed. Anal.*, **8**(8) (1990), 867–870.
38. Roberts, D. W., Lawrence, J. G., Fairweather, I. A., Clemett, C. J. and Saul, C. D. *Tenside Surf. Det.*, **27**(2) (1990), 82–86.
39. ECOSOL *Dialkyl-Tetralins in Linear Alkylbenzene*, method published by ECOSOL, Brussels, August 1992.
40. Bauer, K., Gros, L. and Sauer, W. *Thin Layer Chromatography—An Introduction*, Huthig, Heidelberg, 1991.
41. Yonese, C., Shishido, T., Kaneko, T. and Marayuma, K. *JAOCS*, **59**(2) (1982), 112–116.
42. Matissek, R. *Parfuem. Kosmet.*, **64**(2) (1982), 59–64.
43. Li, P. and Wang, Y. *Sepu*, **5**(3) (1987), 191–193.
44. Hellmann, H. *Fresenius' Z. Anal. Chem.*, **321**(2) (1985), 159–162.
45. Armstrong, D. W. and Stine, G. Y. *J. Liq. Chromatog.*, **6**(1) (1983), 23–33.
46. Takano, S. and Tsuji, K. *JAOCS*, **60**(4) (1983), 870–874.
47. Read, H. *Lipids*, **20**(8) (1985), 510–515.
48. Small, H., Stevens, T. and Bauman, W. *Anal. Chem.*, **47**(9) (1975), 1801–1811.
49. Ianniello, R. M. *J. Liq. Chromatog.*, **11**(11) (1988), 2305–2314.
50. Geissler, P. R. *JAOCS*, **66**(5) (1989), 685–689.
51. Onuska, F. I. and Terry, K. A. *J. High Resolut. Chromatog.*, **11**(12) (1988), 874–877.
52. Lafosse, M., Elfakir, C., Morin-Allory, L. and Dreux, M. *J. High Resolut. Chromatog.*, **15**(5) (1992), 312–318.

10 Infrared spectroscopy

P. E. CLARKE

10.1 What infrared spectroscopy offers the analyst

Infrared (IR) spectroscopy is one of the most versatile and cost-effective instrumental analytical methods available to the analytical chemist. The traditional attractions of infrared are well known:

—used for the confirmation of chemical structure, particularly in raw material analysis;
—enables functional group analysis of unknown samples;
—is familiar and easy to use (the overwhelming majority of chemists will have used infrared spectroscopy at one time during their training);
—can be readily understood without recourse to complex mathematical theories.

The drawbacks of the technique tend to be as familiar:

—awkward sample preparation;
—bewildering number and variety of bands in the spectrum;
—difficulty with aqueous samples.

The development of spectrometers utilising the principle of the Fourier transformation in the early 1970s, and the rapid rise in the availability of cheap, powerful computers, has propelled infrared once more into the forefront of techniques used in the characterisation and measurement of organic and inorganic materials. These developments have done much to offset the traditional disadvantages of the technique outlined above. Types of analysis that are now routine would, only 20 or so years ago, have been thought of as either impossible or else extremely difficult. Into this category fall the following:

—analysis of aqueous solutions;
—quantitative analysis;
—micro-analysis;
—factory gate raw material clearance;
—kinetic studies on the microsecond timescale;
—environmental analysis.

In this chapter we will first introduce the basics of the subject—in terms both of underlining theory and instrumentation—and then investigate specific applications of infrared to the detergents industry. The chapter will conclude with some recommendations on what to look for when buying a spectrometer.

10.2 Theory

Although this chapter is not intended as a treatise on the fundamental theory of the infrared behaviour of molecules, a few basic principles need to be outlined before we move on to consider the more practical aspects of the technique.

10.2.1 Origin of infrared absorptions

The infrared region of the electromagnetic spectrum extends in wavelength from approximately 2.5 to 50 microns. The vast majority of organic compounds absorb radiation in some parts of this region. These absorptions are caused by transitions between different vibrational states of the molecule. In general, the observed transition is from the ground vibrational state to the first excited state. Transitions are quantised, and can only accurately be described using the tools of advanced quantum theory—something well outside the scope of this book. However, for simple diatomic molecules the analogy to a vibrating spring exemplifies many of the basic principles well.

The frequency at which two masses, joined together by a spring, vibrate, is described by the equations of simple harmonic motion:

$$\nu = (1/2\pi)\sqrt{k(1/M_1 + 1/M_2)} \qquad (10.1)$$

where k is the force constant (stiffness of the spring) and M_1 and M_2 the two masses.

As equation (10.1) indicates, the frequency at which the spring vibrates will increase with the strength of the spring, but will decrease with increasing mass of either particle.

The analogy holds well for simple gas-phase molecules. Of course, we cannot directly observe the vibration of a molecule itself, only the transition between one state and another. However, the frequency of the transition between vibrational states is comparable to the fundamental frequency of the vibration itself.

For a transition between vibrational states to be observed it must be infrared active—i.e. it must lead to the absorption or emission of a photon of infrared light. The rules governing whether or not this can occur are rather complicated—especially for larger molecules—and to be applied require a knowledge of the symmetry properties of the species and recourse to established symmetry tables. Nevertheless, one condition is fundamental—the molecule must possess an electric dipole that changes magnitude during the vibration. Consequently, O_2 and N_2, the most common gases in our atmosphere, having no dipole, do not possess significant infrared spectra.

Returning to equation (10.1), we can see that vibrational frequency is governed by the constants k and the magnitudes of the two masses. Thus a 'stronger' bond should be manifested by an absorption band at higher frequencies. For carbon, the triple bond vibration is observed at approximately 2400 cm^{-1}, double bond vibration at 1600 cm^{-1}, and single bond vibration around 1100 cm^{-1}. The effect of heavy substituents can be seen in the stretching vibrations of halocarbons: C–I, C–Br, C–Cl and C–F vibrations occur at approximately 500, 600, 700 and 1200 cm^{-1}, respectively. This 'heavy atom' effect can be useful for identifying molecules such as acids and alcohols by exchanging H with D and observing the marked shift in the frequency of the O–D bands compared to the O–H bands.

10.2.2 Modes of vibration

So far we have deliberately simplified the subject by considering only diatomic species. Most molecules are considerably more complex than this and consequently exhibit a wider range of vibrational behaviour. The simple formula shown in equation (10.2) quantifies the number of modes of vibration, n_{vib}, exhibited by a non-linear molecule:

$$n_{vib} = 3n - 6 \tag{10.2}$$

where n is the number of atoms in the molecule (note that n_{vib} is really the number of degrees of freedom possessed by a molecule minus those modes required for translational and rotational motion—three each).

Fortunately, equation (10.2) is not fully borne out in practice. Many modes of vibration are often identical in energy (degenerate) and others either too weak to be observed or else entirely infrared inactive (for information on the use of character tables in the assessment of infrared and Raman activity see [1]).

Although it rapidly becomes impractical to describe fully the coordinated vibrations of any molecule of significant size (more than 3 or 4 atoms), spectroscopists find it useful to consider certain bonds and functional groups as though they were somehow isolated within the molecule and behaved as much simpler species. For instance, the methyl group of a long-chain hydrocarbon will have bands in the infrared spectrum assigned to it and described in terms such as 'stretching' or 'bending'. Within these two general categories of vibrational motion are further subcategories. 'Symmetric' stretching occurs when all bonds either expand or shrink at the same time, 'asymmetric' stretching occurs when some bonds expand as others shrink. Similarly, bending can be simply a scissoring motion, or more complex behaviour such as the 'umbrella' motion of methyl groups, 'wagging' or 'rocking' of CH_2 groups, and 'in-plane' or 'out-of-plane' deformation of aromatic C–H bonds.

Before ending this section, note that stretching of bonds is generally more energetic (occurs at higher frequencies) than bending, and that the strength of the interaction with infrared light (i.e. intensity of the resulting absorption band in the infrared spectrum) is generally greater for stretches. It is also generally true that the more highly polarised a bond, the stronger the observed band (this partly explains the great success of infrared when used to look at carbonyl species with their highly polar C=O bonds).

10.2.3 Raman scattering

The Raman effect is due to the same vibrations that give rise to the infrared spectrum. Raman scattering describes the inelastic scattering of incident light by certain vibrational transitions (described as Raman active). Note that this is *not* a fluorescence effect. The molecule is not electronically excited and the incident photon interacts with the vibration of the molecule on a time-scale of the order of 10^{-14} seconds. The Raman effect is also *weak*—except for resonant transitions, no more than one photon in a million is inelastically scattered in this way. Hence the need for powerful sources of monochromatic radiation (lasers) and sensitive detectors (photo-multiplier tubes or charge-coupled devices).

Raman spectra are a useful complement to infrared spectra. Some vibrations give rise to identical bands in infrared and Raman spectra, while others are weak or non-existent in one while being strong in the other. *Polarisability* of bonds determines Raman activity, not the strength of the electric dipole, as is the case for infrared. The interested reader is referred to [2] for full descriptions and applications of the technique.

10.2.4 Near-infrared spectroscopy (NIR)

Originally a laboratory curiosity, the applications of NIR in the industrial environment have burgeoned miraculously in the last five years. This expansion owes more to the greater understanding of chemometric techniques for data analysis, and the development of superior optical fibres for the transmission of NIR radiation, than to any breakthrough in NIR science itself.

Essentially, NIR spectra are the result of vibrational overtones. The fundamental vibrations can normally be observed in the mid-infrared region, where they are much more intense. Overtones involving the vibrations of H atoms are generally the strongest. NIR spectra typically appear devoid of sharp bands, but contain extensive, highly overlapped broad curves. Extracting useful information from such data was previously thought a hopeless task. However, multivariate regression techniques have been developed which do not need sharp features for input. The only requirements are for features that are reproducible and possessed of high signal-to-noise. NIR has the virtue that spectra are generally weak (no saturation), but still contain very little noise, thanks to detectors which are orders of magnitude more sensitive than their mid-infrared counterparts.

Recently, NIR has been used for raw material confirmation, quantitative analysis of mixtures, protein assays and structure–activity relationships. A selection of recent work in the field is contained in [3].

10.3 The infrared spectrometer

Until the last twenty years, an infrared spectrometer meant a dispersive instrument; this had a grating or prism to split light into its component frequencies, slits to provide the required resolution, and usually a dual sample beam to provide internal referencing of transmission. Today, Fourier transform infrared has almost completely replaced dispersive machines in the analytical laboratory. There are several reasons for this, which to be understood first require a brief introduction to the principles and technology involved.

10.3.1 The time domain

Fourier transform (FT) instruments, whether used for infrared, NMR, ion cyclotron resonance, etc., all operate in the *time* domain. The intensity of the analytical signal measured by the detector is related to the time at which the measurement was made, rather than the frequency, or energy, of the signal itself. Time-domain spectra are meaningless to the spectroscopist, and must be converted into the *frequency* domain to appear intelligible. This is done using a mathematical, computationally intense, procedure known as Fourier transformation. The Fourier transformation for a continuous function is given in equation (10.3).

$$\mathrm{f}(t) = \int F(\nu) e^{2\pi i \nu t} \mathrm{d}\nu \qquad (10.3)$$

Note that real instruments do not sample continuously but in discrete, equally spaced intervals. The discrete Fourier transformation is similar in form to equation (10.3). Note also that the discrete Fourier transformation converts a series of *complex* time-domain data into another series of *complex* frequency-domain data. Even a time-domain signal containing only real data, such as is provided by a Fourier transform infrared (FTIR) spectrometer, still produces a complex frequency domain on transformation. Only half these data are displayed as the resulting frequency spectrum.

Several mathematical algorithms have been developed to speed up the Fourier transformation of discrete data. The most successful of these has been the fast Fourier transform (FFT) derived by Cooley and Stukey. There are also inverse FFTs (IFFTs) to convert from frequency data to the time domain. IFFTs are mostly used in mathematical manipulations of data, such as data reduction, deconvolutions, derivatives, etc.

Further information on the mathematics and applications of Fourier transforms can be found in [4].

10.3.2 Instrumentation

Figure 10.1 gives a simplified schematic of an FTIR spectrometer. The labelled parts are now discussed.

The heart of most FTIR spectrometers is a Michelson interferometer. This consists of two mirrors, one fixed and the other moving, separated by a *beam splitter*, which reflects 50% of the incident light while transmitting the other 50%. One half hits the fixed mirror, the other the

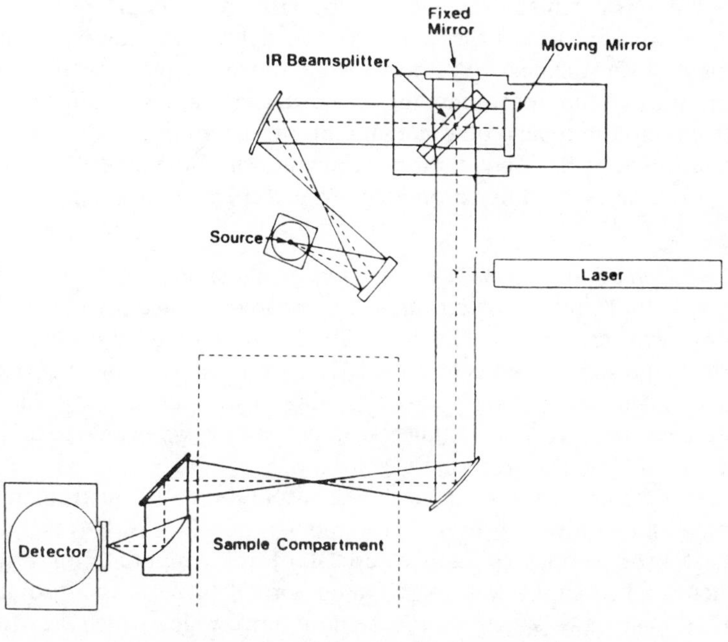

Figure 10.1 Schematic of an FTIR spectrometer.

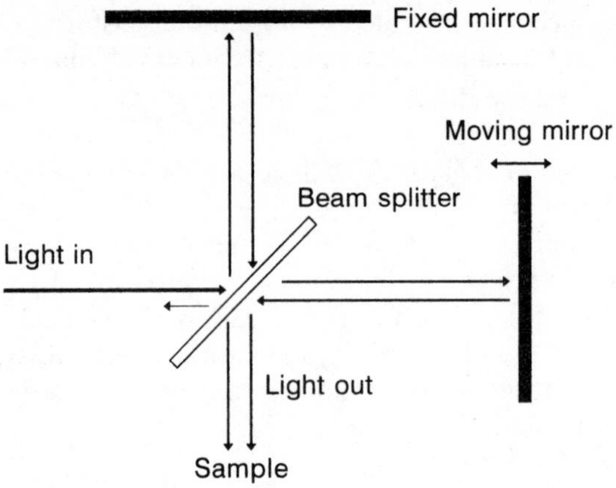

Figure 10.2 Behaviour of a beam of light passing through the beam splitter.

moving mirror. Returning from these mirrors, the light is recombined at the beam splitter (see Figure 10.2) where interference occurs. The nature of the interference depends on the wavelength and optical pathlength difference between the two rays. Out-of-phase rays destructively interfere, while those which are in-phase constructively interfere. As the moving mirror approaches the beam splitter, light of varying wavelength is brought to resonance. As the moving mirror completes its travel between its extreme positions the original, flat intensity (in-time) is modulated into the typical time-domain interferogram displayed in Figure 10.3, equivalent to the frequency spectrum shown in Figure 10.4.

The advantages of interferometry over dispersion are summarised below.

Fellgett (multiplex) advantage. All wavelengths are sampled simultaneously with a Fourier spectrometer, compared to successive sampling in a dispersive instrument. This is by far the most significant advantage that affects the signal-to-noise ratio (SNR). An improvement of 30 times or greater, can be achieved over grating instruments. In terms of measurement *time* this advantage is even more significant—equal to 30×30, i.e. a 900-fold speed increase!

Jacquinot (throughput) advantage. The resolution of FT instruments is determined largely by the distance the moving mirror travels (resolution is in fact proportional to the reciprocal of the amplitude of mirror movement). The beam size is largely unimportant (at least for resolution $>2\ cm^{-1}$). Dispersive spectrometers require narrow slits which inevitably reduce the amount of light reaching the detector.

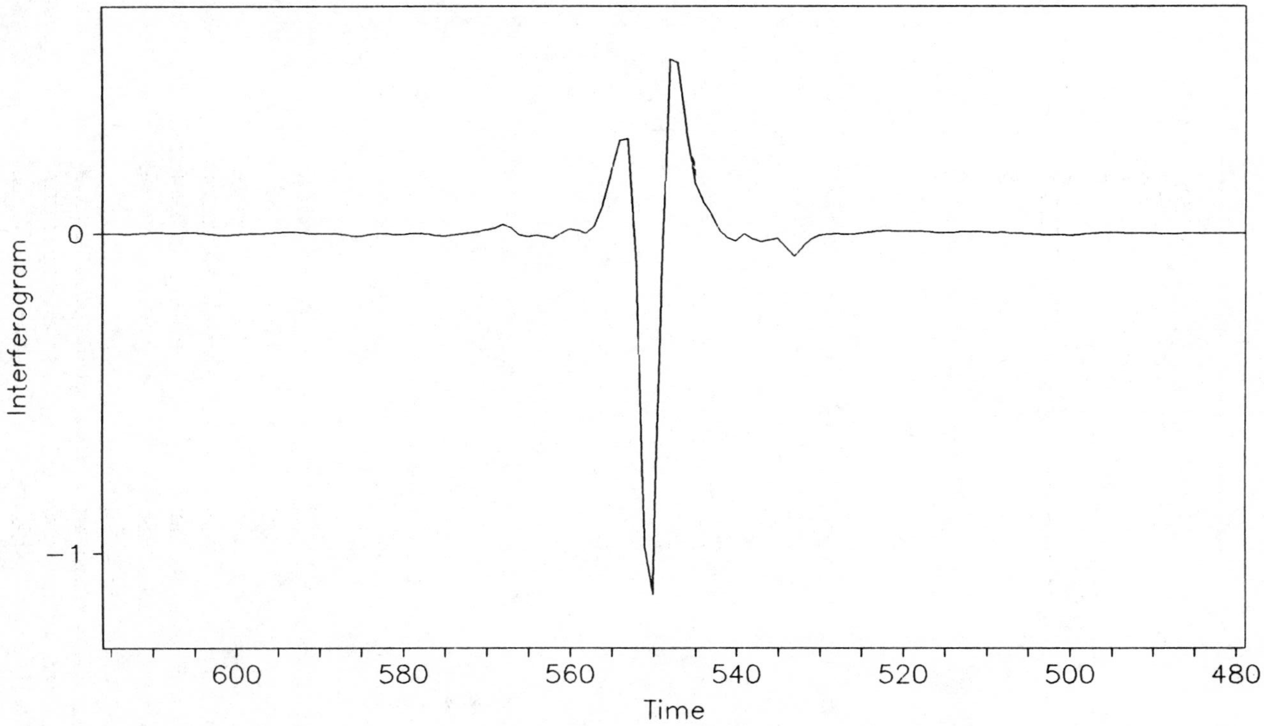

Figure 10.3 Single beam spectrum—time domain.

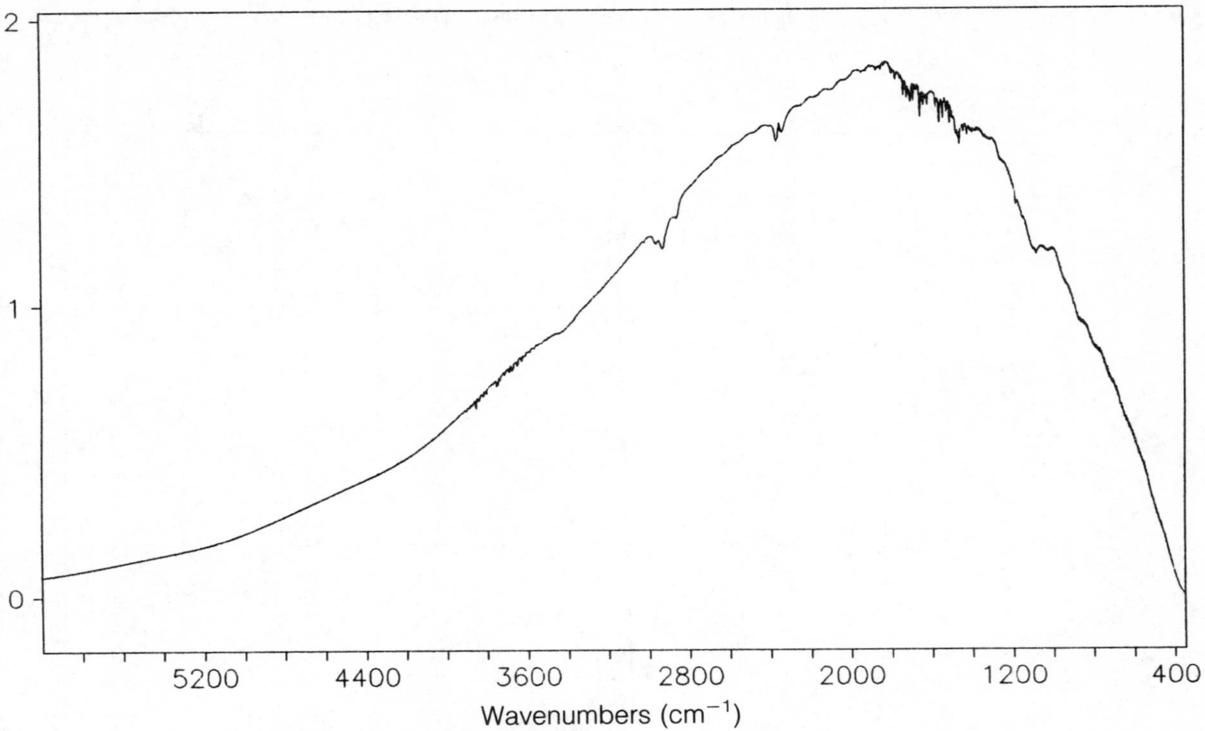

Figure 10.4 Single beam spectrum—frequency domain.

Connes advantage. All FTIR instruments make use of a He–Ne laser beam coaxial with the infrared beam. A separate laser detector triggers data collection at every zero-crossing of this signal—i.e. when the interferometer has introduced the maximum destructive interference into the modulated laser beam.

This internal registration of wavelength provides FTIR instruments with unparalleled wavelength accuracy and repeatability, enabling mathematical manipulations such as spectral subtraction to be performed. Note, however, that if laser and infrared radiation differ in their pathlengths after modulation by the interferometer, accuracy is compromised.

The following terms are commonly encountered in conjunction with FTIR processes and instruments.

Apodisation. The moving mirror of the Michelson interferometer must stop at some point. At the extremes of its motion the interferogram is truncated. If no treatment is applied to the interferogram before Fourier transformation, the resulting infrared spectrum would exhibit peaks with distorted line shapes (so-called 'sinc-ringing'). In order to prevent this, the interferogram must be truncated gradually. The easiest way of doing this is to apply an 'apodisation' function to the interferogram. The simplest apodisation, and probably the most common, is to use a simple triangular function which linearly reduces the interferogram to zero at the turning point. Other functions are also available which are more curved, such as Happ–Genzel, Norton–Beer, etc. All these functions serve to subtly alter the line shape of bands in the spectrum, but the cost of each is the same—a small but unavoidable loss in resolution.

Nyquist frequency. We do not 'know' what the real shape of the time-domain spectrum is when it is only sampled at discrete intervals. If we sample too infrequently, high-frequency signals will be incorrectly analysed. The question of how many data points are sufficient is answered by the Nyquist criterion, which directs that we must sample a signal at least twice per cycle to know its true frequency (a cycle representing successive resonances brought about by the mirror travel). Undersampling can lead to high-frequency signals giving ghost peaks at half, or quarter, their true frequency (known as 'aliasing' or 'foldover'). Unless the optical arrangement of the spectrometer blocks these frequencies, an electronic low-pass filter must be employed to remove them.

Digitisation. In this digital age, all electrical signals issuing from the detector must be digitised prior to Fourier transformation. The typical analogue-to-digital converter is 16 bit, giving a theoretical signal-to-noise ratio of the order of 10 000:1. Some manufacturers claim a higher effective number of bits by applying a non-linear preamplification to the detector output—so-called gain-ranging. This can only be done when it is known that the centre-burst of the interferogram is much stronger in intensity than the wings, so that increased amplification of the wings will therefore not saturate the AD converter. For Fourier transformation, the interferogram must, of course, be unweighted. The data points acquired at higher gain must, therefore, be reduced by dropping the appropriate number of least significant bits.

Signal averaging. A single, complete time-domain scan at moderate resolution (4 cm^{-1}), with a fast detector, will take considerably less than a second. Sometimes the signal will be strong enough that no more scans are necessary. Usually, however, several scans are collected successively and averaged before Fourier transformation. The improvement in signal-to-noise thus achieved is proportional to the square root of the number of scans co-added and averaged. Therefore, using 25 scans rather than a single scan brings about a fivefold reduction in noise.

Zero filling. The fast Fourier transform requires that the number of data points in the time domain be equal to an integral power of two. If insufficient points exist, either the data must be reduced, which sacrifices resolution, or else enough zeros are added to the end of the interferogram so that the magic number is reached. Zero filling also makes the frequency spectrum appear significantly smoother, although this can sometimes give a misleading smoothness to the data.

10.3.3 Hardware

The basic configuration of the FTIR spectrometer illustrated in Figure 10.1 contains fundamental components which need to be briefly described.

Source. This usually consists of a filament or rod of some refractory material, heated to a temperature of around 1500 K so as to emit infrared radiation. The Globar is probably the most common source of mid-infrared radiation, consisting of synthetic silicon carbide. This usually has to be water-cooled, however. Filament (Nernst), and nichrome wires are also popular—and may not require water cooling. Water-cooled sources should deliver a higher and more stable output, which is better suited to quantitative applications.

Far-infrared (FIR) experiments, at least above 100 cm^{-1}, can also utilise the Globar source. Below 100 cm^{-1}, however, a mercury discharge lamp must be used. This emits considerable UV radiation as well as FIR, and care must be taken that the sensitive electronic components of the instrument are not exposed to this ionising radiation.

Table 10.1 Typical beam splitter materials and their useful spectral ranges

Material	Spectral range (cm^{-1})
Ge on KBr	7400–350 (can be extended to 10 000)
Ge on CsI	6400–225
Quartz	15 250–2000
Si on CaF_2	10 000–1200
Mylar	500–10 (requires several different thicknesses of Mylar to span whole range)
Wire grid	650–50

Near-infrared experiments may also require their own source, typically a quartz halogen lamp.

Beam splitters. The splitting of the infrared beam into reflected and transmitted beams is usually performed by a thin coating of germanium supported on an infrared transmitting material. Typical beam splitter materials are shown in Table 10.1, which indicates the wavelength range over which these materials can be used. KBr is by far the most popular choice for the mid-infrared region, and can be used at wavelengths up to 10 000 cm^{-1} if a wide-range KBr beam splitter is used. The drawback with this material is that it is hygroscopic. If the air around the beam splitter becomes too humid, irreparable damage will occur.

Detectors. In mid-infrared spectroscopy, there are really only two detectors which are commonly used. These are the deuterated triglycine sulphate (DTGS) or the mercury cadmium telluride (MCT) detector.

The DTGS is a pyro-electric detector—infrared radiation falling on a DTGS element causes a rise in temperature, which in turn changes the polarisation of the material, inducing a voltage on attached electrodes. The detector does not need to be cooled, returns a linear response with intensity, and furthermore has a very wide operating frequency range—4000 cm^{-1} to below 50 cm^{-1} (with a suitable window over the element—CaF_2 for mid-infrared, polyethylene for far-infrared).

The MCT detector operates from a photoelectric effect. Infrared radiation promotes electrons from the valence to the conducting band and hence generates a current. Because the energy of infrared radiation is low (<1 ev), thermal electrons will spontaneously populate the conduction band unless the detecting element is cooled. For this reason, MCT detectors must be kept at liquid nitrogen temperature (77 K). Even so, the operating range is limited, and must be specified in advance of purchasing the instrument. Narrow-band MCT detectors are the most sensitive, offering around an order of magnitude greater sensitivity than DTGS—at a cost of cutting off around 720 cm^{-1}. Intermediate- and wide-range MCTs extend the lower limit to 600 and 450 cm^{-1}, respectively, but at great cost to sensitivity in the mid-infrared region (around 1000 cm^{-1}).

A DTGS detector would normally be preferred for measurements below 700 cm^{-1}, unless the greater speed of the MCT, in experiments such as kinetic or time-resolved work, is required (MCT detectors can operate at much higher mirror velocities—modulation frequencies—than can DTGS—by up to an order of magnitude).

Care must also be taken when using MCT detectors for quantitative analyses. At high photon flux, MCTs exhibit serious non-linear responses. An aperture may be required to reduce the amount of light reaching the detector in this case.

Amplification and data handling. The signal from the detector is first filtered, if aliasing is to be avoided, amplified and then digitised. The digitised interferogram is usually then fed into a micro- or mini-computer through an appropriate data-acquisition board, and the resulting raw data is ready for pretreatment before Fourier transformation.

Early FTIR instruments required specialist computing resources, which partly explains their high cost. Today, the electronics which required a console the size of a small room to house them, can be contained within a PC of moderate performance and low cost, bolstered by fast data-acquisition cards, transputers or coprocessors.

A series of scans, each containing a few thousand data points, can now be averaged, digitised, apodised, phase-corrected and Fourier transformed in much less than a second—even using computers of only moderate performance. The frequency-domain spectrum produced will still, however, contain unwanted spectral artefacts caused by the distinctive spectral signatures of source, mirrors, beam splitter, sampling accessory and detector—as well as the desired information from the sample itself. The spectrum in this form is referred to as a single beam spectrum, as it has not yet been referenced to the background spectrum obtained through the same optics but without the sample. Dividing the sample single beam by this background single beam successfully removes unwanted artefacts, leaving the typical transmission spectra displayed later in this chapter.

10.3.4 A few tips on running spectra

Before we consider the thorny issue of how best to present the sample to the infrared radiation, it is worth considering briefly the general approach to use when recording spectra.

Traditionally, infrared spectra are displayed with abscissa units of wavenumber and ordinate units of percent transmission. The infrared active bands of the sample are consequently revealed as the dips in the spectrum (peaks pointing downwards). However, transmission units are not appropriate for quantitative measurements, as the relationship between quantity (concentration) and transmission intensity is not linear. However, if we take the negative log of the transmission intensity, the resulting units (called absorbance) are now proportional to concentration, at least for systems obeying the well-known Lambert-Beer law. Note that absorbance peaks point upwards.

Ideally, if the sample has been prepared correctly and the appropriate means selected for presentation, the infrared spectrum should contain bands that are sharp, where the transmission of infrared light through the strongest band is around 10% (and certainly not less than 1%) and where weak peaks are still easily discernible above the inherent noise in the spectrum. In addition, the spectrum should be free of artefacts caused by reflection or scattering of the incident radiation (specular reflectance from a sample lends a 'differentiated' aspect to peak shapes, with positive and negative components present), and possess a flat baseline. Such perfect spectra are, alas, none too common, and allowances have to be made when tight schedules demand fast turnover. If a quick confirmation of a particular band is all that is required, sloping baselines and noisy signals can be tolerated. Quantitative analysis does, however, demand more in terms of the quality of the data. In particular, the Lambert–Beer law cannot be relied upon when absorbance values exceed one absorbance unit, or when the noise is a significant fraction of the peak intensity. Clearly the former case requires a change in the sampling conditions, as no mathematical manipulation will compensate for poor sampling. Improving the signal-to-noise ratio is more straightforward. Noise is inversely proportional to the square root of the number of scans recorded before averaging and Fourier transformation—by simply doubling the number the noise is reduced by approximately 40%. A useful trick to save time is to record fewer scans for the background than for the sample. If the optical throughput (total integrated intensity, equal to the height of the centreburst of the interferogram) of the sample is much less than that of the background, the signal-to-noise becomes dominated by the noise in the sample single beam. Recording lots of scans in the background achieves little, and recording half as many as for the sample will probably suffice.

Typically, 20 or 30 scans will do for background and sample single beams. Recording hundreds of scans should be avoided, except for specialist applications (such as very weakly absorbing samples, e.g. monolayers, or in areas where high signal-to-noise is essential, deconvolution, for instance).

Another way of boosting signal-to-noise is to reduce the resolution. Although 4 cm^{-1} is largely the standard in FTIR analysis, bands are never that narrow, except for gases. Indeed, a peak with an intrinsic FWHM (full width, half-maximum) of $12\,cm^{-1}$ is remarkably sharp. Reducing the instrument resolution to $8\,cm^{-1}$ will not cause much deterioration in spectrum quality. Although in theory there should be little relationship between resolution and signal-to-noise, for the same number of scans, in practice most interferometers perform better if the shortest mirror travel is used (corresponding to the 'lowest', i.e. $8\,cm^{-1}$, resolution). Even a perfectly aligned mirror with a completely friction-free bearing will still record an $8\,cm^{-1}$ spectrum in half the time needed for $4\,cm^{-1}$

10.4 Sample presentation

Samples for infrared analysis can be solid, liquid or gas; they can be analysed in solution or following separation. With the appropriate technique, environments can range from *in situ* measurements to the analysis of picogram quantities of material for forensic analysis. This versatility and power are unrivalled amongst other instrumental analytical techniques.

We will now consider some of the more popular methods for sample preparation and presentation to the infrared beam.

10.4.1 Pressed halide disc method

The most common means of recording the spectrum of solid samples, the pressed halide disc method, relies on containing the material to be analysed in a self-supporting matrix of infrared transparent material. KBr is by far the most popular salt to use, although other metal halides, such as CsI or KCl, have their own particular advantages. CsI, for instance, transmits down to $180\,cm^{-1}$ whereas KBr cuts off at $350\,cm^{-1}$.

To press a disc, the following is required: infrared-quality KBr, a press (hand-held or hydraulic) capable of delivering from 15 000 to 25 000 psi pressure, a die, and a mortar and pestle (usually made of agate). The KBr should be kept in a clean, dry environment—preferably a desiccator or a drying oven maintained at 105°C. The storage unit should not be used for anything which might contaminate the KBr by sublimation or evaporation. The die is a simple infrared accessory which holds the powder between two platens while pressure is applied.

The procedure for preparing a KBr disc is as follows:

> Place 1 to 10 mg of sample in the mortar and grind to as fine a consistency as possible. Grinding reduces particle size and thereby helps prevent unnecessary scattering of the incident radiation. Add approximately 0.5 g of KBr to the sample and mix together. Further grinding is not required. Place the resulting mixture on the lower

platen, which should already be in place in the die. Typical platen size is 16 mm. Place the second platen on top, together with the plunger, and place the whole assembly in the press. Apply pressure for a minute or two, remove the die and mount it on a KBr disc holder. The disc/disc holder assembly can now be slipped into the bracket within the sample compartment of the spectrometer and a spectrum recorded.

It is sometimes advisable to record the background through a pure KBr disc rather than an empty compartment. This may help to correct for baseline slopes or water accumulation.

Ideally, the die should be evacuated before being compressed. This removes air from the matrix and helps prevent micro-cracks appearing in the disc during scanning. The effect is not usually too damaging, however, and this stage is frequently omitted.

10.4.2 Mulls

The principle of mulling is that a powdered sample is held in a viscous liquid with a similar refractive index to the sample, supported by an infrared transmitting window such as KBr or NaCl. The method is similar to the early stages of making a pressed disc—grind a small quantity of sample and add the mulling agent, mix well together, remove the mull from the mortar with a spatula and place on the window. A second window can then be placed over the first and the mull squeezed to the desired thickness. The appropriate window mount is then used to introduce the sample to the spectrometer.

Typical mulling agents are mineral oil ('nujol') or perfluorinated hydrocarbon. The latter is used if there is an interest in recording the CH bands of the sample. The region from 4000 to 1350 cm^{-1} can be recorded using perfluorinated hydrocarbon, and the region from 1350 to 400 cm^{-1} using nujol. These two can then be combined to show a complete spectrum exhibiting few effects due to mulling agent.

10.4.3 Liquids

Non-volatile, viscous liquids can be analysed as films supported on, or between, infrared windows. Samples that are difficult to spread can be taken up in a volatile solvent (such as methanol or ether), and 'cast' onto the window by applying a drop of solution to the window. A heat lamp will quickly drive off the solvent leaving a thin film.

Thicker liquids will behave like mulls and can be trapped between two windows. To obtain good-quality spectra, the separation of the plates should be around 25 microns, although non-polar materials may need much greater pathlengths for adequate spectra.

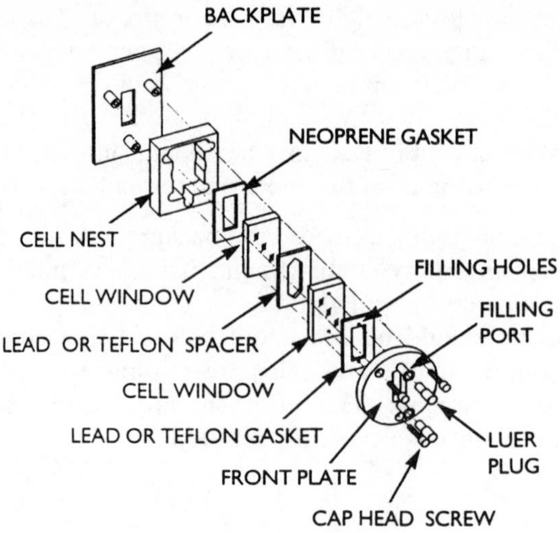

Figure 10.5 Schematic of a transmission cell (reproduced by kind permission of Graseby Specac Ltd).

10.4.4 Solutions

Solutions are generally used for precise quantitative measurements. Simply squeezing some solution between two plates is not an acceptable method for obtaining quantitative results. Transmission cells and internal reflection accessories are the best methods for obtaining quantitative data. A schematic of a transmission cell is shown in Figure 10.5. A spacer of known thickness (usually PTFE, 0.01–0.1 mm thick) maintains a constant separation between two infrared windows. The top window contains holes at the top and bottom which allow the solution to be injected between the plates using a syringe. The sample spectrum should be referenced to a background obtained by passing light through a cell containing the solvent only. Do not record a background through the empty cell as interference fringes will distort the spectrum.

If the same cell is used for calibration purposes, cleaned, dried and then reused for the test sample, no compensation is needed for the pathlength between the plates. However, if fresh calibration solutions are not to hand, historical data can be used as long as any difference in pathlength is taken into account. The best way to measure the pathlength is to record a single beam spectrum through an empty cell and measure the distance (in cm^{-1}) between consecutive peaks or troughs of the interference fringes. The pathlength, in centimetres, equals half of the reciprocal of this.

Remember that absorbance values must be used for all quantitative

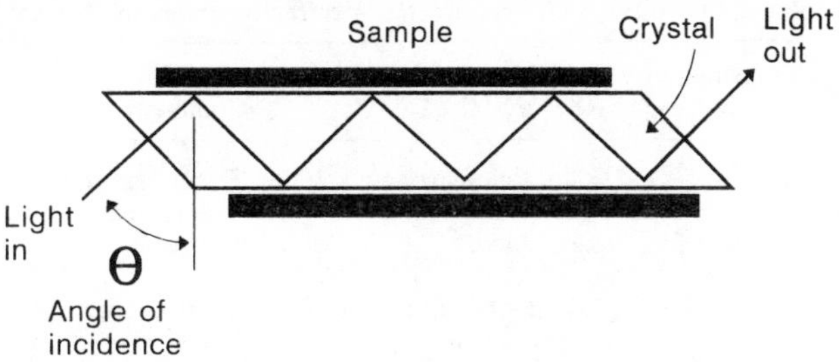

Figure 10.6 Behaviour of light through an ATR crystal.

work. Concentration is then proportional to the inherent band strength, the absorbance of the selected peak, and the pathlength of the cell.

10.4.5 Attenuated total reflectance (ATR)

Analysis of low-viscosity, volatile liquids is greatly facilitated using a technique based on the principle of total internal reflection. Accessories which make use of this principle are usually referred to as attenuated total reflection (ATR) or frustrated multiple internal reflection (FMIR) accessories. The principle behind the method is illustrated in Figure 10.6. Briefly, infrared radiation is trapped within an infrared transmitting material of high refractive index. This crystal can take several forms, notably a flat plate, or a cylinder (circle or tunnel cells). Light introduced into the crystal subtends an angle of incidence to the interface between crystal and surrounding medium that is greater than the critical angle. Rays are reflected back at the interface, but only after they have penetrated the surrounding medium to a distance that is a fraction of the light wavelength. This 'evanescent' wave samples a distance *d* into the surrounding medium, where *d* is described by equation (10.4):

$$d = \frac{\lambda}{2\pi n_1} \sqrt{(\sin^2 \theta - n_2/n_1)} \qquad (10.4)$$

where n_1 and n_2 are the refractive indices of crystal and surrounding medium, respectively, at wavelength λ, and θ is the angle the infrared radiation subtends at the crystal surface.

Generally each ray is forced to make several internal reflections before exiting the crystal. As the penetration depth at each reflection is around a micron, a crystal allowing 10 internal reflections will generate a spectrum that, in transmission, would require a cell 10 microns thick.

The simplicity of sample presentation in such a cell has generated a rapid rise in the popularity of the technique. It is ideal for quantitative

Table 10.2 Properties of typical ATR crystals (RI = refractive index at 1000 cm^{-1})

Material	Transmission range	RI	Liquids which attack crystal	Comments
ZnSe	20 000–450	2.4	Acids/strong alkali	Easily cracked/fogs in detergents
ZnS	17 000–833	2.2	Acids	Robust
Ge	5500–600	4.0	Hot H_2SO_4, aqua regia	Brittle, low transmittance
AMTIR	11 000–625	2.5	Alkalis	High transmission, resists detergents
KRS-5	20 000–250	2.4	Slightly soluble in water	Toxic

analyses, as the sampling depth does not depend on the quantity of sample poured onto the crystal. In addition, the technique is not seriously affected by scattering from particulates, or by artefacts caused by the generation of standing waves set up in transmission cells, which can smother a spectrum with sine waves. Aqueous solutions are particularly suited to the sampling depth that ATR provides, and multicomponent calibrations for aqueous detergent mixtures are not difficult [5].

ATR crystals may be made of a variety of materials, and offer varying sampling depths. Control of the incident angle at which light rays impinge on the interface will alter both the number of internal reflections *and* the penetration at each incidence. Incident angles of 45° are the most common variety. However, a higher incident angle will lessen the sampling depth. Aqueous solutions, in particular, will benefit from the use of a 55° internal reflection element if high quantitative accuracy is required and bands lying over the 1640 cm^{-1} water peak must be used. Angles less than 45° are less common, as the critical angle does not permit much exploitation of smaller angles.

Table 10.2 lists the names and properties of some common types of ATR crystals. ZnSe is perhaps the most common material, as its wavelength range is wide and it is not easily attacked by solvents. Germanium is particularly resistant to highly corrosive chemicals, and AMTIR particularly well suited to long-term exposure to detergents (ZnSe has a tendency to fog if left in contact with anionic detergents for extended periods).

The appearance of an ATR spectrum is very subtly different to a transmission spectrum. Individual band shapes are slightly unsymmetrical, but, more importantly, the intensity of the high-wavenumber spectrum (O–H and C–H stretches for instance) is significantly reduced. This cannot easily be avoided, as the ATR effect (equation (10.4)) shows that sampling depth is proportional to wavelength. However, this is not a problem for quantitative analyses, only library searching may be hindered by it. Searching of non-ATR libraries for a match to an unknown spectrum recorded with an ATR accessory requires a correction to be made for this high-frequency insensitivity.

ATR may also be used to study solids which cannot easily be examined using mulls or KBr discs. The sample should ideally be quite plastic so that contact with the crystal is better effected. Carbon-filled rubber is an example of a material with a spectrum which is almost imposssible to record using conventional transmission. Fibres and films, such as wrappers, are usually easier to examine with ATR than with transmission. A weight, or sprung pressure pad attached to the ATR assembly, will help ensure good contact.

10.4.6 Other reflection techniques

External reflection. This is not as well developed a technique as internal reflection: the physics of reflection of light from surfaces is less accommodating to the infrared spectroscopist. Smooth or shiny surfaces are particular problems. 'Specular' reflection from the surface itself is governed by Fresnel's equations—the reflectance depends on a complicated combination of refractive index, sample absorbance and polarisation. Consequently, samples where the reflectance is mainly from the surface give rise to spectra which bear little relation to conventional transmission spectra. A transformation known as the Kramers–Kronig transformation does exist which attempts to convert a 'specular' reflectance spectrum into a conventional-looking one. It is not 100% successful, and also very computer-intensive. For these reasons, specular reflectance is not commonly used by the analyst.

Diffuse reflectance (DRIFT). This is increasing in popularity as more accessories become available at affordable prices. Accessories are usually configured so as to reduce the proportion of surface (specular) to sub-surface sampling. The theory of diffuse reflectance holds that a ray penetrates a rough surface, is scattered by the sub-surface region, and finally re-emerges to be collected by the optics of the system. Sometimes devices are employed to prevent specular light from being collected. A 'blocker' may be used to physically prevent surface reflections escaping, or the optics can be arranged in such a way that diffuse and specular rays are brought to different focal points. Figure 10.7 shows a schematic of a diffuse reflectance accessory.

The uses for DRIFT often overlap with KBr discs or mulls. However, much less sample preparation is required and interference by the matrix (KBr or the mulling agent) is avoided. This may be essential if the sample

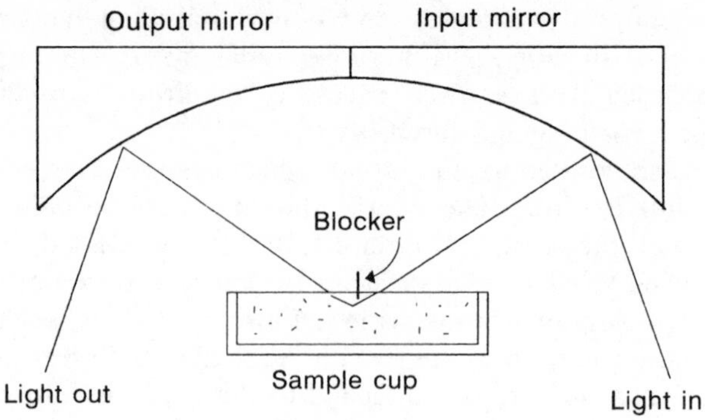

Figure 10.7 Behaviour of light through a DRIFT accessory.

reacts with these materials (strong oxidising agents, such as peroxy acids, are known to react with KBr, liberating bromine). Quantitative analysis is also possible using the Kubelka–Munk function:

$$k/s = \sqrt{(1 - R\infty)/2R\infty} \tag{10.5}$$

where $R\infty$ is the reflectance of the sample, k is the absorption coefficient, and s is the scattering factor. Note that as the scattering coefficient is rarely known, it is generally ignored by ensuring reproducible particle size distribution in samples and standards. This may require consistent grinding and compaction of the samples before analysis. We also still need to record, and divide by, a background spectrum. This is usually KBr.

The reflectance spectra of many materials are very often too strong if the sample is used neat. Mixing with excess KBr may be required to reduce intensity.

10.4.7 Other techniques

Other methods are available, but are somewhat specialist.

Emission. The sample acts as its own light source. The emission spectrum will depend on sample temperature, transmittance and reflectance. Most instruments offer an external port for use in emission experiments. There is normally little point in measuring an emission rather than a transmission spectrum, as considerable reconfiguration of the spectrometer is required and the noise in the spectrum is inherently greater. However, certain measurements may demand such an approach.

For instance, measuring the spectrum of a distant object (on-line analysis of hot processes, analysis of chimney stack exhaust or remote temperature measurement) [6].

Gas cells. Long-pathlength cells are required for gases. A typical arrangement uses a series of mirrors to send the infrared beam through a cell several times. In such a way, an effective pathlength of tens of metres may be achieved. With this pathlength, infrared can be used to measure gas concentrations at ppm levels, for use in factory safety monitoring or car exhaust analysis [7].

Photoacoustic spectroscopy. The sample is placed in a sealed cell where it is exposed to the modulated infrared beam. As the sample absorbs radiation, it heats up, and a thermal wave travels to the surface where it creates a weak acoustic wave in the surrounding medium. This sound wave is 'picked up' by a sensitive microphone and amplified.

In contrast to most other techniques, the strongest absorbers give the greatest detector signal—thus charcoal is often used to provide a background spectrum. Photoacoustic accessories are expensive, but can provide information unattainable otherwise. Intractable samples—strong absorbers or odd shapes (e.g. carpet fibres)—can give quality spectra. Also, more advanced applications make use of the depth-sampling possibilities inherent in the technique. By varying the modulation, the detector can be sensitised to acoustic waves emanating from different depths within the sample. Multilayer polymers can thus be analysed for their different constituents simply by varying modulation. This application is not trivial, and requires high-quality instrumentation to be successful.

Microscopes. Another expensive accessory. An external infrared microscope, fitted onto the side of an existing spectrometer, may well cost more than the spectrometer itself. The usefulness of the accessory for forensic analysis cannot be understated, however. Samples as small as 10 microns in size can be analysed. Particular strengths of infrared microscopy include the analysis of individual fibres, occlusions in polymers, individual crystallites within mixtures, contaminants, layers within paint flakes, multilayer polymers and trace amounts of organic deposits on metal surfaces. The instrument is extremely versatile—some operators report never having to use KBr discs again once the microscope is installed. The wavenumber range is limited, however, as the most sensitive detectors have to be used (generally narrow-range MCT). The small-area detector required is an inherent part of the microscope, it cannot easily be changed. See Messerschmidt and Harthcock [8] for applications and principles of the technique.

10.5 Data manipulations

Having recorded our raw data we will usually have to perform some arithmetical manipulations. These are briefly outlined in the following sections.

10.5.1 Baseline corrections

Elastic scattering often leads to a sloping baseline across the spectrum. Baselines can easily be removed, and instrument manufacturers always provide some means of doing this. Normally, the spectrum is converted into absorbance units before correction, and if necessary reconverted into transmission afterwards. The simplest correction procedure is to take a straight line between the two end-points of the baseline and subtract this from the spectrum. Semi-automatic methods are available on some instruments, although, of course, they should be used with great care as a broad peak may be mistaken for a curving baseline.

10.5.2 Subtractions

The accurate wavelength registration of the FT technique enables the subtraction of one spectrum from another with, at times, remarkably good results. The procedure is really quite trivial: usually the subtraction is done interactively—the operator sweeps the subtraction coefficient through a range of values to find the one that looks the best. Beware programs that place the resulting spectrum in the original data file—you will lose your original data.

One of the most common, and successful, subtractions is the removal of water vapour. The physical removal of this from the spectrometer takes time and expensive purging equipment. However, equally good results can sometimes be achieved by keeping on disc an absorbance spectrum of water vapour and subtracting it whenever necessary. Bands between 1800 and 1450 cm^{-1} are particularly vulnerable to water vapour. Figure 10.8 shows the results of a subtraction on the amide I band of haemoglobin.

A water vapour spectrum can be produced by recording a sample single beam under 'wet' conditions (perhaps immediately after the sample compartment has been opened) and dividing by a background recorded under the driest conditions. Aim to get the absorbance of the strongest water vapour band (usually at 1560 cm^{-1}) above 0.05 absorbance units; a weaker water vapour spectrum may introduce undesirable extra noise into the spectrum following subtraction.

Subtraction is also used to remove the contribution of known chemical constituents when combined with other (unknown) compounds. The success of the subtraction is really determined by the amount of interaction between the components of the mixture—sometimes little can be gained. Remember that no subtraction is perfect, and artefacts (caused by band shifts) can be introduced if the process is taken too far. Avoid making any part of the spectrum negative in intensity.

10.5.3 Peak-picking

Most operating software offers the facility to pick out the peak maxima in the spectrum and store the results in a file and/or label the peaks on the paper hardcopy. This is a very useful facility and greatly eases the task of interpretation. Coupled with peak-picking, there is usually a facility to store peak heights and areas in the same data file. This may be of use for quantitative work, although it remains preferable to analyse peaks personally when quantitation is required. Peak areas are to be preferred for such analyses, as areas better compensate for matrix effects (interactions).

10.5.4 Smoothing

'Noisy' data can be made to appear more attractive by 'ironing out' the spectrum using a smoothing algorithm. Smoothing can indeed greatly improve appearance and enable the positions of peak maxima and their heights and areas to be calculated with greater precision.

The Savitky–Golay algorithm is perhaps the most commonly used device for smoothing spectra. It works by effectively convolving the data with a cubic function; the intensity of each data point is adjusted by an amount determined by a weighted average of the intensities of the points before and after. The smoothing operation simply requires the operator to enter the number of data points to be used—'5' means the centre datum point plus two either side. The more points that are selected, the greater the smoothing. The best choice will depend on the width of the peaks and the resolution at which the data were recorded. It is best not to use more points than there are data points across the narrowest peak in the spectrum. At resolution $4\,cm^{-1}$, a data point is usually recorded every $2\,cm^{-1}$. A peak $20\,cm^{-1}$ wide can be smoothed with a Savitky–Golay algorithm using around 9 data points. Note that a 9-point smooth will inevitably truncate the data slightly by losing 4 points at each end of the spectrum.

Although other techniques have been developed for smoothing, notably using Fourier or maximum entropy techniques, they do share one common feature—noise can only be reduced at the expense of resolution. Smoothing can easily result in loss of information by removing shoulders from bands or erasing small, narrow peaks situated near large ones. It can

(a)

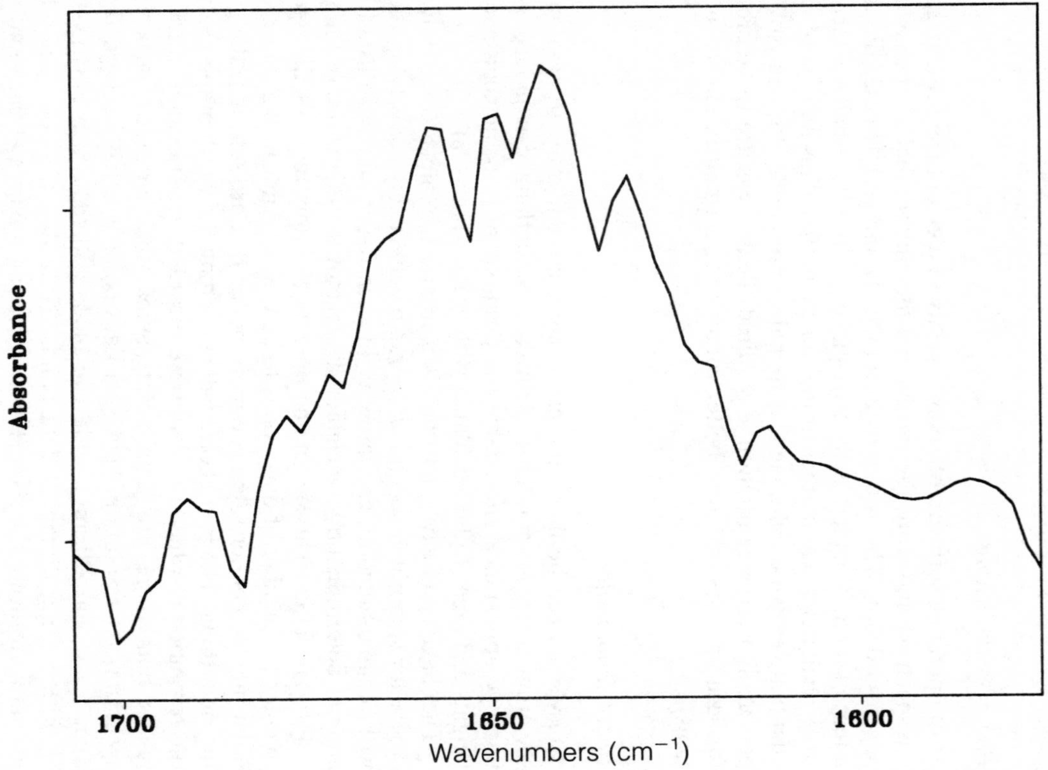

(b)

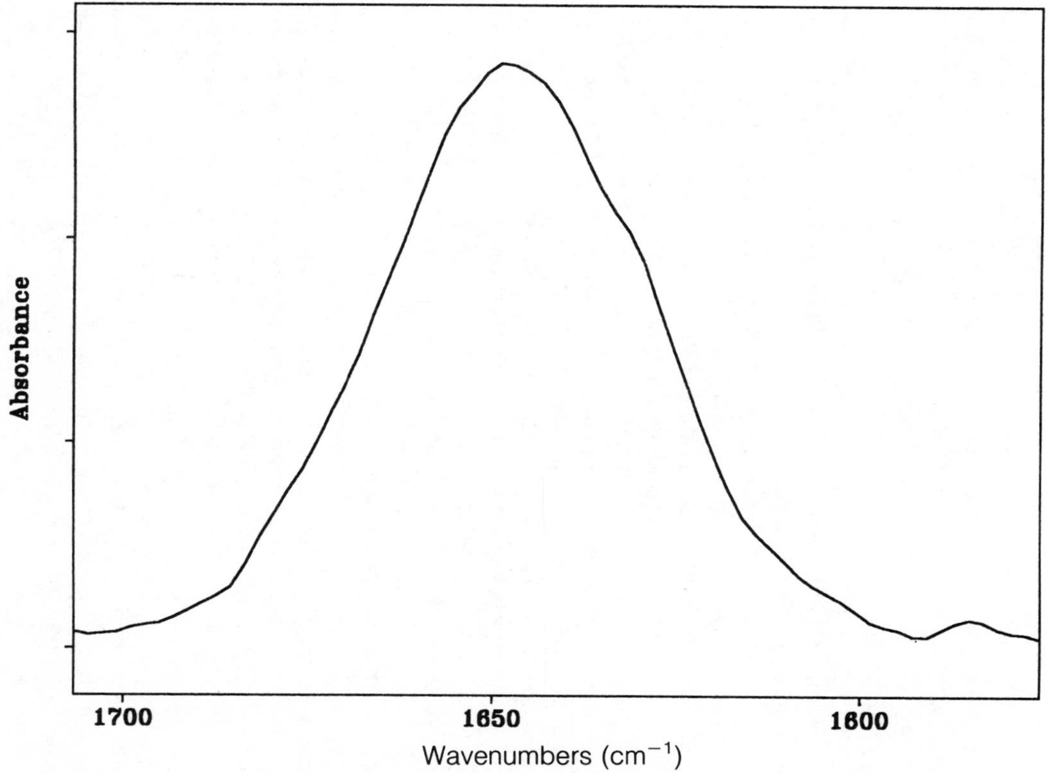

Figure 10.8 Amide I band shape before (a) and after (b) subtraction of water vapour.

be seen that smoothing is conceptually the reverse of deconvolution—a price always has to be made; this price may be resolution for the sake of noise or noise for the sake of resolution.

10.5.5 Differentiation

Differentiation offers perhaps the simplest, and certainly the most primitive, form of band deconvolution. A differentiated spectrum contains sharper features than the original, and this may be useful for pinpointing the positions of individual peaks that combine to form overlapped bands.

An exact differentiation can only be done for a *continuous function*; digital spectrometers do not produce such data, and therefore approximations have to be made.

Taking the simple difference between successive data points is the crudest means of 'differentiating' a spectrum. This method is not popular as it leads to a tremendous corruption in the signal-to-noise ratio of the spectrum. Other methods, including a second Savitky–Golay algorithm, are based on fitting a function, usually a cubic, to a series of data points. The function can then be differentiated accordingly. Fourier transforms may also be used to achieve even-order differentials.

Infrared spectroscopists do not generally use first-order differentials, as such curves do not help in pinpointing the positions of individual peaks present in the original spectra. First-order curves may be useful for quantitative methods, however, and in other chemometrics techniques which benefit from the enhancement which differentiation brings in the prominence of small, sharp bands over broad, sloping baselines.

Second-order spectra are more useful as an interpretative tool. Peaks in the original (i.e. zero-order) spectrum are revealed as sharper, negative-going peaks after the second differentiation. Second-order differentiation therefore offers a crude form of deconvolution.

Figure 10.9 shows zero-, first- and second-order spectra of the amide I band of haemoglobin shown in Figure 10.8. This band is made up of several narrower peaks which strongly overlap each other. The individual peaks which contribute to the overlapped band can reveal important information on the secondary structure of the protein [10]. Differentiation helps in pinpointing the positions of these individual peaks.

10.5.6 Curve-fitting

We sometimes need to calculate the heights or areas of a series of peaks that severely overlap each other. One way to do this is to fit model functions to overlapped bands and allow the computer to calculate the optimum individual area of each. This approach requires some knowledge

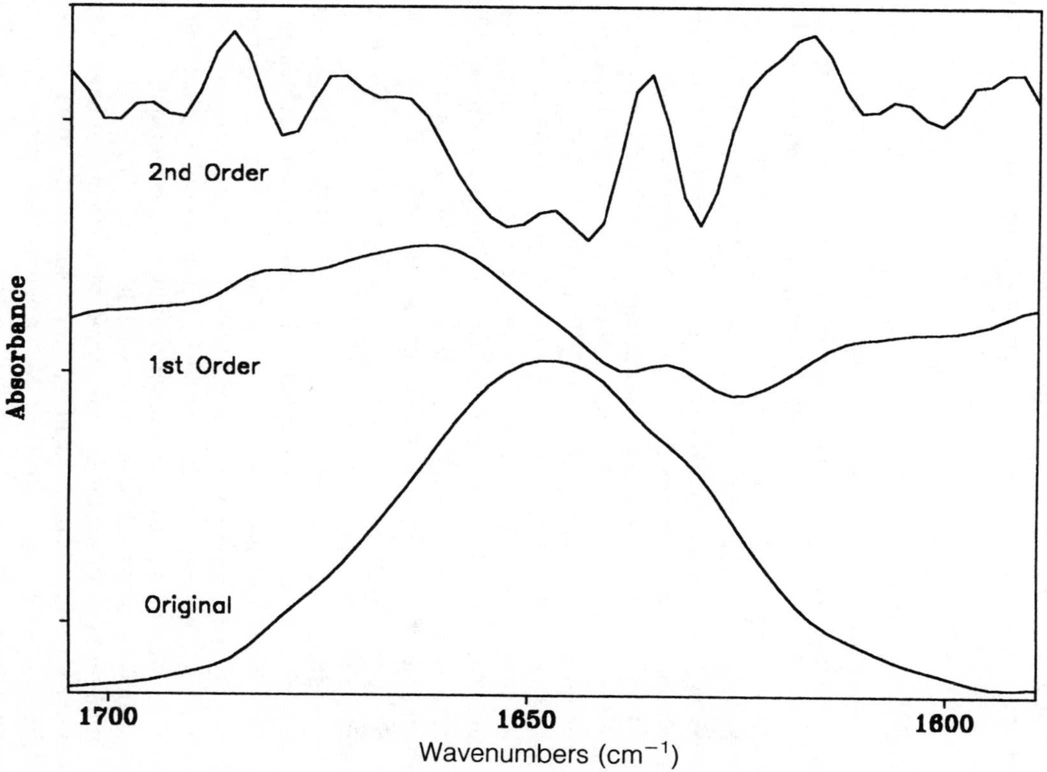

Figure 10.9 Differentiated spectra of the amide I band shape shown in Figure 10.8(b).

of the band shape of the peaks, together with the peak width at FWHM (full width, half-maximum). Unfortunately, infrared spectra do not consistently contain the same band shapes or FWHMs. Indeed, we often encounter unsymmetrical line shapes which do not correspond well to typical Gaussian, Lorenzian or Voigt (combination of both) curves. When the line shape is not known beforehand, the best compromise is generally a Gaussian curve.

10.5.7 Deconvolution

There is an alternative to curve-fitting when it comes to the analysis of highly overlapped bands, and that is spectral deconvolution. Deconvolution describes the process of artificially increasing the apparent resolution of the spectrum so that the FWHM of peaks is significantly reduced.

The most common form of deconvolution is Fourier self-deconvolution, which, as the name implies, takes place in the Fourier (time) domain. Deconvolution in the frequency domain is extremely arduous and lengthy. In the time domain, however, deconvolution equates to multiplication—and is therefore much quicker, even after taking into account the extra processing required for the FFT and IFFT. The procedure requires a knowledge of the band shape and full width, half-maximum resolution of the peaks involved. It is also common to enter a resolution enhancement factor—i.e. how much we want to improve the resolution by. As deconvolution is tremendously sensitive to noise, do not attempt it if the signal-to-noise ratio is less than a hundred. Also, enhancement factors tend not to be much greater than 2.5, owing to the noise amplification that occurs.

Other methods are available for deconvolution. One of the most successful of these is maximum likelihood deconvolution. Although a description of this subject is outside the scope of this text, maximum likelihood can give results that are generally superior to Fourier self-deconvolution. The maximum likelihood algorithm works by searching a response surface of all selected variables until it finds the most likely deconvolution based on the information content of the data.

Maximum entropy is similar in some respects to maximum likelihood, although the technique is perhaps a little more esoteric and requires more expensive hardware and software. Nevertheless, the proponents of maximum entropy argue that it is the best *possible* solution to a problem such as deconvolution, as probability calculus is employed in the calculation of the most likely (maximum entropy) solution.

All the methods outlined, with the exception of peak-picking and baseline correction, will lead to a reduction in the signal-to-noise quality of the specrum. This should be considered *before* the raw data are recorded, as signal-to-noise cannot be increased after the event. Bear in mind that a respectable-looking zero-order spectrum may yet be too noisy following several sequences of subtraction, followed by deconvolution and differentiation.

10.5.8 Spectral search

Having recorded a spectrum from an unknown compound, and smoothed any sloping baselines or removed water-vapour contamination, we are then faced with the difficult job: namely identifying the structure of the unknown from its spectrum.

The examples given in section 10.6 are designed to aid in the task of interpretation. Although it is becoming increasingly common to let computers shoulder some of the burden of interpretation and identification, the current state of the art in computerised interpretation is not yet sufficiently advanced that human experts need feel endangered by them. However, digital libraries have made an impact in the identification of routine samples, such as raw materials, and in certain forensic analyses, particularly the analysis of polymers and plastics (particularly copolymers), and of gaseous products. The danger in relying on libraries is obvious: the analysis can only be successful if the right spectrum is present in the library. There are tens of thousands of chemicals in everyday use, but few companies can afford to buy all the commercial library spectra available. Also, consider that a computer cannot distinguish a mixture of two or more components, whereas a good infrared spectroscopist can normally pick out several components present in one mixture spectrum.

Polymers are a special case: the range of polymers used commercially is such that even a modest library of around a thousand spectra (such as the Hummel–Scholl library [11]) can contain enough pure polymers, copolymers and common mixtures to ensure a reasonably high probability that a computer spectral search will correctly reveal the identity of the unknown material.

GC–IR, where an FTIR spectrometer is interfaced to a GC column, is a second area where spectral searching has a good track record. The number of possible library spectra are again limited (not that many chemicals successfully negotiate the column) and one is confident that the unknown spectrum contains one compound only.

Raw material analysis can be based on a limited spectral library containing only those compounds used as feedstocks. It is better to make one's own libraries for such an application: not only will the results be better but it is much cheaper as well.

There are several methods and algorithms for performing spectral searches. The oldest relies on comparisons of the peak maxima in the unknown spectrum with library peak tables. This is still a useful method for searching large databases quickly, although a large quantity of the useful information in the spectrum is lost.

The speed of modern computers has allowed full-spectrum comparisons to become practical. The spectra are compared point by point, and a coefficient corresponding to the quality of the match is computed. The library spectra are then ranked in order of the top ten or so best matches. The coefficient is usually computed by scaling the unknown spectrum in some way (either by normalising the total integrated intensity or adjusting the largest peak to 100% intensity) and calculating a sum of the squares of the difference between library and test, or computing a correlation coefficient by projecting one spectrum onto the other. Most commercial search software packages allow a variety of approaches to be taken, such as searching differential spectra, peak tables or restricting the wavenumber region. Increasingly, suppliers are adopting principles used in relational databases, and allowing searches to include information other than simple spectral data, such as chemical name, class, physical properties, etc. This is being extended to allow databases of higher dimensions of complexity to be constructed, where, for instance, NMR, GC, mass spectra and infrared spectra would all be included in the same database on a network of computers. GRAMS386 Level 3 (Galactic Inc.) is one such database currently available.

10.6 Examples and applications

Before considering surfactants in detail, it is appropriate to consider those features which the vast majority of compounds to be discussed have in common—namely, a hydrocarbon skeleton or an aromatic ring. Infrared spectra contain information on chain length, branching, isomerism, the position of aromatic ring substituents and other fine details relating to the structure of the material.

Table 10.3 shows hydrocarbon bands commonly found in the spectra of organic materials. The table reveals how infrared spectra can be used to gauge the degree of branching of the compound by comparing the methyl to methylene (CH_2) intensities, the positions of aromatic substitution, and the *cis* or *trans* structure of alkenes.

A confusing aspect to these spectra is the activity in the region 1350–900 cm^{-1}—the so-called fingerprint region. Many of the bands present in this region are strongly dependent on the physical structure of the material, such as its crystallinity or phase behaviour. As such, the region can be useful for providing information on glass transition temperatures [9] or amorphous to crystalline ratios of, for instance, polyethylene [12]. Because of this complexity, the region can be of great

Table 10.3 Typical alkane, alkene and aromatic vibrations (Key: sym—symmetric; asym—asymmetric; str—stretch; def—deformation; R,R^1,R^2—alkyl chain).

Group	Vibration	Wavenumber	Comments
Saturated species			
$-CH_2-$	C–H asym str	2940–2915	
	C–H sym str	2870–2840	
	Scissor	1480–1440	
	Rock	725–720	Observed for longer chains ($>C_5$).
$-CH_3$	C–H asym str	2975–2950	Frequency raised by electronegative substituents.
	C–H sym str	2885–2865	
	Asym def	1465–1440	
	Sym def (umbrella)	1390–370	
C–C	Skeletal vibrations	1260–700	Complex series of bands, usually very weak. Very sensitive to physical structure. Not useful for interpretation.
Unsaturated species			
$>C{=}C<$	Double bond str	1680–1620	Medium-weak, absent for sym species.
$=C-H$	C–H str	3100–3000	
	Out-of-plane def	995–980	Vinyl groups ($R-CH{=}CH_2$).
		895–885	$R^1R^2C{=}CH_2$ groups.
		980–955	*Trans* disubstituted.
		730–665	*Cis* disubstituted.
		850–790	Trisubstituted.
$R^1HC{=}CHR^2$	skeletal vibrations	670–455	Two bands typical of *cis*.
		455–370	Strong band typical of *trans*.
Aromatic species (Ar–H out-of-plane deformations are indicative of degree and position of substitution. These bands are little affected by the nature of the substituents).			
Ar–H	C–H str	3080–3010	
	In-plane def	1290–1000	
	Out-of-plane def	770–730 (strong)	Two strong bands indicative of monosubstitution.
		710–690 (strong)	
		770–735 (strong)	Characteristic of *ortho*.
		960–900 (medium)	*Meta* disubstitution.
		880–830 (medium)	
		820–765 (weak)	
		710–680 (strong)	
		860–800 (strong)	*Para* disubstitution.
$C{=}C$	'Double' bond str	1625–1430	Usually two or three bands present.
	In-plane ring def	650–500	Position and intensity dependent on substitution.
	Out-of-plane def	560–415	

value in identifying the specific nature of a compound by library searching (generic identification is historically all an infrared specialist aspired to). However, this complexity also makes the region a minefield for human interpretation, and for this reason even highly experienced spectroscopists treat this region with great caution when making their interpretations.

10.6.1 Fatty acids and soaps

Figure 10.10 shows the spectrum of a typical fatty acid, Figure 10.11 the spectrum of a fatty acid salt. Clearly infrared is an excellent method for revealing whether the acid has been neutralised or not.

In most environments, fatty acids form dimers with strong hydrogen bonds between acidic O–H and carbonyl oxygen. This H-bonding gives rise to a broad O–H stretch extending over an extremely wide wavenumber range (3300–2500 cm^{-1}). The formation of dimers allows both symmetric and asymmetric C=O stretches to occur—however, only the latter is infrared active. The carbonyl band typically appears at 1710 cm^{-1} and is quite sharp. The C–O stretch occurs at 1320–1280 cm^{-1}, also quite strong, and there is a weak, but significant 940 cm^{-1} band due to O–H out-of-plane bending. This latter will disappear if dimerisation is prevented. Dilute solutions of fatty acids in CCl_4 remain as monomers, and their spectra lack this peak at 940 cm^{-1}. Also, the C=O stretch shifts to higher frequency (about 1740 cm^{-1}).

Dissociated fatty acids (soaps) have quite different features. Delocalisation of the C=O double bond across the CO_2^- group depresses the frequency of the stretching vibration. Stretching of the two C–O bonds is coupled (i.e. the vibration of one influences the vibration of the other). Both asymmetric and symmetric vibrations occur and are infrared active. Bands appear between 1600 and 1560 cm^{-1} (asymmetric), and 1420 and 1340 cm^{-1} (symmetric), with short-chain carboxylates absorbing at the higher end of the range.

The nature of the counter-ion can also affect the nature of the spectrum. Calcium palmitate, for instance, shows a doublet due to asymmetric stretch at 1540 and 1580 cm^{-1}, whereas sodium palmitate only displays a singlet at 1560 cm^{-1}.

Although the foregoing applies to aliphatic and aromatic acids and soaps, there are differences between the two other than those expected from the nature of the hydrocarbon components. In general, the effect of the aromatic ring is to lower the C=O frequency. Typically, aryl acids absorb between 1700 and 1680 cm^{-1}, although the presence of other electronegative substituents in the ring can lower this by a further 10–30 cm^{-1}. Rocking of the CO_2 group also generates a prominent band at 570–540 cm^{-1}.

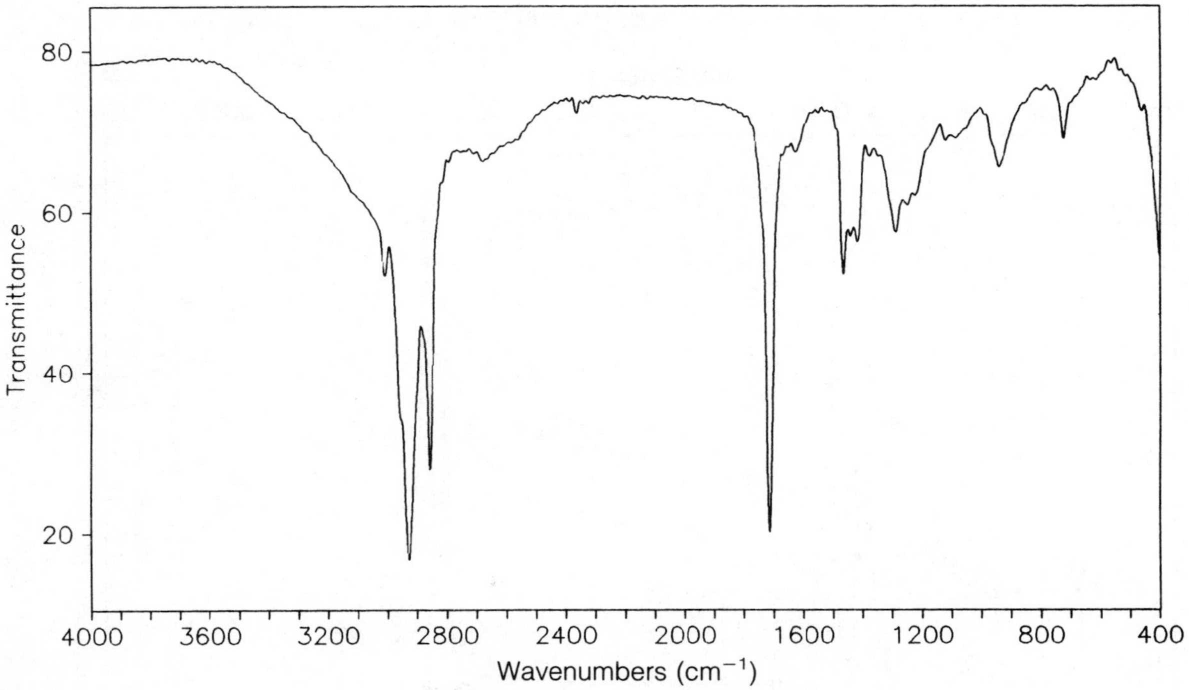

Figure 10.10 Oleic acid.

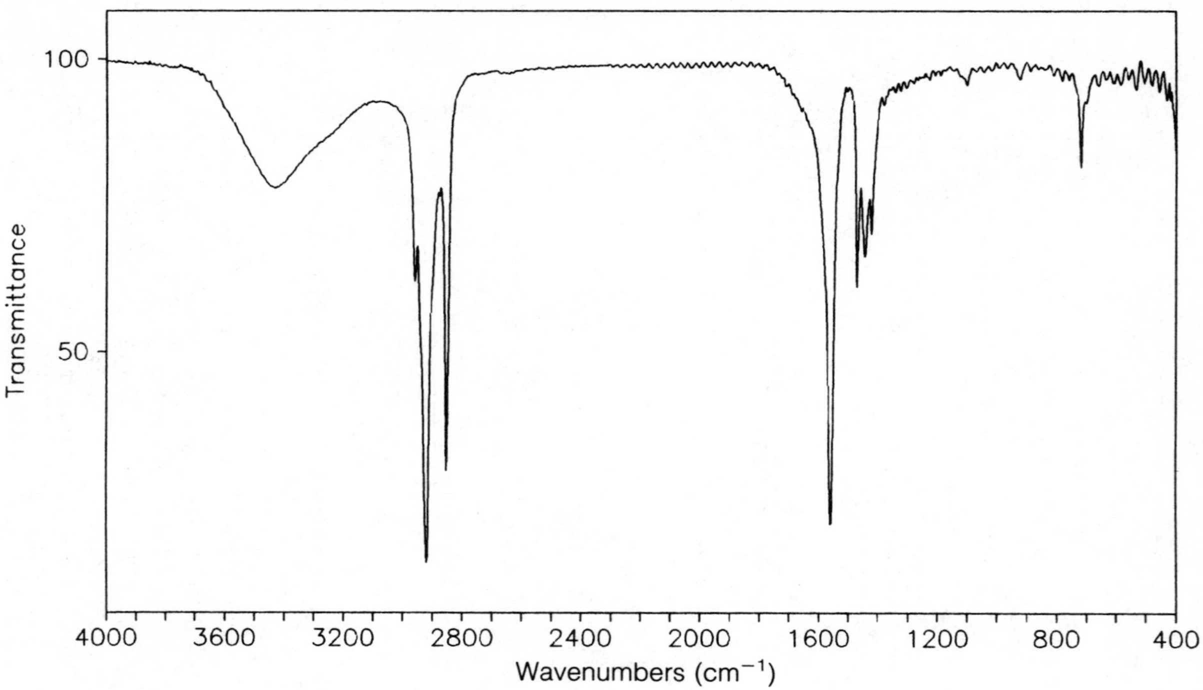

Figure 10.11 Sodium stearate.

10.6.2 Alcohols and alkoxylates

Alcohols are very common surfactants, acting alone, as a solvent for other surfactants, or in combinations forming, amongst others, esters and ethers.

Simple alcohols are often used as solvents—for instance iso-propanol (see Figure 10.12). The main features of such spectra are due to either O–H or C–O bonds. The O–H stretch is usually broad, again due to hydrogen bonding. The band centre is generally at a higher frequency than fatty acids (around 3300 cm^{-1}), and is also much less broad. O–H in-plane bending may sometimes be seen in the region 1400–1260 cm^{-1}, and out-of-plane bending at around 700–600 cm^{-1}. The C–O stretch is highly diagnostic of the nature of the alcohol—whether primary (1085–1030 cm^{-1}), secondary (1125–1085 cm^{-1}) or tertiary (1205–1125 cm^{-1}). Polymeric alcohols, and carbohydrates, also show broad bands centred around 1100 cm^{-1}. Glycerol, an important raw material in many industrial applications as well as an ingredient in products such as high-quality soaps, has a characteristic spectrum with important peaks at 1110, 1040, 990, 920 and 850 cm^{-1} (see Figure 10.13). Phenols exhibit complex spectra which are very sensitive to the nature of any substituents and to the degree of hydrogen bonding. Alkyl phenols generally exhibit medium to strong bands in the regions 1255–1240 cm^{-1}, 1175–1150 cm^{-1} and 835–745 cm^{-1}, and hydrogen bonding gives rise to a weak, though characteristic, band at 450–375, due to in-plane C-OH bending.

Polyethoxylates and polypropoxylates are extremely common nonionic surfactants. The repeating unit in the former is $-CH_2CH_2O-$(EO), and in the latter $-CH_2-CH(CH_3)O-$ (PO). Usually there are between 1 and 7 such ether units per surfactant molecule, however, higher molecular-weight species are used and may contain in excess of 10 such units. Low molecular-weight polyethoxylates have the usual O–H stretch at 3000–3500 cm^{-1}, and strong ether vibrations at around 1110 cm^{-1}. A characteristic series of sidebands is observed at 1450, 1345, 1290 and 1245 cm^{-1}, as well as at 940, 890 and 740 cm^{-1} (Figure 10.14). Infrared is a very powerful tool for identifying the precise nature of polyols, as both generic and fingerprint interpretations can be performed.

Polypropoxylates show many of the same bands as ethoxylates. Polymers made from 1,2-propanediol contain significant numbers of methyl groups, which give characteristic bands at 2980 and 1380 cm^{-1}. An EO/PO mix can be semi-quantitatively analysed by comparing intensities of 1350 cm^{-1} (EO) to 1380 cm^{-1} (PO). Figure 10.15 shows a typical EO/PO condensate.

Alkyl phenol ethoxylates, which generally possess an EO chain *para*-substituted into phenol, exhibit features common to both phenol and polyethylene oxides (Figure 10.16). A distinguishing feature is the

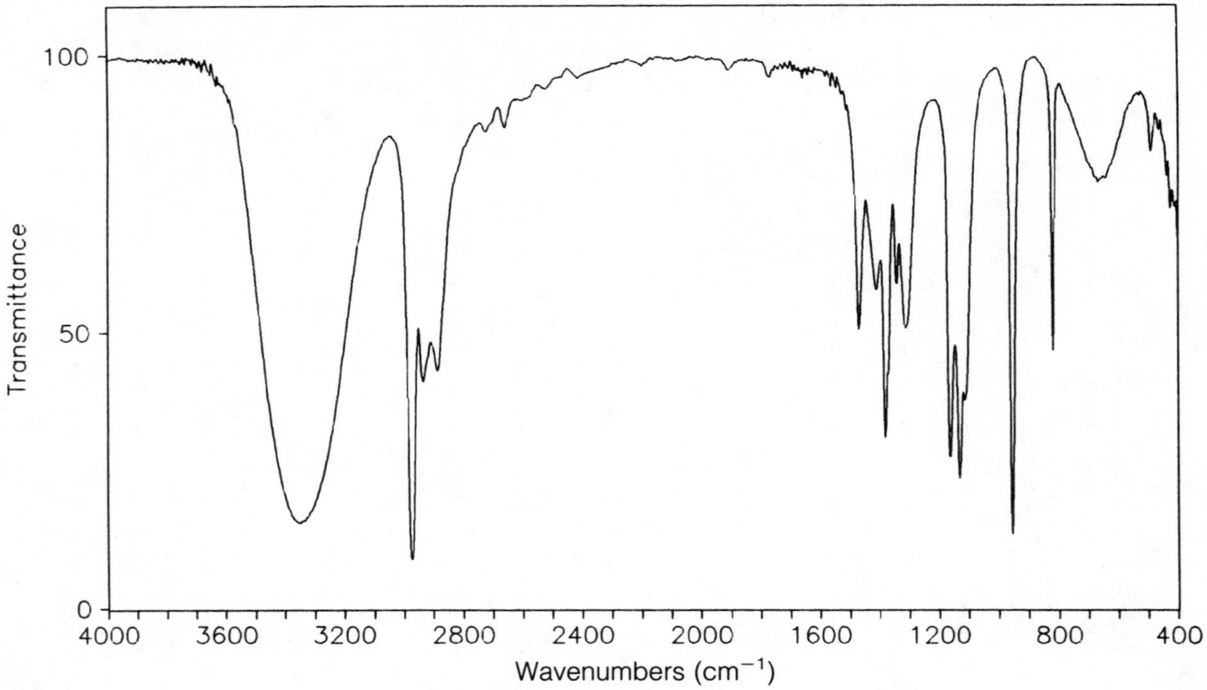

Figure 10.12 Iso-propanol.

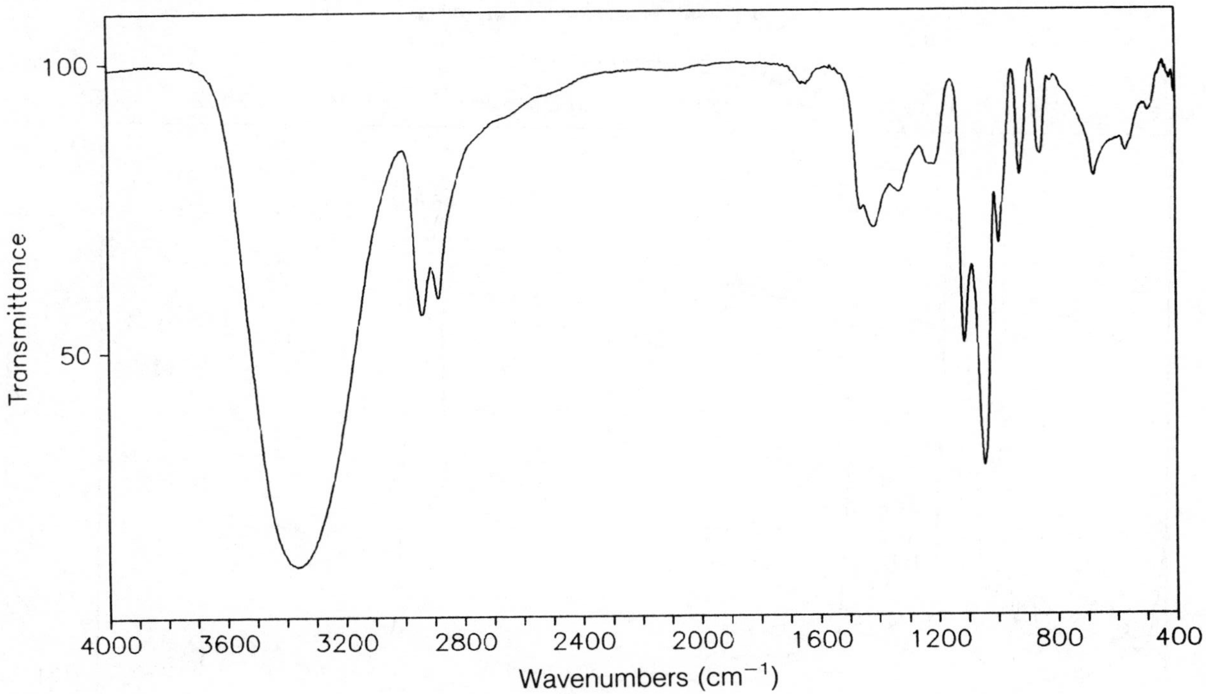

Figure 10.13 Glycerol.

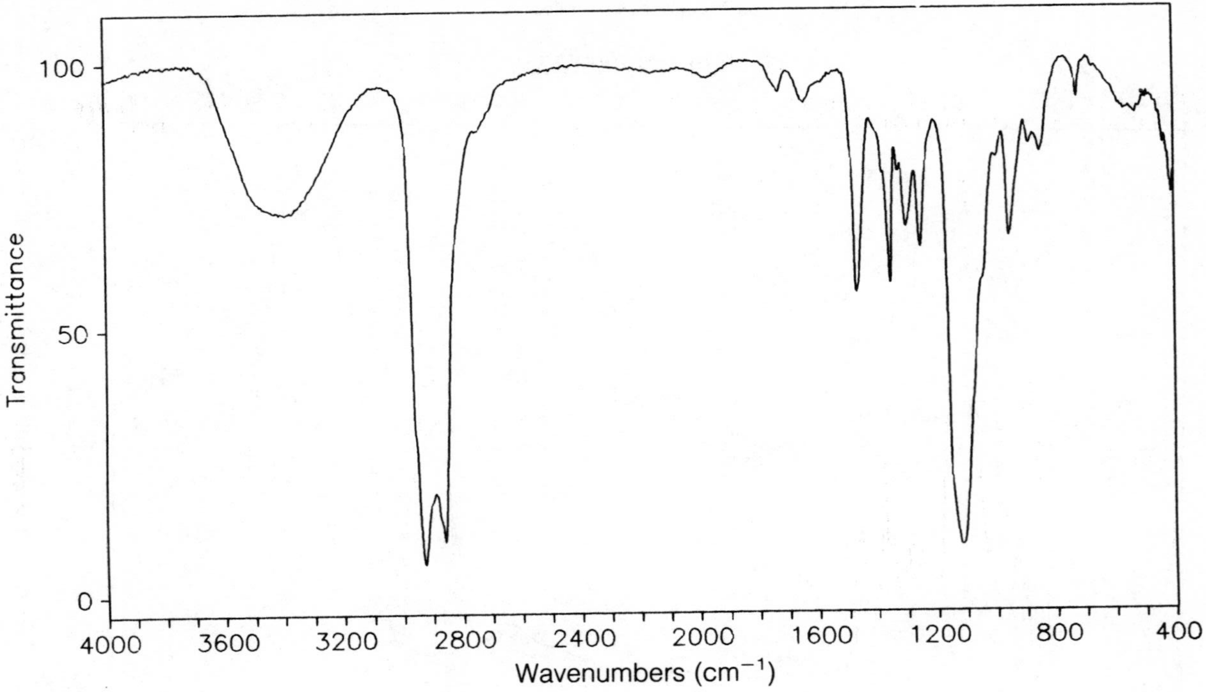

Figure 10.14 Fatty alcohol ethoxylate.

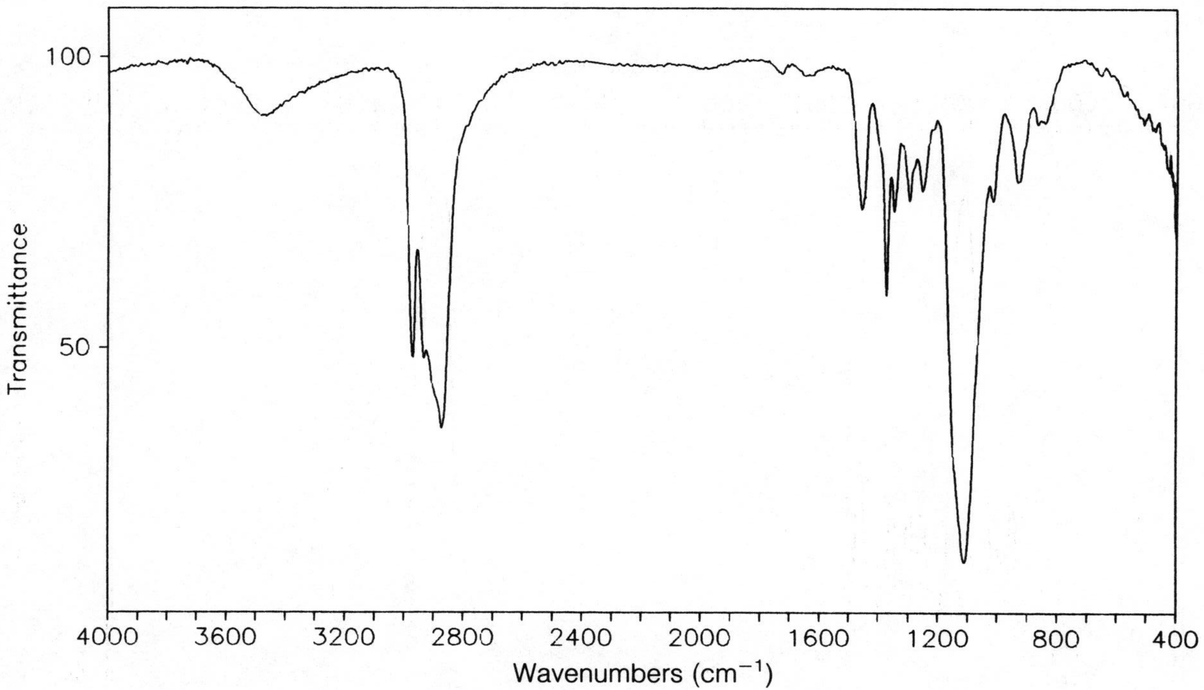

Figure 10.15 Ethoxylate–propoxylate copolymer.

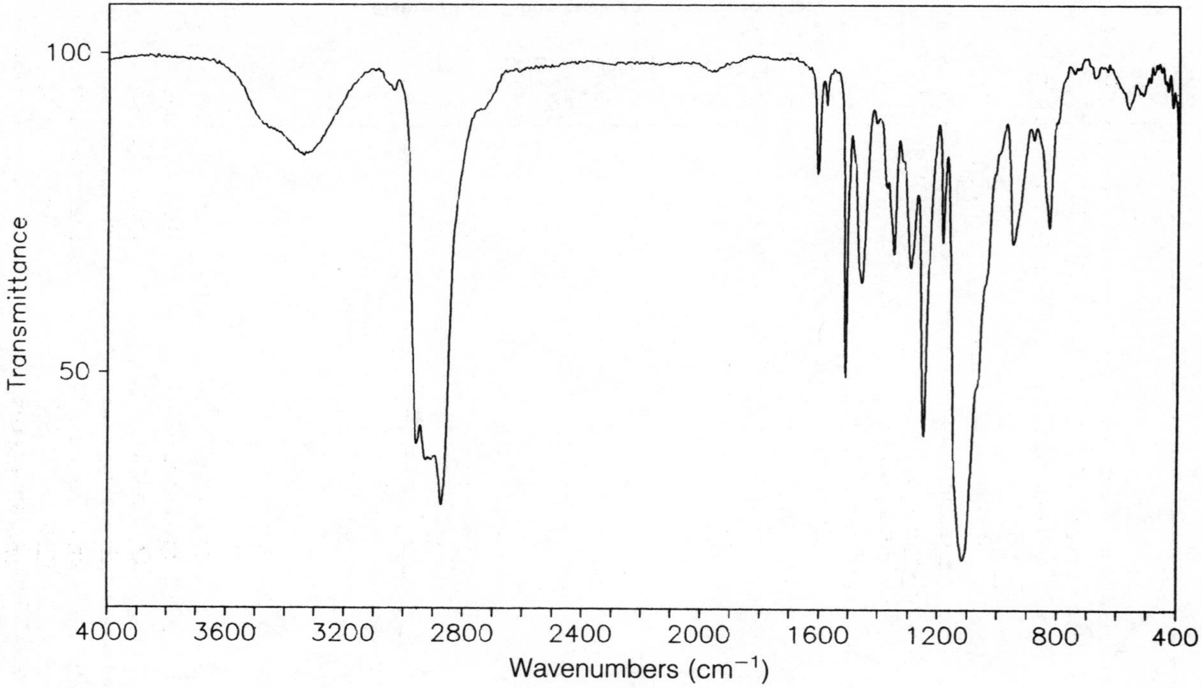

Figure 10.16 Alkyl phenol ethoxylate.

strength of the band at 1510 cm^{-1}—compare its intensity here with that in alkyl benzene sulphonate (see Figure 10.20).

10.6.3 Esters

Simple straight-chain alkyl esters, such as stearyl stearate, exhibit a carbonyl stretch between 1750 and 1730 cm^{-1}, and C–O stretches at 1250 and 1170 cm^{-1}. In aryl esters the carbonyl stretch is reduced in frequency to 1720 cm^{-1}.

Esters based on polyols include the family of glycerides. Triglycerides, the raw material for soapmaking, have no O–H band at 3400, mono- and diglycerides exhibit this and other C–O bands at 1050 and 1100 cm^{-1} Figure 10.17 shows triolein—a triglyceride based on unsaturated C_{18} oleic acid, and the major ingredient in olive oil.

Polyethoxylates and polypropoxylates also form surface-active esters. Their spectra exhibit the classic EO or PO bands described in section 10.6.2. The alkoxy component may or may not contain terminal O–H groups, which should be evident from the spectrum.

10.6.4 Alkanolamides

Used as foam boosters in conjunction with anionic actives, these compounds contain one or two ethyl alcohol groups attached to the nitrogen atom of a long-chain alkyl amide. Monoethanolamides are secondary amides and therefore exhibit typical amide I and amide II bands at 1650 and 1560 cm^{-1}, as well as an N–H stretch at 3090 cm^{-1} (Figure 10.18). The spectrum also contains a strong, broad O–H stretch centred near 3400 cm^{-1}, and C–O vibrations at 1070 cm^{-1}. Diethanolamides do not exhibit any N–H features—no 3100 cm^{-1} band and only one peak between 1700 and 1500 cm^{-1} (Figure 10.19).

10.6.5 Anionic surfactants

Alkyl and aryl sulphonates. Probably the most common detergent molecule, at least for fabric washing, is still alkyl benzene sulphonate (ABS), where an SO_3^- functional group is attached at the *para* position to a benzene ring which also supports a long alkyl chain. The spectrum of a typical ABS is shown in Figure 10.20. The characteristic bands are mainly SO_3 stretching at 1200 cm^{-1}, and sulphonate deformation at 1135, 1045 and 1015 cm^{-1}. Aromatic ring vibrations can be seen at 3060, 1600, 1495 and 830 cm^{-1} (*para*-substituted benzene). Alkyl sulphonates may be distinguished from aryls by the absence of aromatic ring vibrations, and by the presence of one strong band at 1060 cm^{-1} rather than two sharp ones at 1045 and 1015 cm^{-1}.

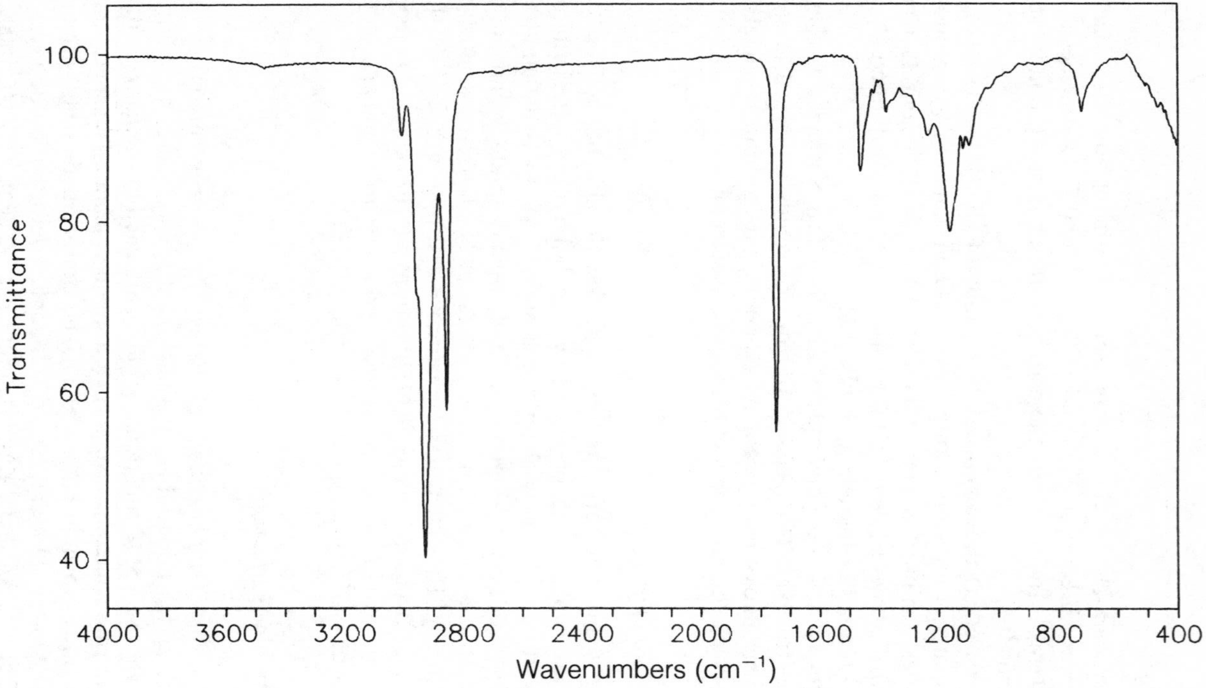

Figure 10.17 Glycerol trioleate.

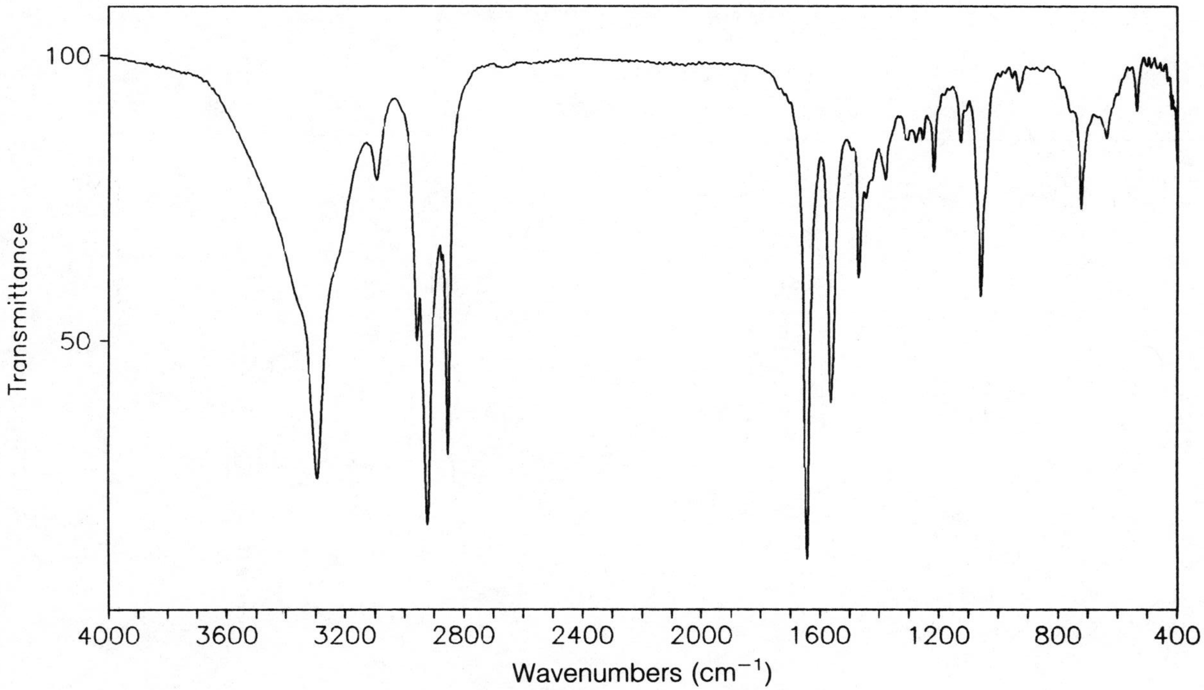

Figure 10.18 Alkyl monoethanolamide.

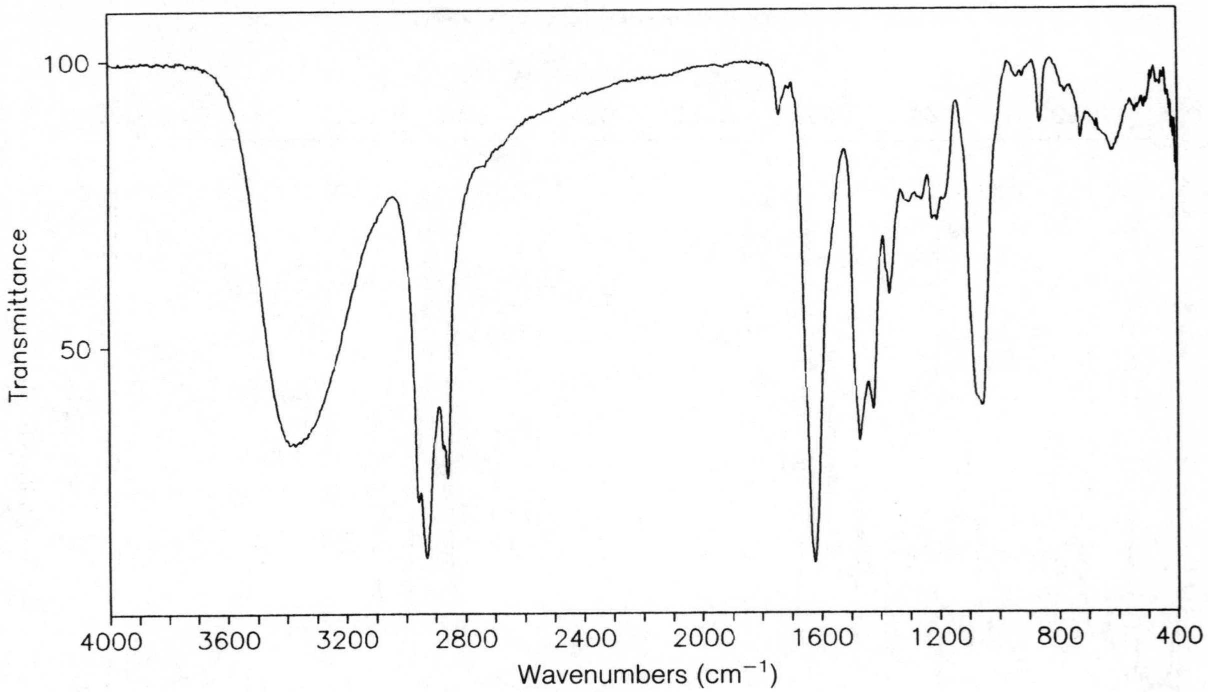

Figure 10.19 Alkyl diethanolamide.

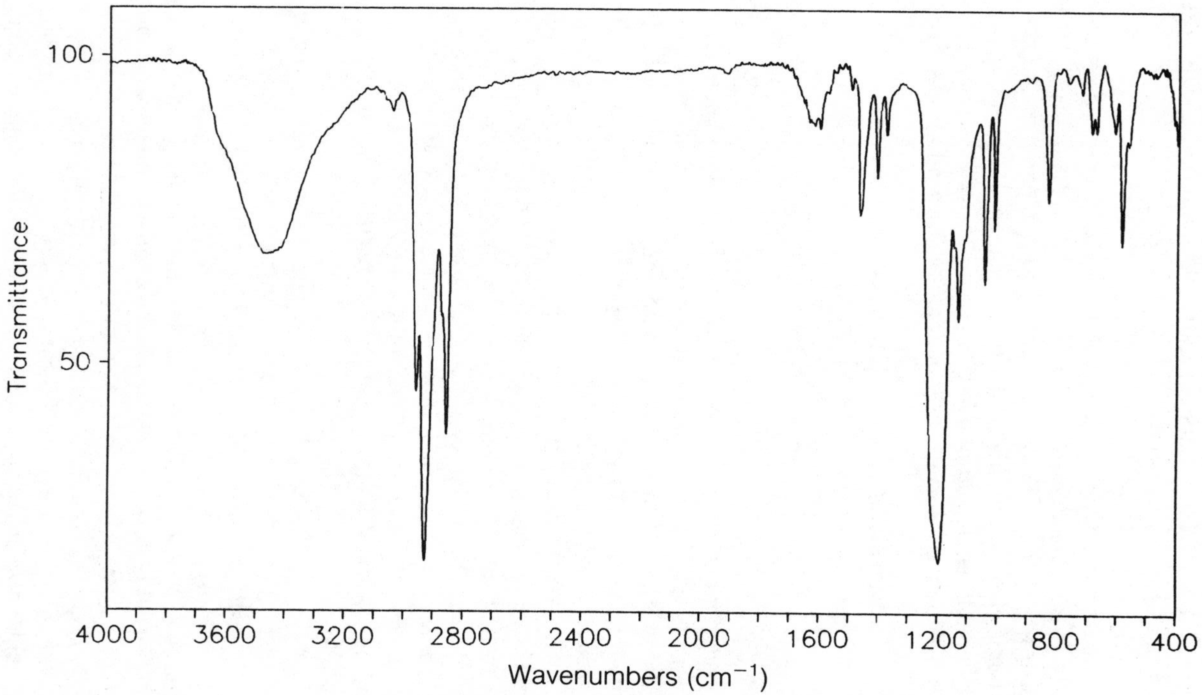

Figure 10.20 Alkyl benzene sulphonate.

The degree of branching of the alkyl chain can be analysed from the intensity of CH_3 bands at 2960 and 1380 cm^{-1}. 'Hard' ABS—highly branched and therefore less biodegradable—exhibits more intense CH_3 bands than does 'soft' linear ABS.

Alkyl and alkoxy sulphates. Alkyl sulphates can be distinguished from sulphonates by the position of the S=O stretch; this is above 1200 cm^{-1} for sulphates but just below 1200 cm^{-1} for sulphonates. The SO_4 deformation is also shifted higher (to approximately 1080 cm^{-1}).

Alkyl ether sulphates feature bands characteristic of ethoxy groups (see section 10.6.2) and alkyl sulphate functionality, with additional bands at 1030 and 780 cm^{-1}.

Figures 10.21 to 23 show some typical spectra.

10.6.6 Cationic surfactants

Cationic surfactants generally comprise quaternary ammonium compounds. Traditional cationics would contain two long alkyl chains and two methyl groups attached to the charged nitrogen atom, with typically Cl^- as the counter ion. The spectra of such species are not grossly different from branched long-chain hydrocarbons. It is usually possible, however, to observe subtle indications of the presence of these quaternary ammonium compounds from the slight shifts in band positions of the alkyl groups attached to N^+. The N^+–CH_3 groups may be observed at 3005 cm^{-1} rather than the customary 2960 cm^{-1}, and a shoulder at 1490 cm^{-1} may be seen on the 1460 cm^{-1} CH_2 band (Figure 10.24).

Amine oxides are also very difficult to identify from their infrared spectra alone. A small band at 930 cm^{-1} is perhaps the best indication (Figure 10.25).

Alkyl betaines ($RCH_2N^+(CH_3)_2CO_2^-$), popular personal product surfactants owing to their mildness, show a characteristic doublet at 1640 and 1610 cm^{-1} due to the carboxylate group (Figure 10.26). There are also shifts in the CH_3 stretching frequency owing to N^+.

Amidopropyl betaines exhibit features that include all the bands referred to for alkyl betaines, plus typical secondary amide bands at 1640 and 1550 cm^{-1} (amide II). The spectrum also contains N–H stretches at 3300 and 3100 cm^{-1} (Figure 10.27).

Imidazolinium salts are also difficult to identify from their infrared spectra, as features due to the charged five-membered ring are generally swamped by other functional groups which are generally present, such as an amido group or $CH_3SO_4^-$ counter-ion. However, *N*-methyl groups can sometimes be spotted at 3005 cm^{-1}, and ring vibrations at 1610 cm^{-1} (Figure 10.28).

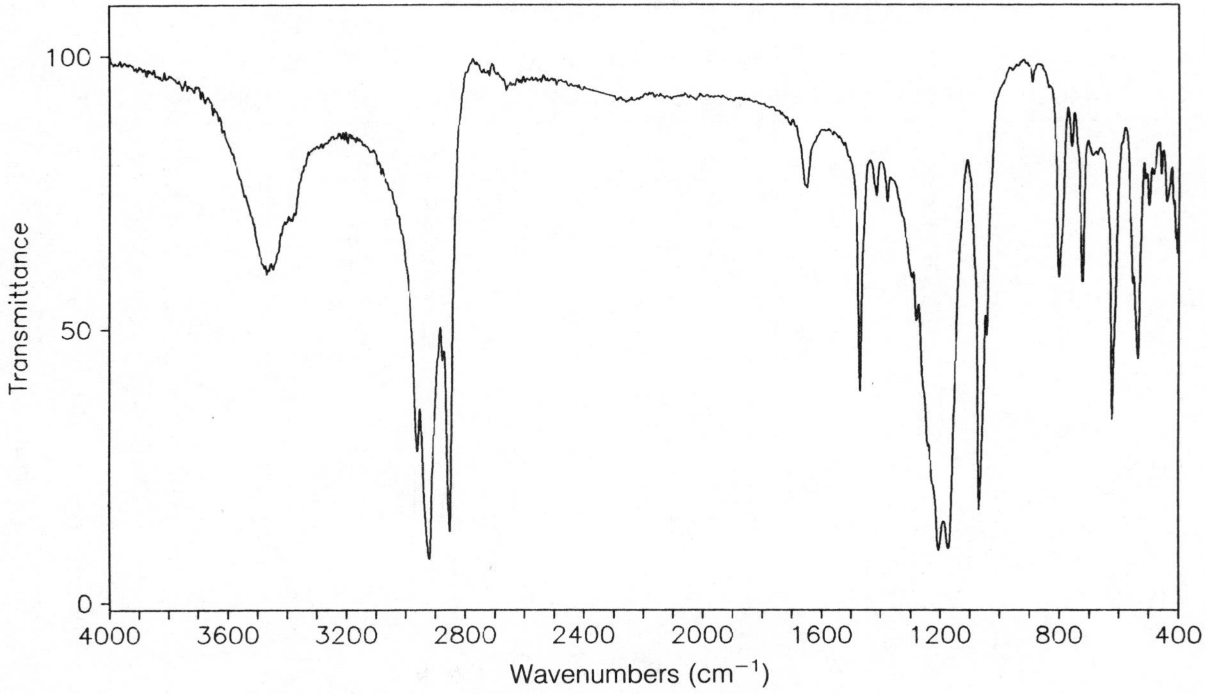

Figure 10.21 Alkyl sulphonate.

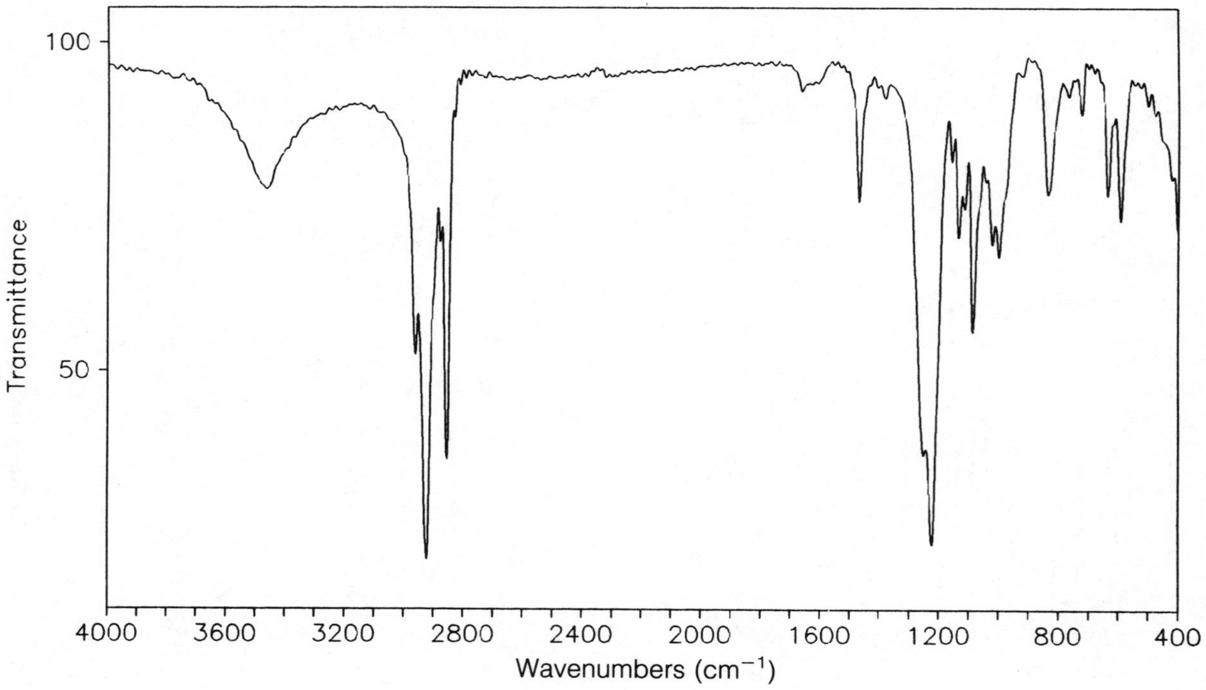

Figure 10.22 Alkyl sulphate.

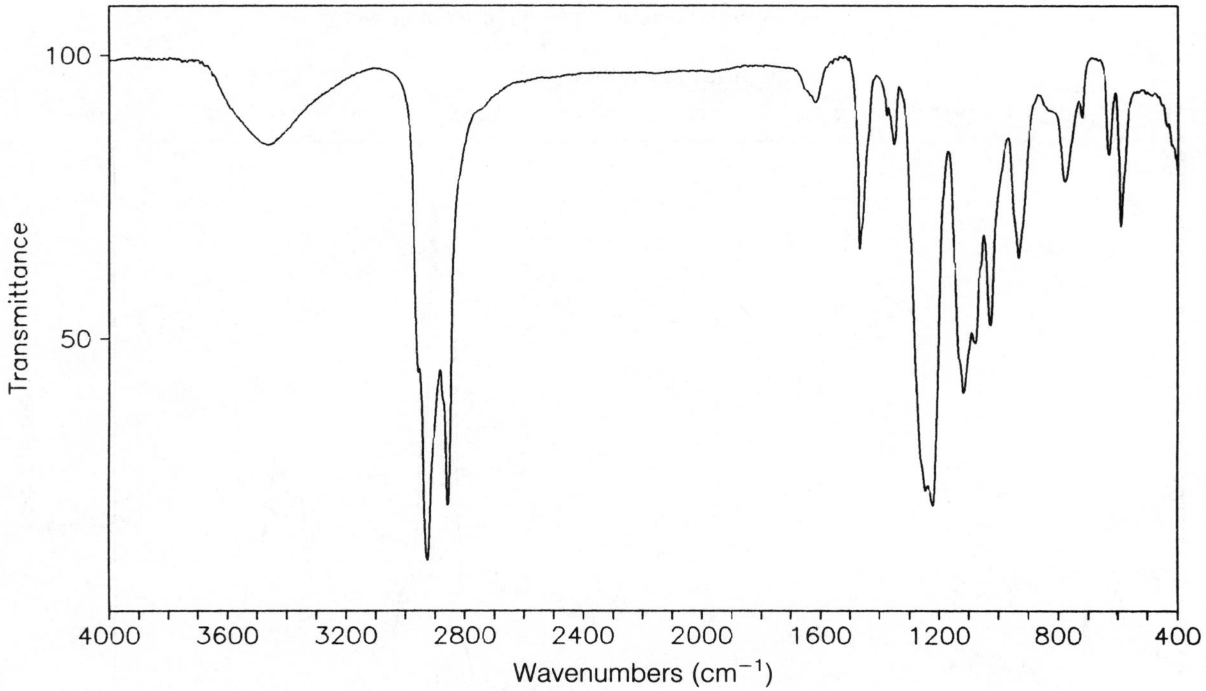

Figure 10.23 Alkyl ether sulphate.

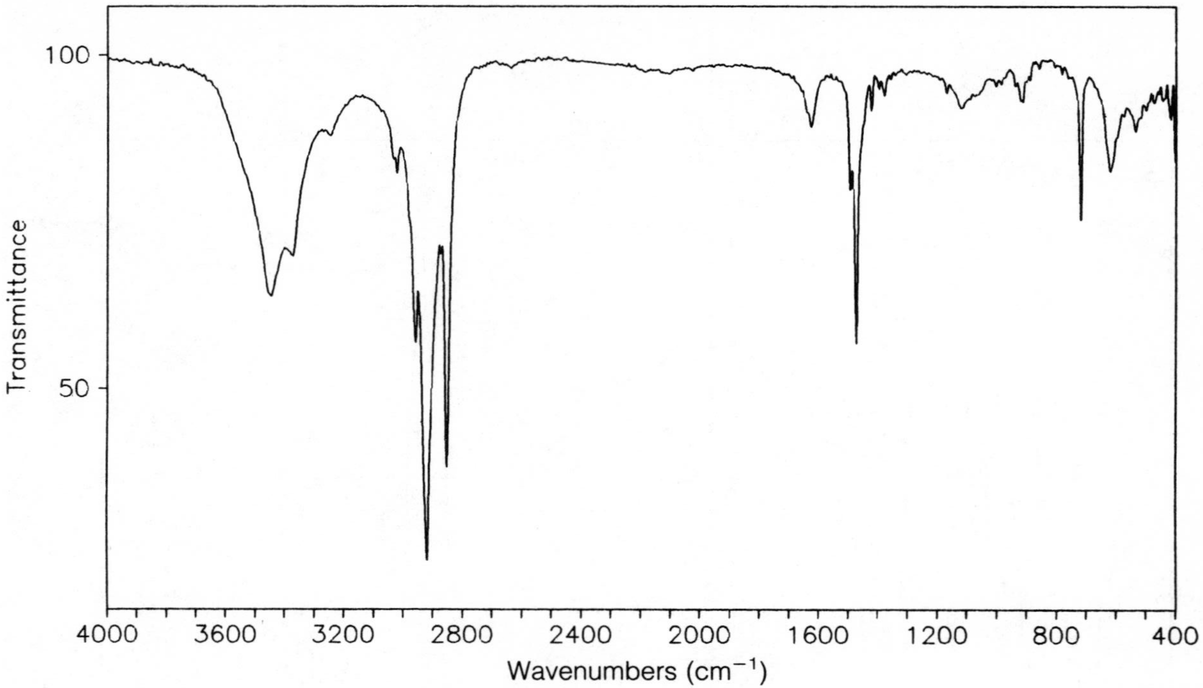

Figure 10.24 Dialkyl dimethyl ammonium chloride.

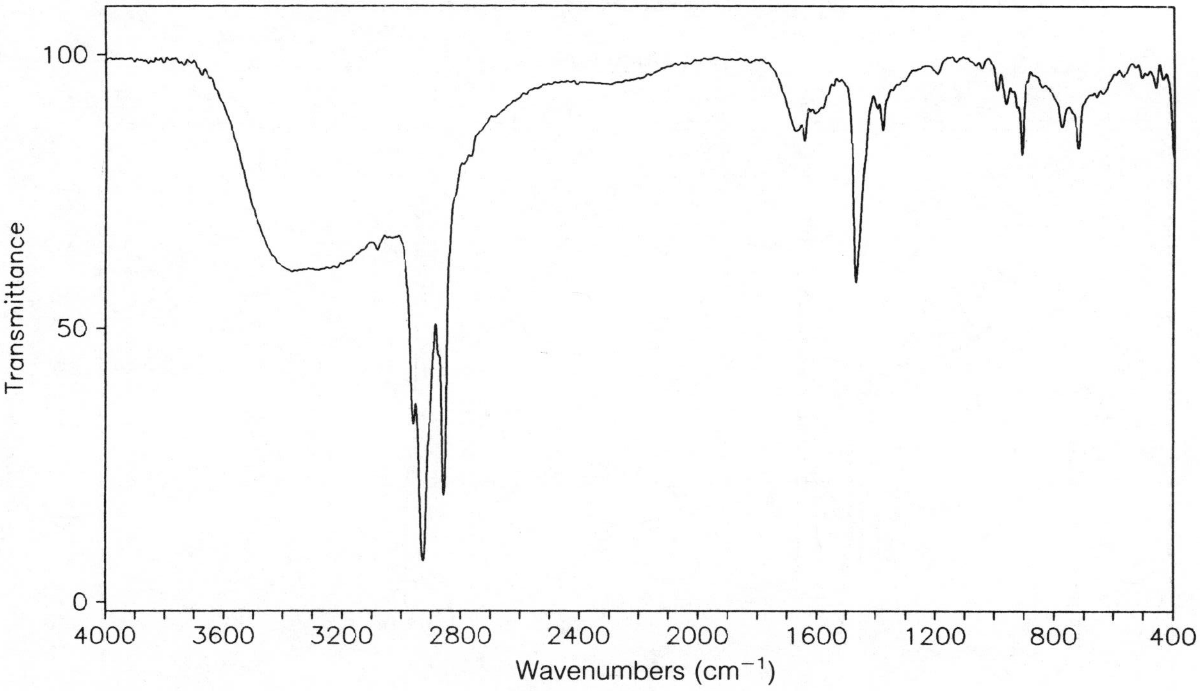

Figure 10.25 Amine oxide.

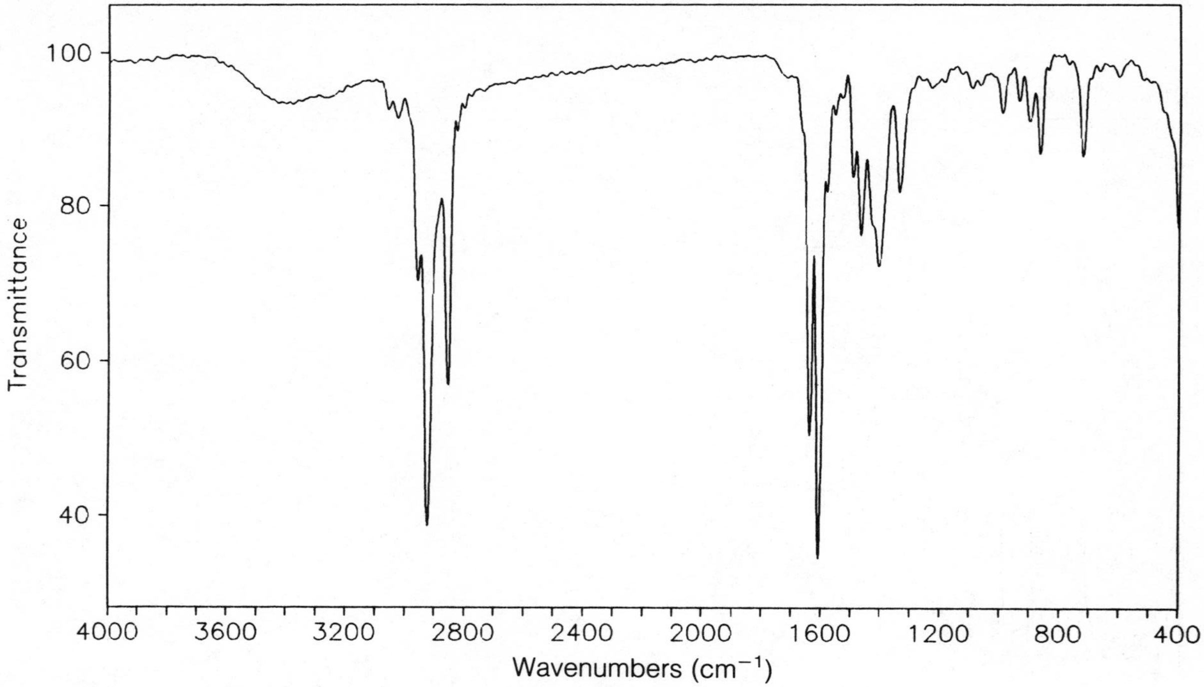

Figure 10.26 Alkyldimethyl betaine.

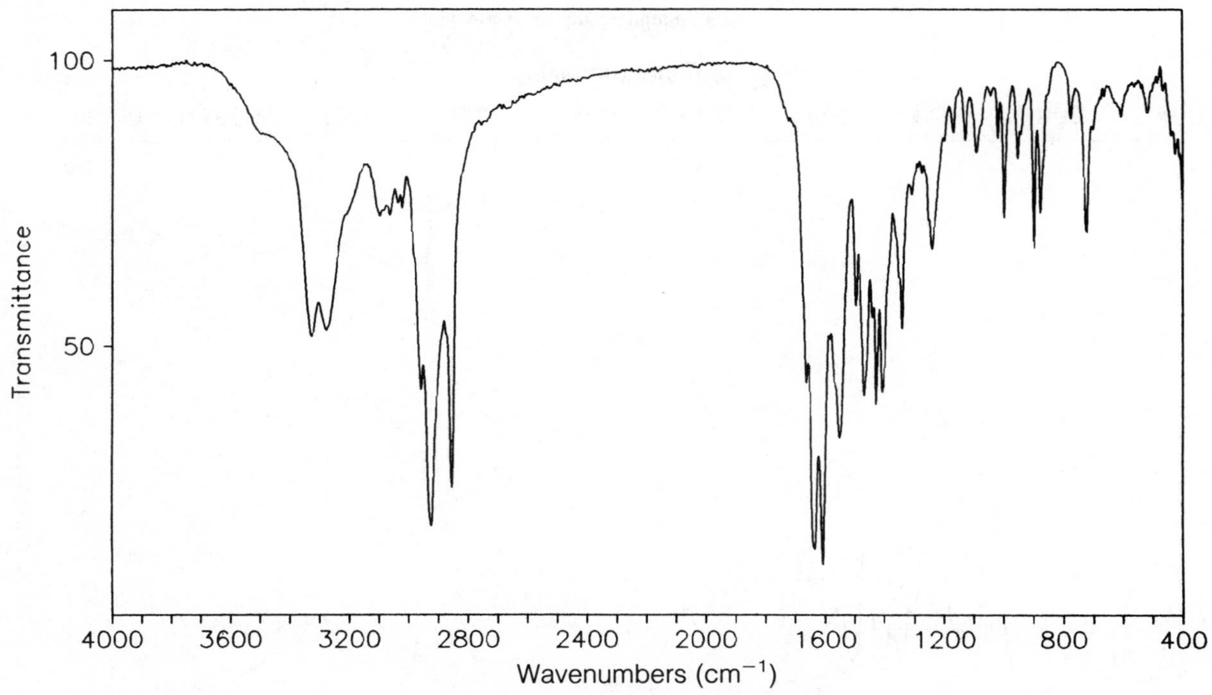

Figure 10.27 Alkyl amidopropyl betaine.

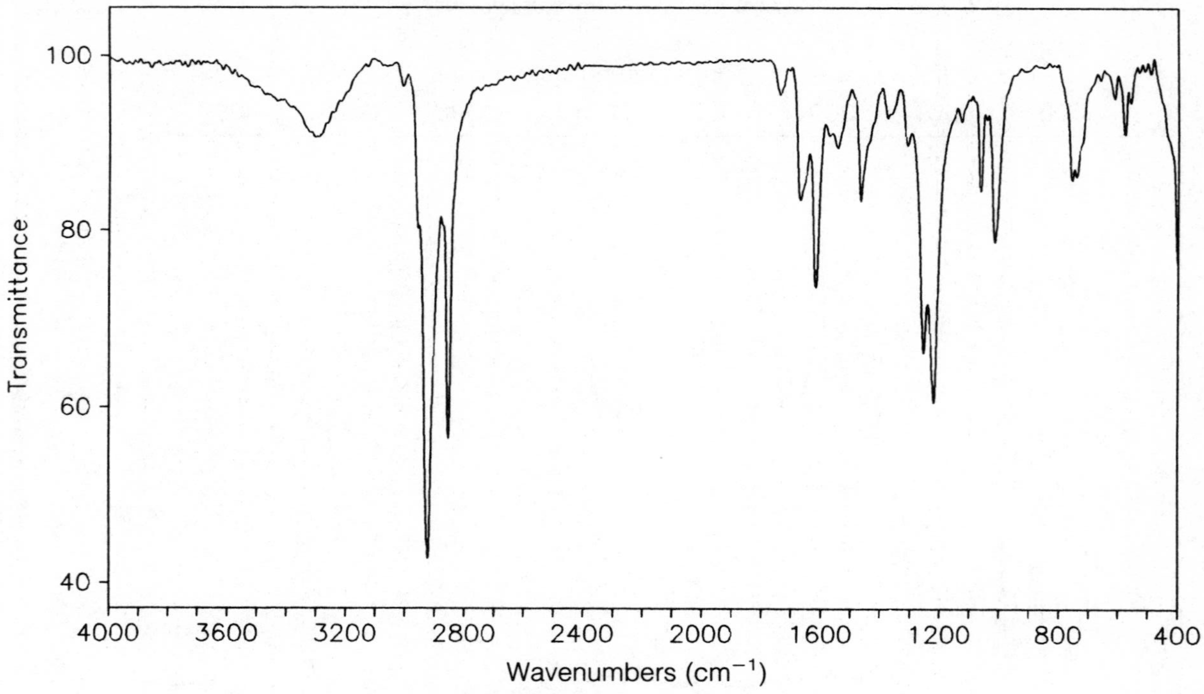

Figure 10.28 Imidazolinium salt.

10.7 Multicomponent analysis

The use of infrared spectroscopy to provide quantitative information on multicomponent systems has blossomed in recent years, spurred on by the development of chemometric techniques which have revitalised NIR spectroscopy. The principle of multicomponent analysis is outlined below, but see the books by Martens and Naes [13] and Massart *et al.* [14] for more thorough developments of these ideas.

In the same way that the absorbance at a single peak maximum for a single material is related to concentration through Beer's law, intensity across the whole spectrum is related to the concentrations of all infrared active ingredients in a mixture through the equation:

$$\mathbf{A} = \mathbf{kC} \tag{10.6}$$

where **A** is a matrix of absorbances ($wn \times 1$), **k** is a matrix of regression coefficients ($wn * n$) and **C** is a matrix of concentrations ($n * 1$) (n is the number of pure components, wn the number of different wavenumbers).

The above relation can be rearranged for the prediction of concentration from absorbance. However, we first need **k**:

$$\mathbf{k} = \mathbf{AC}'(\mathbf{CC}')^{-1} \tag{10.7}$$

where **C**′ is the transpose of **C**.

Concentrations are then given by:

$$\mathbf{C} = (\mathbf{k}'\mathbf{k})^{-1}\mathbf{k}'\mathbf{A} \tag{10.8}$$

Calculation of concentrations therefore requires two matrix inverse calculations. This method is often referred to in the literature as the **k** matrix method of multilinear regression. The advantage of the technique is that it uses more variables (absorbances) than there are reference samples. In this sense, the solution is said to over-determined.

Perhaps a superior method of multilinear regression is the so-called **p** matrix method. Here, we assume that concentration is a function of absorbance, rather than vice versa:

$$\mathbf{C} = \mathbf{pA} \tag{10.9}$$

p is calculated by regression:

$$\mathbf{p} = \mathbf{CA}'(\mathbf{AA}')^{-1} \tag{10.10}$$

Only one inversion is now necesary. However, the drawback is that we need more references (calibration samples) than variables. Modern spectrometers supply hundreds of absorbances without any difficulty, but preparing so many samples is simply not viable.

The **p** and **k** matrix methods are two classical least squares approaches to multicomponent calibration. There are techniques based on factor analysis, however, that are increasingly popular; these include the methods of principal components regression (PCR) and partial least squares analysis (PLS). We shall describe these as data-reduction techniques.

10.7.1 Data-reduction techniques

Although a spectrum contains hundreds of data points, the 'dimensionality' of the data is not generally as great as this. By dimensionality, we mean the number of inherent variables which comprise a data set. Take a mixture containing five components, say. We can vary the concentration of each component between certain limits, and record spectra of any combination we choose. Each spectrum contains, perhaps, a thousand data points. However, there can be no more than five genuine variables, as this is the number of components (ignoring any interactions between them in the mixture). The object of data reduction is to extract these five variables, which we can visualise at this stage as the pure-component spectra.

Principal components analysis, factor analysis, singular value decomposition, etc. are all techniques used in data reduction. The aim of the overall process is to reduce the data set **X** into the product of two matrices, **T** and **P**, with residual error matrix **E**:

$$\mathbf{X} = \mathbf{T}\mathbf{P}' + \mathbf{E} \tag{10.11}$$

The so-called factor loadings **P** are calculated from the matrix by singular value decomposition. **T** is called the scores matrix. The inner workings of this algorithm are not of interest here. Suffice to say that, if **X** has *n* variables, **P** will contain less than *n* linear combinations of variables which yet retain the information present in the original data.

In calibration, where a matrix of dependent data, **Y** (e.g. concentration), is related to a matrix of independent data, **X** (e.g. absorbances), the requirement is to relate the scores of **Y** to the scores of **X**:

$$\mathbf{T}_{\mathbf{Y}} = \mathbf{T}_{\mathbf{X}}\mathbf{R} \tag{10.12}$$

where **R** is the matrix that translates **X** scores into **Y** scores. In prediction, concentrations can be calculated from:

$$\mathbf{Y} = \mathbf{T}_{\mathbf{Y}}\mathbf{Q}' \tag{10.13}$$

where **Q** is the factor loadings from the calibration matrix of **Y**. Note that **R**, **P** and **Q** are all calculated in the calibration step. $\mathbf{T}_{\mathbf{X}}$ is calculated by projection of the test spectrum onto the factor space **P**.

Decomposition of data matrices into principal components has other advantages. Relationships within the set of variables, or within the set of samples, become clear. Scores may be depicted as points on a graph of orthogonal principal components, where clustering of samples reveals the existence of natural groupings (this is used in discriminant analysis, for instance). The association of particular variables to particular samples may be indicated by a biplot of loadings and factors.

However, multicomponent quantitative analysis is the area we are concerned with here. Regression on principle components, by PCR or PLS, normally gives better results than the classical least squares method in equation (10.8), where collinearity in the data can cause problems in the matrix arithmetic. Furthermore, PLS or PCR enable a significant part of the noise to be filtered out of the data, by relegating it to minor components which play no further role in the analysis. Additionally, interactions between components can be modelled if the composition of the calibration samples has been well thought out; these interactions will be included in the significant components.

10.7.2 Performing multicomponent analyses

The following is a list of the steps involved when performing multicomponent analyses.

(1) *Decide how many components you want to measure.* Remember that precision is strongly dependent on the sensitivity of infrared to the presence of each component. In the author's experience, there is little point in trying to calibrate for ingredients present in a mixture at less than 1% concentration. You will also need to prescribe limits on the allowable variation in concentration of each component.

(2) *Design of the calibration.* You will need at least twice as many calibration samples as there are ingredients. Each major component must be varied across the calibration set in such a way that its concentration is correlated with no other component.

An example of an experimental design for a three-component mixture, where each component is varied independently, is based around a cubic design. This cube has points at each corner (8 of them) at the centre of each face (6) and in the centre (1). Each point corresponds to a unique combination of concentrations. The point in the middle is the specified concentration, all the others contain one or more components present in either higher or lower than average concentrations. If *H* is higher than the specified concentration, *L* is lower and *M* the average ('correct' concentration), then the other 14 points can be described as follows:

H H H,
H H L, *H L H*, *L H H* Corner points
H L L, *L H L*, *L L H*
L L L

M M L, *M M H*, *M L M*, *M H M*, *L M M*, *H M M* Face points

This design defines the three-dimensional space of the analysis. Ideally, all future test samples will fall inside this cube. Samples falling outside may not be so well modelled, as this involves extrapolation of the model.

(3) *Manufacture of samples.* In an ideal world, the calibration samples will be made in the same way as future test samples. If an analysis for quality control purposes is desired, we should use the same mixing devices as are used in production. However this is rarely possible. We normally have two alternatives:

(i) Take what samples production gives us, analyse them in the conventional way, and use these data for calibration;
(ii) Use a smaller-scale production process—a pilot plant, perhaps, or even test tube mixing.

The analyst should include all the components of the mixture, even if he doesn't want to calibrate for them all. Minor ingredients can usually be combined in their nominal concentrations and added in one dose to the samples.

If there are other variables which may affect the calibration, such as temperature for instance, it is wise to record spectra twice and include both sets of data in the calibration. If possible, use genuine replicates (same concentrations, but made up separately).

(4) *Run the samples and extract the useful data.* We may want to use the whole spectrum, or simplify by discarding data on all but the intensities at peak maxima, plus one or two points on the baseline.

(5) *Construct the calibration model.* Use PLS, PCR or both, depending on the software available. Select the desired number of principal components, which will be at least equal to the number of ingredients, and may be several more if interactions occur. Cross validation—the removal of the data from one sample and its prediction from the model calculated from the remainder—is commonly used to select the required number of principal components.

The importance of validating the multivariate model cannot be over-emphasised. In particular, the data should be checked for 'outliers', that is, samples whose properties are different from the rest of the calibration set. If outliers are not detected and either removed or corrected, serious errors may be built into the model. Check for outliers by plotting actual compositions (Y data) against predicted compositions. In a good model, all the samples will lie close to the line of best fit. Outliers will be isolated and associated with poor predictive accuracy.

(6) *Test the accuracy of the multivariate model by running test samples.* This may involve a conflict if we want to use all of our samples in

calibration, as testing involves withholding some data from the model. If this is the case, we can perform a 'jack-knife' test by removing one sample from the data, recalibrating and testing the removed sample, repeating this procedure for as many samples as we care to choose.

To give an idea of the accuracy of multicomponent calibrations, this author has constructed calibrations for a mixture of three detergents in water, using ATR for recording spectra and PLS for the mathematical analysis. The accuracy of this analysis was as follows:

major component (anionic surfactant, concentration 12%) precision +/− 0.15%
second component (zwitterionic surfactant, concentration 2%) precision +/− 0.06%
third component (nonionic surfactant, concentration 1%) precision +/− 0.05%

The precision compares favourably to analysis by titration.

10.8 Buying a spectrometer

The following companies supply FTIR spectrometers. Inclusion in or omission from this list does not, of course, imply any endorsement or criticism of a particular product.

Bio-Rad, Bomem, Bruker, Laser-Precision (Analect), Mattson, Midac, Nicolet, Perkin-Elmer, Shimadzu.

A prospective purchaser of a low-cost, worktop FTIR has a bewildering variety of makes and models to choose from. The following are some recommendations which may be useful in deciding which spectrometer is most suitable.

Performance. Modern FTIR spectrometers, even the least expensive types, generally offer a level of performance that is more than adequate for simple sampling techniques—such as KBr discs or mulls—and will normally give excellent results with more sophisticated accessories such as ATR or diffuse reflectance. Only the most demanding accessories (microscopes or photoacoustic devices) or technically difficult experiments (step-scanning or high-speed and time-resolved work) require more expensive machines. The prospective customer who requires FTIR for raw material confirmation and occasional forensic analysis can be confident that whatever machine is chosen should be up to the job.

However, there are differences between products and suppliers, and these differences can be grouped into three categories:

1. Hardware
2. Software
3. Support

Hardware. Most low-cost, worktop FTIR spectrometers contain sealed optics. This reduces the effect of water vapour on spectra and protects hygroscopic KBr beam splitters from damage that would occur in high-humidity environments. Sealed optics remove the need for a separate supply of dry air, although the disadvantage is that the sample compartment itself cannot be kept dry and some water vapour will inevitably be manifest in spectra.

The moving mirror assembly has traditionally been based on an air-bearing system, and this still gives the best performance in demanding applications. Low-cost machines seldom use air-bearings, though, but employ other means of moving the mirror, such as pivots, moving wedges or low-friction bearings. These alternatives inevitably compromise the instrument to some extent, particularly at the limit of system performance.

Some simple tests will reveal the quality of the bearing:

Record a 100% line and examine the spectrum for linearity and noise. Do this for medium resolution (8 or 4 cm^{-1}) and the highest resolution (1 or 2 cm^{-1}). The peak–peak noise in the region 2100–1900 cm^{-1} (normally the optimum for signal strength) should be similar at both resolutions, provided no additional aperturing was required at high resolution. Unstable mirror travel will degrade the high-resolution performance. The presence of water-vapour bands in a 100% line recorded in an environment where water should not have varied may also indicate poor stability, especially if these peaks have a pronounced 'derivative' shape—i.e. extend both above and below the 100% line.

The beam splitter, and its alignment, is perhaps the other key factor affecting the quality of spectral data. This can be assessed by looking for drifting in the high-frequency region of the 100% line, and also by comparing the ratio of intensities at 4000 cm^{-1} and 1300 cm^{-1}—ideally this should be around 10%, and certainly greater than 5%.

Other simple tests that can be performed are described by Johnston [15].

Software. For many users, software is now the most important factor in choosing a machine. There is clearly no substitute for lengthy, hands-on evaluation of a system's software. This is best done on-site, and many suppliers will agree to the loan of a demonstration machine for a limited period. Almost universally now, the platform for the software is an IBM PC, usually with a fast data-acquisition board added on. Evaluation of software is inevitably a very personal and subjective process, but bear in mind some essentials. Software is expensive—how much do you get with the initial purchase of the machine? Are search and quantitative analysis packages included or available as add-ons? Does the software crash periodically, or when you try to do two things at once? Is the software multitasking—i.e. can you collect and plot in the background while manipulating a spectrum in foreground? Can you easily write macros to perform frequently repeated sets of commands? If you require Windows programs, can the software take advantage of genuine Windows capabilities, such as data and graphics exchange with other programs? Finally, are you entitled to free upgrades when they become available—remember that while the technology has perhaps now reached a plateau of sophistication, software is evolving all the time.

Support. It is in the nature of mechanical things that they periodically break down. FTIRs are now much less prone to breakdown, and with luck a few years will pass before anything goes wrong. However, at some point you will need assistance of one kind or another, and this is where the support offered by the manufacturer is vital. Many spectroscopists will always buy a new machine from a trusted manufacturer, even when a rival company's machine offers greater capability or a cheaper price. For those looking to buy for the first time, the best advice is to seek out existing users of FTIRs for their impressions of rival suppliers. Geographical considerations also apply—if you have to call out an engineer to repair your instrument the charge will cover travelling time as well as parts and labour.

Your instrument supplier should also be a resource to be used for his/her advice concerning the analysis of intractable samples, and help in training, interpretation and macro-programming. Insist on talking to the application scientist rather than just the salesperson.

10.9 Further information

A good practical guide to infrared spectroscopy is contained in the book by Willis and van der Maas [16]. The books by Colthup *et al.* [17] and Griffiths and De Haseth [18] provide a thorough introduction to theory and mechanics of the subject, as does the *Analytical Chemistry by Open Learning* publication [19], although at a less challenging level. Socrates [20] provides extensive correlation charts for use in interpretation;

Hummel and Scholl [11] is essential for the analysis of polymers and polymer additives.

Libraries of spectra can still be obtained in paper form, although electronic versions are rapidly gaining the ascendancy. Apart from the well-known collections by Sadtler, a useful single volume item is the IR/Raman atlas compiled by Schrader and Meier [21].

Many universities now offer short courses in practical infrared spectroscopy and/or interpretation. Some instrument manufacturers also hold regular tutorials on aspects of data collection or interpretation.

This chapter represents the view of the author and is not necessarily the view of Unilever Research.

References

1. Fadini, A. and Schnepel, F.-M. *Vibrational Spectroscopy—Methods and Applications*, Ellis Horwood, Chichester, 1989.
2. Hendra, P., Jones, C. and Warnes, G. *Fourier Transform Raman Spectroscopy—Instrumentation and Chemical Applications*, Ellis Horwood, New York, 1991.
3. Murray, I. and Cowan, I. *Making Light Work: Advances in Near Infrared Spectroscopy*, VCH Weinheim, 1992.
4. Marshall, A. G. and Verdun, F. R. *Fourier Transforms in NMR, Optical and Mass Spectrometry—A User's Handbook*, Elsevier, Amsterdam, 1990.
5. Fuller, M. P., Ritter, G. L. and Draper, C. S. *Appl. Spectrosc.*, **42**(2) (1988), 228–236.
6. Beaven, H. and Ricketts, R. E. *J. Sci. Instr.*, **44** (1967), 1048.
7. Bishop, G. A., Starkey, J. R., Ihlenfeldt, A., Williams, W. J. and Stedman, D. H. *Anal. Chem.*, **61** (1989), 671A–674A.
8. Messerschmidt, R. G. and Harthcock, M. A. *Infrared Microspectroscopy—Theory and Applications*, Marcel Dekker, New York, 1988.
9. Cesteros, L. C., Meaurio, E. and Katime, I. *Macromol.*, **29**(9) (1993), 2323–2330.
10. Surewicz, W. K., Mantsch, H. H. and Chapman, D. *Biochem.*, **32**(2) (1993), 389–394.
11. Hummel, D. O. and Scholl, F. *Infrared Analysis of Polymers, Resins and Additives. An Atlas*, Vol. 1, Parts I and II, Wiley-Interscience, 1971.
12. Mochola, K. K. and Bell, J. P. *J. Polym. Sci., Polym. Phys.*, **11**(9) (1973), 1779–1791.
13. Martens, H. and Naes, T. *Multivariate Calibration*, Wiley, Chichester, 1989.
14. Massart, D. L., Vandeginste, B. G. M., Deming, S. N., Michotte, Y. and Kaufman, L. *Chemometrics: A Textbook*, Elsevier, Amsterdam, 1988.
15. Johnston, S. *Fourier Transform Infrared—A Constantly Evolving Technology*, Ellis Horwood, New York, 1991.
16. Willis, H. A. and Van der Maas, J. H. *Laboratory Methods in Vibrational Spectroscopy*, 3rd edn. Wiley, Chichester, 1987.
17. Colthup, N. B., Daly, L. H. and Wiberley, S. E. *Introduction to Infrared and Raman Spectroscopy*, 2nd edn, Academic Press, New York, 1975.
18. Griffiths, P. R. and De Haseth, J. A. *Fourier Transform Infrared Spectrometry*, Wiley, Chichester, 1986.
19. George, W. O. and McIntyre, P. S. *Infrared Spectroscopy, Analytical Chemistry by Open Learning*, Wiley, Chichester, 1987.
20. Socrates, G. *Infrared Characteristic Group Frequencies*, Wiley, Chichester, 1980.
21. Schrader, B. and Meier, W. *Raman/IR Atlas of Organic Compounds*, Vols 1–3, Verlag Chemie, Weinheim, 1974/75/76.

11 Nuclear magnetic resonance (NMR) spectroscopy

T. I. YOUSAF

11.1 Introduction

NMR is a spectroscopic technique that uses radiation in the radio frequency region of the electromagnetic spectrum. Its use in the analysis of surfactants can be broken down into three main categories:

1. Determination of molecular structure, e.g. characterising unknowns in raw materials, or unexpected by-products.
2. Measurement of molecular parameters, e.g. alkyl chain length, ethoxylate chain length, etc., of surfactants.
3. Quantitation of mixtures, e.g. determining molar ratios in mixtures of carboxylic acids/esters or of mono-, di- and trialkyl phosphates, either in raw materials or in formulations.

11.1.1 Limitations of NMR

When only a minute quantity of sample is available (e.g. <1 mg) MS (mass spectrometry) wins hands down over NMR as the analytical method of choice for molecular structure elucidation of either pure compounds or complex mixtures. Not surprisingly, therefore, it is the technique preferred by analytical chemists when analysing for trace impurities in surfactants. This fact highlights the first drawback of NMR as an analytical technique: it is insensitive, and may require anywhere from *c.* 5 mg of sample for a basic ^1H NMR spectrum to *c.* 0.5 g for the more sophisticated two-dimensional NMR experiments.

The second drawback of NMR is that it cannot be easily interfaced with modern chromatographic separation techniques. Thus GC (gas chromatography), HPLC (high performance liquid chromatography), SFC (supercritical fluid chromatography), and CZE (capillary zone electrophoresis) are all routinely interfaced with MS but not with NMR, although the major instrument manufacturers JEOL (Japan), Varian (USA), and Bruker (Germany) have now developed NMR probes compatible with HPLC columns. If NMR spectra of individual components of a mixture are required then the mixture usually has to be separated beforehand, although examples where this is not necessary are given later (section 11.6).

11.2 Theory of NMR

All nuclei that have an odd atomic number (number of protons) or odd mass number (protons + neutrons) possess a permanent magnetic dipole moment, and therefore behave like tiny bar magnets. This is because such neuclei are not perfect spheres, but are either oblate or prolate spheroids. When placed in a magnetic field, such nuclei absorb electromagnetic radiation in the radio-frequency (RF) region of the spectrum. The exact frequency depends upon various factors, but of critical importance to the chemist is the fact that this frequency (known as the frequency of resonance) varies with the chemical environment of the nucleus, and hence each nucleus in a molecule gives a characteristic signal. NMR therefore allows us to 'see' the individual atoms that molecules are made of, and hence, to determine molecular structure.

What atoms can we see by NMR? It will be quite clear from the above definition that the nuclei of certain commonly occurring isotopes are not NMR active, e.g., ^{12}C, ^{16}O, ^{32}S. For the case of carbon and oxygen, this is a pity because almost all molecules of interest to organic chemists (in academia and industry) contain these two key elements. A second problem is that the nuclei of most atoms of the periodic table that have a dipole moment also possess a nuclear quadrupole moment. This makes them hard (and in many cases, impossible) to detect by NMR. Thus NMR spectroscopy, for all its versatility, is incapable of answering a question as simple as 'Does this molecule contain a chlorine atom?' or 'Does it contain a sulphate group?' Indeed, elemental sulphur is invisible by NMR.

The dominant isotope of hydrogen, ^{1}H, is, however, NMR-active and ^{1}H-NMR has been used routinely in molecular structure elucidation from the time that NMR spectrometers first became commercially available (1956), up until the early 1970s when improvements in NMR technology and in computing allowed ^{13}C, the rarer isotope of carbon (present at 1.1% natural isotopic abundance), to be detected with ease. ^{13}C-NMR was then immediately embraced by organic chemists the world over because it gave them a rapid and unambiguous way of 'seeing' the carbon skeleton of a molecule, thereby determining its structure directly, rather than implicitly as was always the case with ^{1}H-NMR. Other spectroscopic techniques, e.g. IR, UV, do not provide this direct information about molecular structure; they only confirm the presence of certain functional groups (e.g. NO_2, CO, CN) within a molecule.

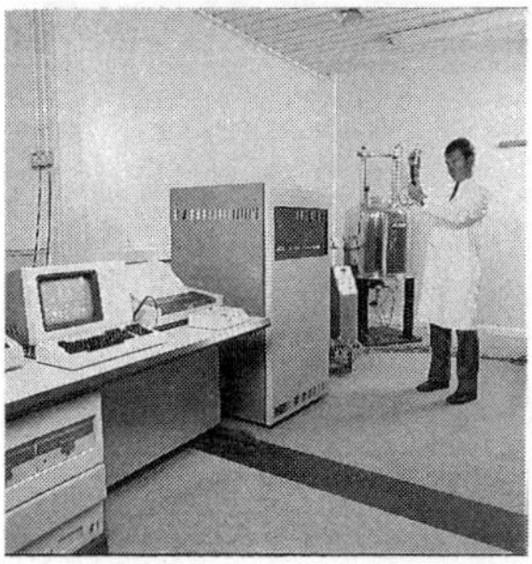

Figure 11.1 A typical FT–NMR spectrometer. The technician in the picture is gauging the sample depth prior to insertion of the NMR tube into the magnet.

In conclusion, ^{1}H- and ^{13}C are the only two nuclei that are routinely used in the analysis of surfactant molecules by NMR.

11.3 The NMR spectrometer

Figure 11.1 is a photograph of a typical NMR spectrometer; in fact it is the JEOL GX270 MHz that has been operating in the Procter & Gamble Analytical Department since 1984. There are three sections to a modern-day NMR spectrometer: the console (foreground), the spectrometer itself (middle), and the most important part, the magnet (background).

Returning to Figure 11.1, the large 'black box' in the middle of the picture is the Fourier transform (FT) NMR spectrometer itself. Figure 11.2 is a close-up of the magnet. A block diagram of a typical non-superconducting electromagnet and of the corresponding spectrometer is given as Figure 11.3. Samples are analysed as solutions in narrow NMR tubes which are then placed inside the magnet. The tubes are spun (at typically 15–20 Hz) using compressed air to ensure that all parts of the sample 'see' the same magnetic flux: this makes up for any minor inhomogeneities either in the sample, or in the tube itself. Any remaining magnetic inhomogeneities are 'ironed out' by the shim coils.

Details of the FT process are given elsewhere [1], but briefly, an NMR spectrometer generates pulsed RF radiation (typically of microsecond duration) which excites nuclei in the sample to non-Boltzmann energy levels; the resulting decay to the ground state generates RF radiation

Figure 11.2 A picture of the author about to insert a 10 mm NMR tube (complete with spinner) into the magnet.

which is picked up by an NMR probe, essentially a coil of wire inserted into the bottom of the magnet (cf. Figure 11.3), amplified, and displayed on the monitor at the console. This signal, known for historical reasons as the free induction decay [2], is a plot of intensity against time, and is Fourier-transformed into a plot of intensity against frequency, which is, in fact, the NMR spectrum. This can be stored electronically and/or recorded on chart paper. For a proton resonating at 300 MHz, the wavelength of the radiation 'emitted' is 1 m (taking $c = 3 \times 10^8\ \mathrm{m\,s^{-1}}$). How can a coil which is typically 5 mm in diameter (for most ^{1}H-NMR work) pick up such long-wavelength radiation 'emitted' by the sample under analysis? The answer is that our model of 'emission' has to be refined: the nuclear dipoles of the sample are best thought of as rotating inside the coil of the NMR probe, generating an AC current in much the same way that a bar magnet would if rotated inside a coil of wire [3].

The resolving power of a spectrometer (and the cost, see Table 11.1) increases with the field strength of the magnet [4, 5]. The SI unit of magnetic flux is the tesla (T = 10 000 gauss); for historical reasons [6] the resolving power of an NMR spectrometer is quoted, not in tesla, but as the frequency at which a proton would resonate at that magnetic field strength (see Table 11.1). Thus, most industrial laboratories and university undergraduate laboratories have 90–200 MHz NMR spectrometers which are provided on an open-access basis to all, requiring little training to take ^{1}H/^{13}C-NMR spectra. Middle of the range spectrometers (270–400 MHz) are also widely used, but are usually operated/managed by PhD spectroscopists.

At the other extreme, there are only five 600 MHz NMR spectrometers in the UK: they are at the Universities of Edinburgh, Oxford and Leicester, with the remaining two at the laboratories of Glaxo and

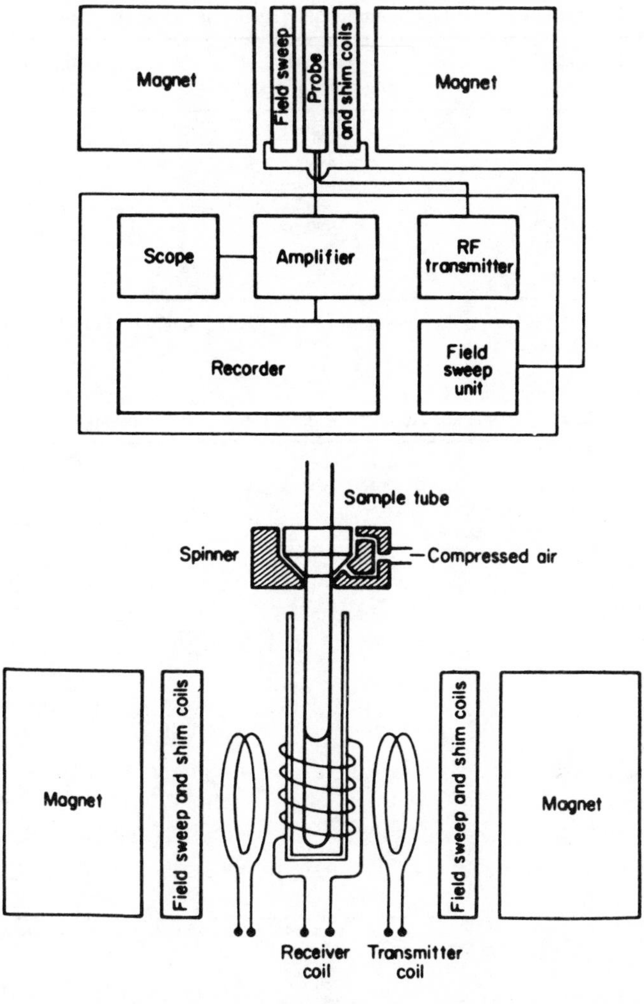

Figure 11.3 Block diagram of a high-resolution NMR spectrometer showing cross-section through the magnet (reproduced by kind permission of Professor R. J. Abraham of the University of Liverpool).

Wellcome. Both Oxford and Wellcome have also recently bought 750 MHz NMR spectrometers.

11.4 The NMR sample

All samples for NMR analysis must be soluble in one of the commercially available deuterated NMR solvents (see Table 11.2). When this is not

Table 11.1 Approximate costs of commercially available NMR spectrometers in 1994

Field strength[a] (tesla)	^{1}H-NMR frequency (MHz)	Approximate cost ($\times 10^3$ £)
2.1	90	73
4.7	200	100
6.3	270	150
9.4	400	180
11.7	500	350
14.1	600	660
17.5	750	1500

[a]For comparison, the horizontal component of the Earth's magnetic field (at the surface) is *c.* 2×10^{-5} T

Table 11.2 Approximate costs of deuterated NMR solvents in 1994

Solvent[a]	£/gram
Acetic acid-d_4	2.4
Acetone-d_6	1.5
Acetonitrile-d_3	2.5
Benzene-d_6	2.2
Chloroform-d	0.2
Cyclohexane-d_{12}	11.9
Dichloromethane-d_2	8.5
Dimethylsulphoxide-d_6	1.1
Dioxane-d_8	21.1
Methanol-d_4	6.2
Nitrobenzene-d_5	3.2
Nitromethane-d_3	3.8
Pyridine-d_5	3.4
Tetrahydrofuran-d_8	21.7
Toluene-d_8	2.6
Trifluoroacetic Acid-d	1.1
Water-d_2	0.4

[a]Purity >99.5 atom %

possible, one has to resort to using solid-state NMR spectroscopy. This is outside the scope of the present review, but is reviewed elsewhere [7, 8]. Table 11.2 also shows that the cheapest polar solvent is D_2O, and the cheapest non-polar solvent is deuterochloroform: these are in fact the two most widely used solvents in university NMR laboratories.

^{1}H-NMR spectroscopy gives a molar response (i.e. the signal intensity of a particular proton is proportional to the molar concentration of the proton), and so all sample concentrations should be expressed in molar units; however, as a rule of thumb, we can say that for most organic compounds (MW 100–500 D), as little as 10–50 mg of sample is enough to produce a ^{1}H-NMR spectrum with a good signal-to-noise ratio within *c.* 1 h.

For ^{13}C-NMR, because of the 1.1% natural abundance, the same sample would require an overnight run to produce a reasonable spectrum. Clearly, for ^{13}C-NMR, the more concentrated the solution, the shorter the analysis time, e.g. 0.3 g of neat pyridine (with just a few drops of deuterated solvent added) produces an excellent ^{13}C-NMR spectrum within 10 min.

Why do we need a deuterated solvent? If an ordinary solvent were used, its own signals would give such a strong response that they would mask signals from the sample under analysis; secondly, and more importantly, the deuterium atoms of the solvent are constantly irradiated during the NMR experiment and their resulting signal is used as an internal calibrant known as an electronic 'field/frequency lock'. This 'lock' ensures constant magnetic-field homogeneity across the sample during the experiment and hence optimum resolution; for this reason deuterated NMR solvents are often known as 'lock' solvents, and are always used irrespective of the nucleus under observation.

Samples are made up in 5 mm (overall diameter) NMR tubes (typical for $^{1}H/^{13}C$ work) or 10 mm tubes (typical for multinuclear NMR) and placed in a spinner, and the whole is then inserted into the top of the magnet (see Figure 11.2). The NMR tube is never completely filled: the total depth of solvent is typically 5 cm, and of that, only the middle part (*c.* 2 cm) will sit inside the coil of the probe, generating the NMR signal (Figure 11.3). The sample depth is therefore always gauged prior to insertion of the tube into the magnet (see Figure 11.1).

Advances in NMR software have now made the process of actually taking an NMR spectrum so simple that just a few simple electronic commands will nowadays produce a basic $^{1}H/^{13}C$-NMR spectrum.

11.5 The NMR spectrum

11.5.1 One-dimensional (1D) NMR analysis of an anionic surfactant

Figure 11.4 shows the ^{1}H-NMR spectrum of a typical anionic surfactant taken at 270 MHz. The molecule is an 'end-capped' (i.e. carboxymethylated) fatty alcohol polyethoxylate. It finds use in industrial cleaning where low foaming is required [9a]. In Figure 11.4 intensity of signal is plotted on the *y* axis and spectral frequency on the *x* axis. As was pointed out earlier (section 11.2), each proton in the molecule has a characteristic chemical environment and therefore will resonate at a unique frequency. It is the convention in NMR (^{1}H, ^{13}C and multinuclear) not to express this frequency (or wavelength) directly, but to quote the difference relative to an internal standard. The one used for ^{1}H and ^{13}C NMR is tetramethylsilane (TMS). Why? Because the signals from most organic

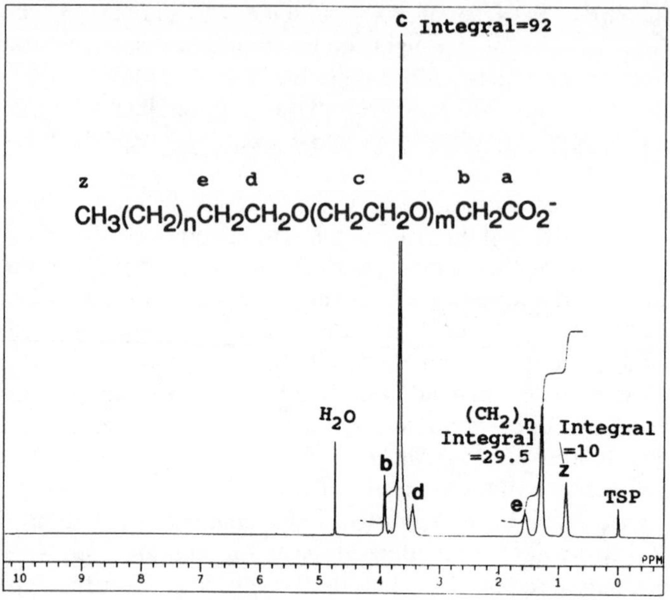

Figure 11.4 ^{1}H-NMR spectrum of an anionic surfactant in D_2O.

compounds resonate at higher frequencies than those of TMS for both ^{1}H- and ^{13}C-NMR. They are therefore given positive numbers known as 'chemical shift' (expressed in ppm relative to TMS) and the corresponding nuclei are said to be deshielded relative to TMS. One notable exception to this rule is methyl iodide, where the carbon atom is actually shielded relative to TMS, resonating at −22 ppm.

TMS is not soluble in water and therefore TSP-d_4 (sodium trimethylsilyltetradeuteropropionate), a water-soluble derivative, is used instead in D_2O solutions, as is the case in Figure 11.4.

11.5.2 Two-dimensional (2D) NMR analysis of a betaine

To the uninitiated, the term 'two-dimensional' seems a bit of a misnomer. Surely the spectrum of Figure 11.4 is a two-dimensional spectrum in the sense that chemical shift (and hence frequency) occupies one dimension and intensity the other? However, by definition it is a 1D spectrum because there is only one frequency dimension. Therefore, 2D NMR spectra by the same definition must have two frequency dimensions. These dimensions can, in principle, both refer to ^{1}H chemical shifts (as in the case in Figure 11.5) or one frequency dimension can be ^{1}H and the other ^{13}C (as is the case in Figure 11.6) or any other NMR-active nucleus

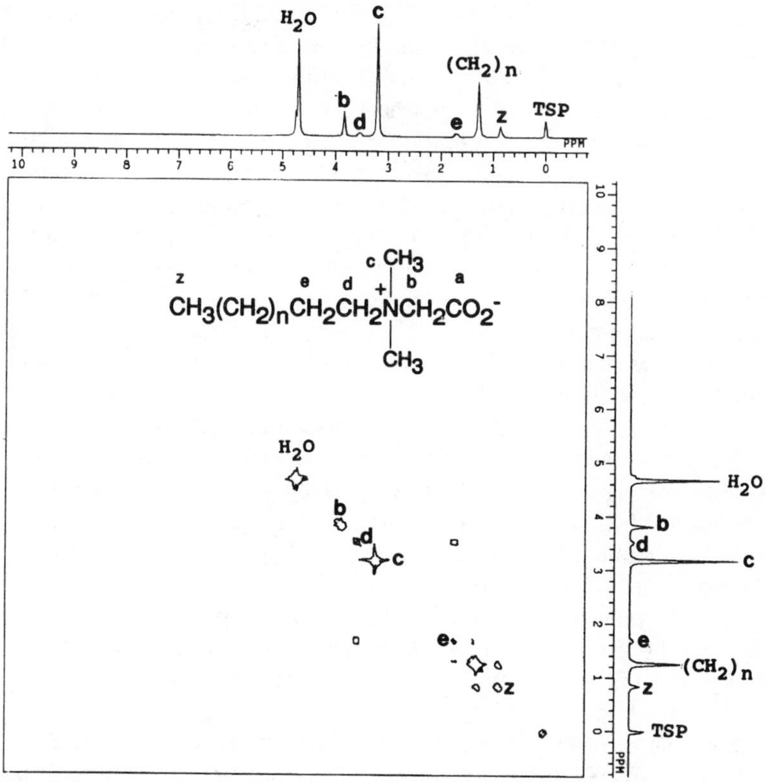

Figure 11.5 ^1H–^1H COSY spectrum of a betaine in D_2O.

(e.g. ^{15}N, ^{29}Si, ^{11}B). Or indeed both frequency dimensions can refer to heteroatoms.

Figures 11.5 and 11.6 illustrate just how useful 2D NMR can be in peak assignment, and therefore in molecular-structure elucidation. The detailed information that they provide may not be necessary in those cases where the 'chemical history' of a molecule is known, but it becomes vital in the analysis of unknowns (e.g. impurities in raw materials, extracts of competitors' products, new surfactant molecules in the development stage, etc.). Indeed, most of the NMR time of modern analytical laboratories is spent on 2D NMR experiments—which is rather fitting because the Nobel Prize for Chemistry in 1991 was awarded to Professor Richard Ernst for developing 2D NMR.

The examples of Figures 11.5 and 11.6 are by no means the only types of 2D spectra that can be taken, and the texts that have already been cited [1–4] are recommended to the interested reader.

Figure 11.5 is a ^1H–^1H COSY (COrrelation SpectroscopY) spectrum of a D_2O solution of *N*-alkyl-*N*,*N*-dimethylbetaine. Such betaines are widely used in the detergent business where 'mild surfactancy' is required [9c]. Also plotted on Figure 11.5 are the spectral projections along the two axes; these formally correspond to the 1D ^1H-NMR spectrum of this compound. Intensity in a 2D NMR spectrum is plotted along the third dimension (i.e. coming out of the plane of the page). It will be immediately obvious therefore that peaks in a 2D NMR spectrum are actually 'mountain ranges' on a three-dimensional surface.

The most popular (but by no means only) way to represent 2D NMR spectra is to take a horizontal 'slice' through this 'mountainous terrain', in which case the peaks will be represented by contours—as is the case in Figures 11.5 and 11.6.

Returning to Figure 11.5, let us go through the 1D spectrum first, and then we will see why the COSY spectrum is so important in characterising this molecule. The three peaks in the region 0.8–1.8 ppm correspond to the fatty alkyl chain, exactly as they do in Figure 11.4. In fact, this pattern of chemical shifts is characteristic of the ^1H-NMR spectra of all surfactants with a linear fatty alkyl chain.

Note also that the spectral resolution is too poor to resolve three-bond ^1H–^1H couplings. For clarification the second methylene of the alkyl chain is labelled 'e' (1.7 ppm); the remaining two methylenes are 'd' (3.6 ppm), and 'b' (3.8 ppm).

The strong peak at 3.2 ppm can be clearly identified from its chemical shift as being the *N*-methyl peak. The only issue is the labelling of peaks 'b' and 'd': how do we know which is which? We would expect *a priori* that 'b' should be deshielded relative to 'd' because it has a carbonyl carbon adjacent to it. But how can we be absolutely sure of this assignment? This is where 2D NMR comes into its own.

If we take each peak and draw an imaginary line down the page we come to a contour that lies on a diagonal going from the top left hand corner of the page to the bottom right hand corner. This diagonal can be thought of as a baseline from which to take correlations. On either side of the diagonal we find other contours; these are known as cross-peaks, and they correlate only those protons that are separated by three chemical bonds, i.e. protons on adjacent methylene carbons (—H_2C—CH_2—).

Let us now see how the cross-peaks give us information about connectivity in the molecule. Starting with the methyl peak ('z') at 0.85 ppm on the diagonal, we find two cross-peaks correlating it to the $(CH_2)_n$ peak at 1.25 ppm. Of course, this tells us what we know anyway, namely that the methyl carbon must be directly bound to one of the —CH_2— carbons in the alkyl chain; the peak at 1.25 ppm in turn has two further cross-peaks to the CH_2 at 1.7 ppm (peak 'e'). This again confirms what we already know, *viz*. the methylene carbon 'e' must be directly

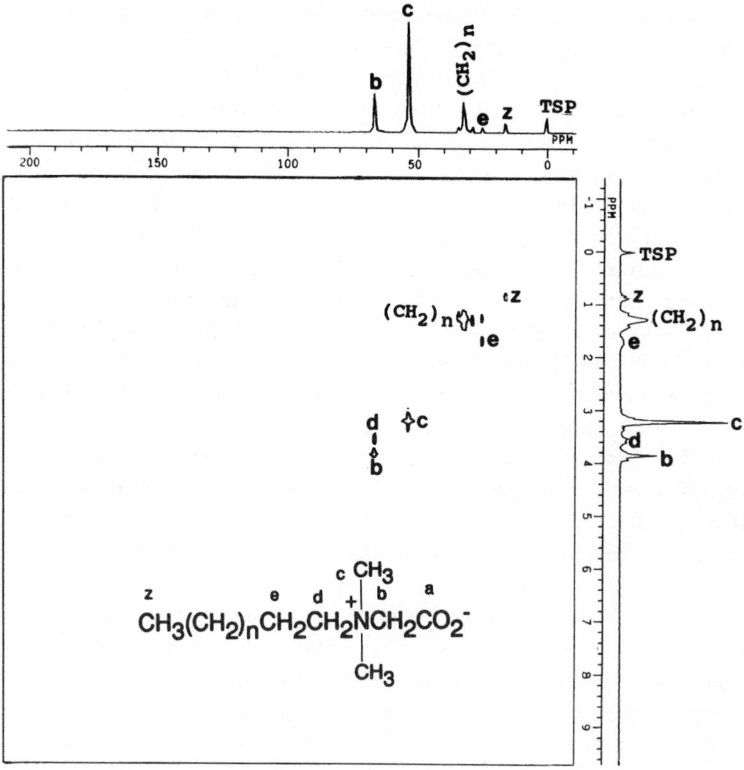

Figure 11.6 ^{13}C–^{1}H HETCOR spectrum of a betaine in D_2O.

bound to one of the methylene carbons in the fatty alkyl chain. Going further out, we find one more pair of cross-peaks that correlate the methylene at 1.7 ppm ('e') with the methylene at 3.6 ppm, which must therefore be peak 'd'. Finally, by default, peak 'b' must be at 3.8 ppm. Note that cross-peaks between 'b' and 'e' are not observed because there are five bonds between these protons, and the spectral parameters have been optimised to correlate only three-bond couplings.

Having now characterised the ^{1}H-NMR spectrum unambiguously, we can use this information to solve a particularly irksome problem in the ^{13}C-NMR spectrum of this molecule. Figure 11.6 is the HETCOR (HETeronuclear CORrelation) spectrum of *N*-alkyl-*N*,*N*-dimethylbetaine. The projection on the *x* and *y* axes correspond formally to the 1D ^{13}C and ^{1}H spectra, respectively.

Once again, this 2D NMR spectrum is represented as a contour plot with peak intensity represented in the third dimension. An obvious difference from the COSY spectrum of Figure 11.5 is that this spectrum does not have a diagonal of symmetry, and therefore all the contours on the spectrum represent cross-peaks.

These cross-peaks result from one-bonded ^{13}C–^{1}H couplings, i.e. each proton peak correlates only with the carbon atom to which it is directly bonded. As before, let us try to interpret the 1D ^{13}C-NMR spectrum first. The most deshielded carbon atom corresponds to the methylene 'b' (67 ppm), by analogy with the ^{1}H-NMR spectrum. The large peak at 53 ppm is due to the two equivalent methyls ('c') directly attached to the nitrogen atom. The remaining peaks in the chemical shift region 15–35 ppm correspond to the carbons of the fatty alkyl chain, as explained later in the analysis of the alkyl polyglycoside spectrum (Figure 11.8). So, where is peak 'd'? Clearly, it must be in the region 50–70 ppm because, like carbon atom 'b', it is directly attached to the nitrogen atom. This problem of 'missing peaks' is often encountered in ^{13}C- and ^{1}H-NMR spectroscopy, and the usual explanation is that the peak in question is overlapped, i.e. 'masked', by another—possibly bigger—peak in the spectrum. And this is indeed the case here: if we draw a vertical line down from the peak at 67 ppm we find that it correlates with two contour peaks; taking the first of these and drawing a horizontal line we see that this contour correlates with peak 'd' (3.6 ppm) in the ^{1}H-NMR spectrum. The second contour correlates with peak 'b' in the ^{1}H-NMR spectrum (3.8 ppm). We can now see that peaks 'b' and 'd' are separated by *c* 0.1 ppm, a difference that cannot be resolved in the 1D ^{13}C-NMR spectrum.

11.5.3 Identification of an unknown surfactant

The principles established in the preceding two sections can now be applied to identifying an unknown surfactant. The example of Figure 11.4 will be used here as the unknown, simply because the information obtained from this exercise will be applied later (section 11.6) to gain quantitative information about alkyl chain length and ethoxylate number for this anionic surfactant.

The carbon atoms in this molecule (and, where appropriate, attached protons) are arbitrarily labelled a–z. Let us scan through the spectrum first and assign the peaks. The peak at 0.9 ppm is due to the terminal methyl group in the molecule ('z'). Rapid rotation along the CH_3—CH_2 bond makes all three methyl protons chemical-shift equivalent; moreover they are coupled to the two protons of the adjacent methylene group, and because of the rules of coupling, this methyl signal should actually be split into a triplet. However, because of the poor resolution in this spectrum (which is often the case with large molecules such as surfactants), only a broad singlet is observed at 0.9 ppm. Nevertheless, it is the coupling pattern, together with the chemical shift, which allows us to say unambiguously that this peak is due to the terminal methyl ('z').

Coupling between protons (and, for that matter, between any two dipolar nuclei) is transmitted through chemical bonds. It diminishes rapidly as the number of bonds between the nuclei increases (for the same reason that the force between two bar magnets decreases with increasing distance), and becomes negligible when the number of bonds is greater than three. Thus we can effectively ignore coupling between the terminal methyl protons and methylene protons further down the alkyl chain.

The large peak at 1.3 ppm, $(CH_2)_n$, is characteristic of a linear fatty alkyl chain. The chemical shifts of these methylenes are virtually identical and so all the peaks (but two) overlap to form one large peak at 1.3 ppm. The exceptions are: the second methylene ('e') of the fatty chain, which resonates at 1.6 ppm, and the first methylene of the fatty chain ('d') which resonates at 3.4 ppm, typical of a proton that is two bonds away from an oxygen atom. This assignment, based purely on chemical-shift grounds, can only be confirmed rigorously from a COSY spectrum of the molecule, as outlined in section 11.5.2. Nevertheless, the example illustrates an important point about ^{1}H (and ^{13}C) chemical shifts: the closer the proton is to an electronegative substituent (e.g. O, N, Cl) the more deshielded it will be relative to TMS, and hence the larger its chemical shift: thus 'd' is more deshielded than 'e'. Tables of chemical shifts are readily available both in textbooks and in the original NMR literature and are helpful in interpreting ^{1}H- and ^{13}C-NMR spectra.

The large peak at 3.7 ppm corresponds to the ethoxylate protons ('c'). The small difference in chemical shift between 'c' and 'd' protons is due to a second-order effect: both protons are deshielded by the electron-withdrawing inductive effect of an oxygen atom two bonds away, but the 'c' protons feel the additional effect of the oxygen atom on the other side, i.e. three bonds away, and therefore have higher chemical shift.

Finally, the 'end cap' methylene ('b') resonates as a singlet at 3.9 ppm—further downfield of the ethoxylate protons because of the carbonyl carbon, which has an additional deshielding effect. Residual water in the solvent (D_2O) can also be seen at 4.7 ppm.

Note that evidence for the presence of a carbonyl carbon atom in the molecule can only come from the ^{13}C-NMR spectrum (see Figure 11.9). The above ideas of chemical shift and connectivity can also be used to assign the ^{13}C-NMR spectrum, giving further proof of its molecular structure. The carbon–hydrogen connectivity information (as obtained from a HETCOR spectrum) can be particularly useful for assigning chemical shifts, and hence functionalities within a molecule, as was outlined in section 11.5.2.

11.6 Applications of NMR to surfactant analysis

11.6.1 Average alkyl chain length by ^{1}H-NMR

Having now assigned all the peaks of Figure 11.4 to the appropriate hydrogen atoms, we can obtain quantitative information from this spectrum. Peak integration, which is readily performed automatically, now affords the value of *n* in the molecule, and hence the average alkyl chain length. The steps in the integrated spectrum are proportional in height to the areas under the peaks, which in turn are proportional to the numbers of nuclei generating them. The calculation is given below.

Area under methyl peak = 3 protons = 10 (in arbitrary units)
Therefore 1 proton = 3.3
Area under alkyl peak at 1.3 ppm = 29.5, i.e. 8.9 protons, i.e. 4.5 methylenes
Therefore n = 4.5 (as defined in Figure 11.4)
Therefore average length of fatty alkyl chain
= 4.5
+ 1 (from the peak at 1.6 ppm)
+ 1 (from the peak at 3.4 ppm)
+ 1 (from the methyl peak at 0.9 ppm)
= 7.5

11.6.2 Ethoxylate number determination by ^{1}H-NMR

The above chemical shift assignments and integrations also allow us to determine the average ethoxylate number for this surfactant. Given that the ethoxylate peak ('c') is at 3.7 ppm:

Area under peak at 3.7 ppm	=92
But 1 proton	= 3.3
Therefore 92	= 27.6 protons
But one ethoxylate group	= 4 protons (—CH_2CH_2O—)
Therefore m	= 27.6/4 = 6.9

It can be clearly seen that the terminal methyl is used as the internal standard in both these calculations, hence the importance of determining its integral with great precision.

Why are *m* and *n* not integers? Because what we have in solution is a complex mixture of molecules, each with its own integral value of *n* and *m*, and the NMR method only gives us an unweighted average of these values. This result can be readily confirmed by HPLC or SFC analysis of the mixture, where the individual oligomers can be seen.

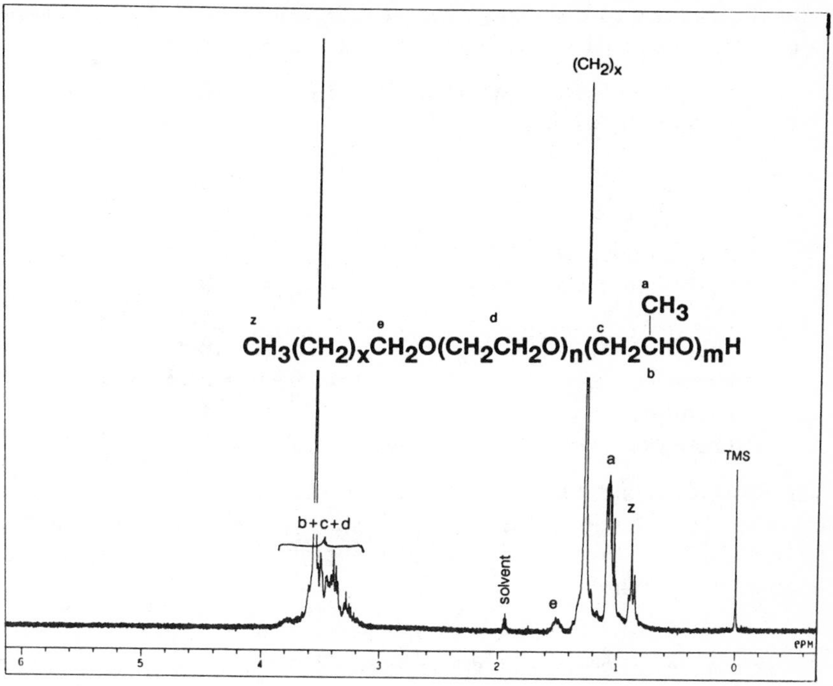

Figure 11.7 ^{1}H-NMR spectrum of an alkyl ethoxylate/propoxylate in CD_3CN.

11.6.3 Ethoxylate/propoxylate ratio by ^{1}H-NMR

Figure 11.7 is the ^{1}H-NMR spectrum of an alkyl ethoxylate/propoxylate. This is made commercially by reaction of the appropriate fatty alcohol with a mixture of ethylene oxide (EO) and propylene oxide (PO). The level of incorporation of the EO/PO units greatly affects the polymer's surfactancy properties. This ratio, i.e. n/m, can be determined by integration, as shown below. The integrals are not shown on this spectrum (for clarity), but they have been calibrated by using the terminal methyl of the alkyl chain (0.9 ppm) as internal standard. The peak assignments on the spectrum have been made using the principles outlined in section 11.5.3.

Protons 'b', 'c' and 'd' are all overlapped in the region 3.2–3.8 ppm. The methyl of the propoxylate group 'a', can, however, be clearly resolved at *c.* 1 ppm. Note that we get a number of overlapping peaks in this region because this is a random copolymer, and so there will be many methyls 'a', each with a slightly different chemical shift. The calculation below applies equally well to random and block copolymers.

Area under methyl triplet ('z') at 0.9 ppm = 3 (by definition)
Area under b + c + d = 26.5
i.e. $4n + 3m = 26.5$
(4 protons from each EO) (3 protons from each PO)
but area under peak 'a' (propoxylate methyl only) = 9.3
Therefore $3m = 9.3$
Therefore $m = 3.1$
Therefore $4n + 9.3 = 26.5$
Therefore $n = 4.3$
Therefore $n/m = 4.3/3.1 = 1.4$

The above calculation shows that there has been greater incorporation of ethylene oxide into the surfactant. The above spectrum was taken in CD_3CN, and therefore a residual solvent multiplet (due to HCD_2CN) can be seen at 1.95 ppm.

11.6.4 Degree of polymerisation by ^{13}C-NMR

^{13}C-NMR can provide qualitative and quantitative information about surfactants—both in raw materials and in consumer products. It is such a powerful analytical technique that entire textbooks have been devoted to it [10, 11]. Figure 11.8 is the ^{13}C-NMR spectrum of an alkyl polyglycoside (APG), a mild nonionic surfactant that is finding increasing use in cosmetic formulations [9b]. Note first that the chemical shift scale of the spectrum is now 0–200 ppm, in contrast to ^{1}H-NMR, where the scale is typically 0–10 ppm. Because of this greater chemical-shift dispersion, the peaks appear much sharper. Note also that each carbon atom resonates as a sharp singlet i.e. ^{13}C–^{1}H coupling has been electronically removed using a powerful decoupling coil: this is the way that ^{13}C-NMR spectra are routinely taken. The analytical question being asked here is not the length of the alkyl chain (which can be easily determined by ^{1}H-NMR), but the degree of polymerisation (DP) of this molecule.

The DP profoundly affects surfactancy properties and skin-penetrating ability of APGs. It is defined as:

$$\text{DP} = \text{No. of glucose units per molecule}$$

The DP cannot be determined by ^{1}H-NMR because the terminal methylene (—CH_2O—) of the alkyl (i.e. 'g') chain overlaps with the glucose ring proton signals, and cannot be resolved from them even at 270 MHz; hence the need for ^{13}C-NMR.

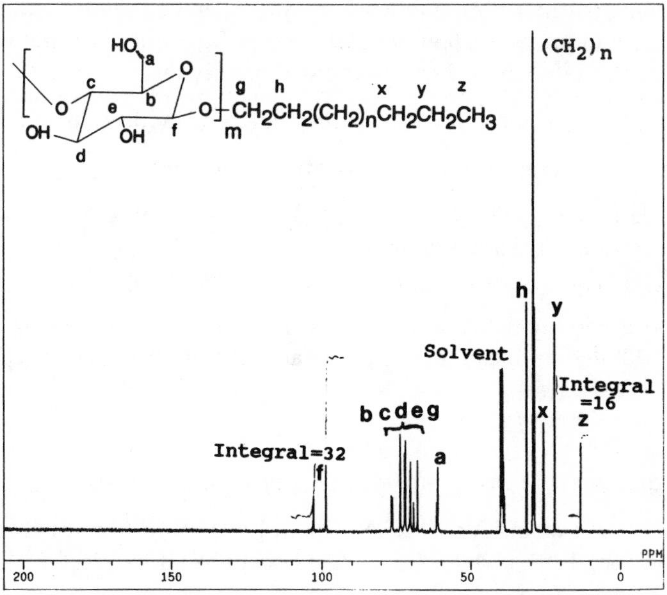

Figure 11.8 ^{13}C-NMR spectrum of a non-ionic surfactant in dmso-d_6.

As before, let us scan through the spectrum (Figure 11.8) before carrying out the necessary integration: the five peaks in the region 10–35 ppm all belong to the fatty alkyl chain. The terminal methyl 'z' resonates at 13 ppm; this time, the last two methylene carbons of the fatty alkyl chain can be clearly discerned, resonating at 22 ppm ('y') and 25 ppm ('x') respectively. Approaching the alkyl chain from the other end, the first methylene (i.e. 'g') resonates at 73 ppm, the second ('h') at 31 ppm, and the remainder of the alkyl methylenes at *c*. 30 ppm. This pattern of ^{13}C chemical shifts is typical of all linear fatty alkyl chains—whatever the surfactant might be—and is seen regularly in the ^{13}C-NMR analysis of detergent formulations. Nowadays, only linear alkyl groups are incorporated into surfactants, because of their ready biodegradation in the environment.

The peak at 61 ppm is 'a' (i.e. —CH_2—OH) of the glucose ring, and the remaining peaks in the region 65–75 ppm are the CH—O carbons of the glucose ring, with the exception of 'f' (the so-called anomeric carbon) which, because it is directly bonded to two oxygen atoms, is more deshielded still, and resonates at 99 ppm and 102 ppm. Why two peaks? The former peak corresponds to the α-anomer, and the latter to the β-anomer [12].

Returning to the *raison d'être* for the spectrum of Figure 11.8, because there is one anomeric carbon per glucose unit, and only one methyl in the molecule, the DP can be expressed most directly as:

$$\text{DP} = \text{area of anomeric carbons/area of methyl carbon}$$

Integration of the appropriate peaks affords the answer:

Area under anomeric carbons ('f')	= 32 (in arbitrary units)
Area under terminal methyl ('z')	= 16
Therefore	DP = 32/16 = 2.0

For a pure alkylmonoglycoside, DP = 1.0, by definition. Therefore the above DP value implies that the raw material is a mixture of alkyl mono- and polyglycosides.

11.6.5 Analysis of a carboxylic acid/ester mixture by ^{13}C-NMR

Figure 11.9 is a quantitative ^{13}C-NMR spectrum of the carbonyl region of the end-capped ethoxylate discussed in sections 11.6.1 and 11.6.2. This carboxylic acid can be made either by reacting the alcohol ethoxylate with chloroacetic acid or by catalytic oxidation of the terminal hydroxyl of the ethoxylate itself. In either event, the product can react with remaining alcohol to form an ester, an undesirable by-product.

Mixtures of acid and an ester can be readily quantitated by integrating the carbonyl resonances in the ^{13}C-NMR spectrum. Note that there are two acid peaks: the minor one at 176.5 ppm is the unethoxylated fatty acid; the major one at 173 ppm is the end-capped alcohol ethoxylate.

Area under ester	= 1.6 (in arbitrary units)
Area under carboxylic acids	= 4.8 + 30.5 = 35.3
Therefore molar % surfactant acid	= 35.3/36.9 = 96
molar % surfactant ester	= 1.6/36.9 = 4

Note that the method does not need an internal standard because the spectral parameters have been adjusted to give a molar response. This type of quantitative ^{13}C-NMR analysis takes up a great deal of NMR time, and is only carried out when the quicker ^{1}H-NMR technique is unapplicable, as is clearly the case here.

11.6.6 Analysis of alkyl phosphate mixtures by ^{31}P-NMR

All modern high-field NMR spectrometers come equipped with broad-band tunable probes that can be used for multinuclear NMR spectroscopy. The phosphorus nucleus is particularly easy to detect, being spin = $\frac{1}{2}$ and 100% abundant. Figure 11.10 is a ^{31}P-NMR spectrum of a

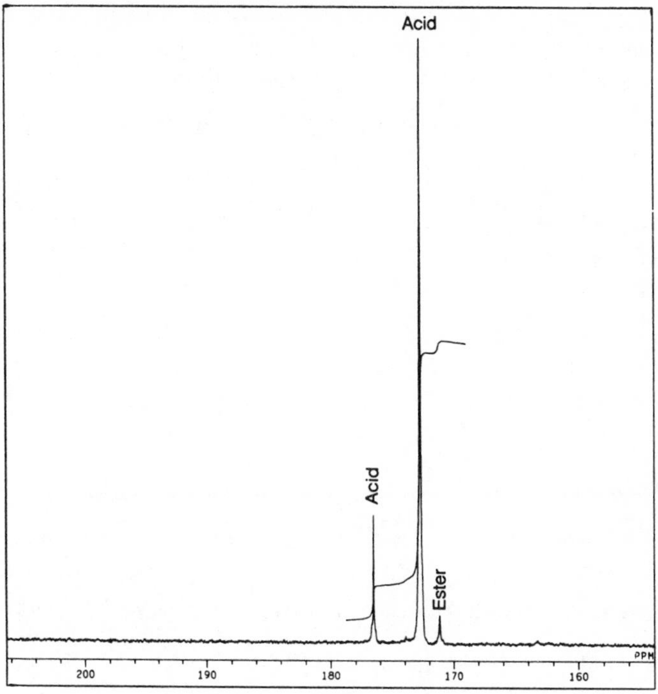

Figure 11.9 Quantitative ^{13}C-NMR spectrum of an acid–ester mixture in D_2O.

mixture of alkyl phosphates ($R = C_{16}H_{33}$) taken in deuterochloroform ($CDCl_3$). The replacement of OH by OR on the phosphorus atom has a characteristic effect on its chemical shift that allows us to resolve, and assign, all these phosphate esters. The peak assignments, referenced to phosphoric acid as external standard, are given on the spectrum. Integration of the peaks, followed by normalisation, affords the molar ratio of each phosphate ester.

Peak area of monoester	= 72 (in arbitrary units)
Peak area of diester	= 15
Therefore: monoester	= 72/87 = 83 mol %
diester	= 15/87 = 17 mol %

Note the absence of phosphoric acid in this sample. This method can also be used to analyse mixtures of alkylether phosphates. The advantage of this method over standard chromatographic procedures is that the phosphorus analysis can be carried out *in situ*; there is no need to extract and subsequently derivatise the surfactants.

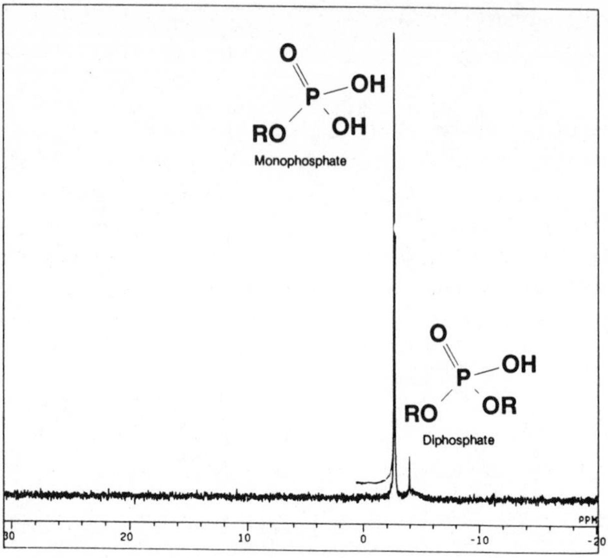

Figure 11.10 ^{31}P-NMR spectrum of an alkyl phosphate mixture in $CDCl_3$.

11.6.7 Determination of surfactant concentration by ^{1}H-NMR

It was mentioned earlier (p. 302) that ^{1}H-NMR always gives a molar response. This fact can be used to determine the absolute molar concentration of a surfactant raw material in solution, as long as a standard is available. If the molar concentration of the standard is known, then that of the surfactant can be calculated from the ratios of the peak integrals. The method is explained below for the anionic surfactant of Figure 11.4. This solution contains 2.4×10^{-2} mol l^{-1} TSP-d_4 (whose structure is given below) as internal standard.

```
      Me
      |
Me—Si—CD2—CD2—CO2⁻ Na⁺
      |
      Me
```

TSP-d_4

Integral of surfactant methyl peak ('z')	= 10 (in arbitrary units)
Integral of three methyls of standard	= 1.7 (integral not shown for clarity)
Therefore, one methyl integral	= 1.7/3 = 0.57

But, concentration of standard $= 2.4 \times 10^{-2}\,\text{mol}\,l^{-1}$
Therefore, concentration of surfactant $= 2.4 \times 10^{-2} \times 10/0.57$
$= 0.42\,\text{mol}\,l^{-1}$

The beauty of the above method is that the concentration of individual components in a complex mixture (e.g. a detergent formulation) can be calculated so long as the corresponding proton peaks can be resolved in the ^{1}H-NMR spectrum.

Note that TMS must never be used in the above method because it is volatile, and so its concentration in solution can never be known accurately.

References

1. Williams, D. H. and Flemming, I. *Spectroscopic Methods in Organic Chemistry*, 4th edn, McGraw-Hill, London, 1989.
2. Kemp, W. *NMR In Chemistry: A Multinuclear Introduction*, Macmillan, Basingstoke, 1986.
3. Derome, A. E. *Modern NMR Techniques for Chemistry Research*, Pergamon, Oxford, 1987.
4. Hunter, B. K. and Sanders, J. K. M. *Modern NMR Spectroscopy: A Guide for Chemists*, Oxford University Press, Oxford and New York, 1987.
5. Brevard, C. and Granger, P. *Handbook of High Resolution Multinuclear NMR*, Wiley-Interscience, New York, 1981.
6. Abraham, R. J. and Loftus, P. *Proton and Carbon-13 NMR Spectroscopy*, John Wiley, Chichester, 1978.
7. Field, L. D. and Sternhill, S. *Analytical NMR*, John Wiley, Chichester, 1989.
8. Fyfe, C. A. *Solid State NMR for Chemists*, CFC Press, Guelph, Canada, 1983.
9. *McCutcheon's Emulsifiers and Detergents*, international edn, vol. 1, 1992; a, p. 9; b, p. 36; c. p. 108.
10. Levy, G. C., Lichter, R. L. and Nelson, G. L. *Carbon-13 NMR Spectroscopy*, 2nd edn, Wiley-Interscience, New York, 1980.
11. Wehrli, F. W., Marchand, A. P. and Wehrli, S. *Interpretation of Carbon-13 NMR Spectra*, 2nd edn, John Wiley, Chichester, 1988.
12. Gurst, J. E. *J. Chem. Ed.*, **12** (1991), 1003.

12 Mass spectrometry

K. LEE

12.1 Introduction

Mass spectrometry (MS) identifies species on the basis of their relative molecular mass, or the relative molecular mass of their fragments. This is achieved by first ionising the sample and then determining the mass-to-charge ratio of the molecular ion and fragment ions. By assuming that the ions are singly charged (which is frequently but not always the case) the resulting mass-to-charge spectrum can be displayed on a mass scale. There are many different types of mass spectrometer. Those applicable to the analysis of surfactants range from small bench-top instruments dedicated to gas chromatography–mass spectrometry (typically quadrupole spectrometers or ion traps), costing several tens of thousands of pounds, up to sector instruments and triple quadrupole instruments costing several hundreds of thousands of pounds. The larger instruments offer higher mass ranges and increased specificity either by high resolution or by use of more involved MS experiments.

A diagram of a typical two-sector instrument is presented in Figure 12.1. Other types of mass spectrometer will be discussed in section 12.4.

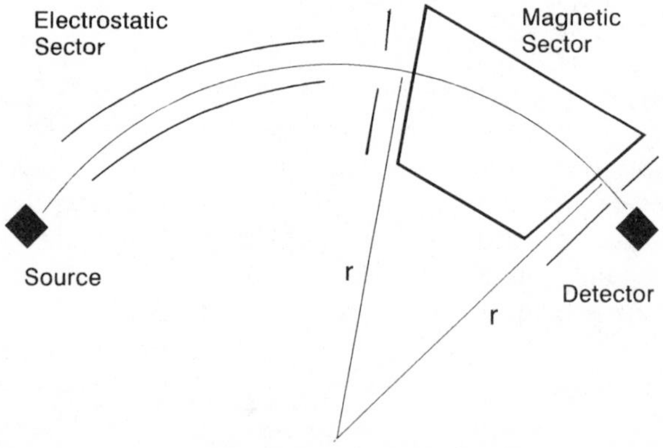

Figure 12.1 Schematic diagram of a two-sector mass spectrometer. The sample is ionised in the source and mass analysed in the magnetic sector. Energy focusing is carried out in the electrostatic sector.

The sample is introduced to the source by one of a variety of inlet systems; here it is ionised and the resulting ions are accelerated through a known potential. Discrimination on the basis of mass-to-charge ratio takes place in the magnetic analyser; the electrostatic analyser aids resolution. Detection is achieved by an electron multiplier, and the resulting signal is analysed by a computer-based data system. The mass spectrum is a plot of the relative abundance of ions versus their relative molecular mass. The standard for molecular mass is a value of exactly 12 for ^{12}C.

While MS is a very sensitive technique, quantities as low as a few picograms (1 pg = 10^{-12} g) being routinely analysed, it is one of the least specific of all spectroscopic techniques. There will be many possible empirical formulae corresponding to any particular peak in a low-resolution mass spectrum. For this reason more sophisticated MS techniques may be required (see section 12.4.4), or MS may be combined with separation techniques such as gas or liquid chromatography.

Areas of application of mass spectrometry in the surfactant industry include the following:

- Characterisation of new synthetic materials
- Raw materials analysis
- Whole product analysis
- Surfactant chain-length distribution determination
- Analysis of factory emissions and trace contaminants
- Determination of trace levels of surfactant-related materials in environmental samples
- Determination of contaminants in products

12.2 Sample introduction

The method used to introduce the sample to the source of the spectrometer will depend on the method of ionisation used (see section 12.3), the physical form of the sample, and any requirement for separation techniques to be applied in line with the MS analysis.

For many routine applications, a solid or liquid sample may be introduced to the source of the spectrometer by direct mechanical insertion, i.e. it is held on the end of metal probe which is inserted into the source via a system of air-locks. All of the sample is presented to the source simultaneously, allowing no separation of the components, and the resulting spectrum is a function of the whole of the sample. For gas-phase ionisation techniques an added level of discrimination may be obtained by selectively vaporising the components of the sample. Volatility discrimination is achieved by acquiring data as a function of time while the probe temperature is increased at a predetermined rate. Spectra can then be related to the corresponding probe temperature.

For most volatile or gaseous species more selectivity may be achieved by combining mass spectrometry with gas chromatography. This is applicable to some nonionic surfactants [1] and derivatives of anionics [2]. The combination of gas chromatography with mass spectrometry (GC–MS) is well established, and relatively cheap dedicated instruments are available for this application.

When less thermally labile compounds require separation prior to analysis, in-line liquid chromatography–mass spectrometry (LC–MS) can be used. Even where liquid chromatographic separation is not required LC–MS interfaces may be used as a convenient method of sample introduction, e.g. for polar, ionic, thermally labile, involatile or high-molecular-weight species. There are several different types of LC–MS interface each with its own characteristics in terms of the solvent systems and sample types that it can handle and the appropriate solvent flow rates; no one interface is applicable to all solvent–solute combinations.

12.3 Methods of ionisation

12.3.1 Introduction

Ionisation techniques that produce little fragmentation are termed 'soft ionisation'; they produce mainly molecular ions or quasimolecular ions. In positive-ion mode the molecular ion is usually denoted M^+, which refers to the radical cation produced by the loss of an electron from the neutral molecule. Quasimolecular ions are the adduct ions formed between neutral molecules and an ionising species, for example the MH^+ species formed by protonation of the molecule M. Molecular ions and quasimolecular ions allow the determination of the relative molecular mass of the analyte. More energetic ionisation processes result in more fragmentation. This is termed 'hard ionisation'. Spectra resulting from hard ionisation techniques may contain only very weak responses for molecular ions, or molecular ions may be absent from the spectra, but strong responses for fragment ions. Fragment ions are important as a source of structural information. Throughout this chapter, fragment ions produced by loss of a group R from a molecular ion M^+ will be denoted as $[M - R]^+$, and similar nomenclature will apply to negative ions.

This chapter will be limited to the following methods of ionisation:

Hard ionisation

Electron impact (EI)

Soft ionisation

Chemical ionisation (CI)
Fast atom bombardment (FAB)
Liquid secondary ionisation MS (LSIMS)
Electrospray (ES)
Thermospray (TS)

The soft ionisation technique of field desorption (FD) will not be covered in this chapter. For a review of this technique see Lattimer and Schulten [3]. In many applications, FAB and LSIMS are more convenient than FD and for this reason FD suffered a fall in popularity in the 1980s. It does, however, have advantages over FAB in certain applications, for example the analysis of surfactants in the presence of inorganic salts [4].

It would be desirable to ionise all samples by both hard and soft techniques, thereby gaining the maximum information. This is not always possible as certain ionisation techniques may be precluded by the nature of the sample, for example most cationic and anionic surfactants cannot be ionised directly by EI. Where species can only be ionised by soft techniques, structural information may be obtained by collisional activation of the molecular ion (or quasimolecular ion, depending on the ionisation mode) to induce fragmentation. This type of technique is denoted as CID in Table 12.1, where ionisation modes appropriate to applications are summarised.

12.3.2 Electron impact ionisation (EI) and chemical ionisation (CI)

These are the most widely used ionisation techniques in mass spectrometry; both take place in the vapour phase.

In electron impact ionisation, the sample is bombarded with a beam of highly energetic electrons. Positively charged molecular ions resulting from this process lose their excess energy by fragmentation. Electron ionisation spectra are very reproducible and may be identified by reference to libraries of previously obtained spectra. Commercial libraries contain tens of thousands of spectra.

Chemical ionisation is a soft ionisation technique in which a reagent gas transfers charge to the sample. Electron impact ionisation of the reagent gas is followed by ion–molecule reactions which produce species capable of cation transfer to the sample to give quasimolecular ions, for example MH^+ ions from proton transfer to the species M. As proton transfer is a much less energetic process than electron impact there is little fragmentation of the quasimolecular ion. Reagent gases include methane, isobutane and ammonia. Methane gives spectra containing more fragments than either isobutane or ammonia.

Since both EI and CI take place in the vapour phase, their application

Table 12.1 Ionisation techniques applicable to different surfactant types

Sample type	Information required			
	Identification in mixtures	Homologue distribution	Structural information	Trace determination
Anionics	−FAB +FAB −ES	−FAB −ES	−FAB/ CID	−FAB +FAB Derivatisation followed by GC–MS
Nonionics	+FAB TS CI +ES	+FAB TS CI SFC/CI	+FAB LC/+FAB SFC/CI	+FAB PB
Cationics	+FAB −FAB	+FAB	+FAB/ CID	+FAB
Amphoterics	+FAB −FAB −ES	+FAB		−ES

CID = Collisionally induced decomposition
PB = Particle beam
GC = Gas chromatography
LC = Liquid chromatography
SFC = Supercritical fluid chromatography

in the analysis of surfactants is limited. They can, however, be applied to nonionic surfactants. They may be used in conjunction with direct insertion probe, GC–MS, LC–MS (using a particle-beam interface) and supercritical fluid chromatography–MS (SFC–MS). Ethoxylated and propoxylated species give CI spectra that are dominated by MH^+ ions, e.g. $RO(CH_2CH_2O)_nHH^+$ for alcohol ethoxylates. These may be used to identify the alkyl group and the degree of ethoxylation, thereby providing information on the distribution of homologues. EI spectra contain weak molecular-ion signals and are dominated by fragment ions.

12.3.3 Fast atom bombardment (FAB) and liquid secondary ion mass spectrometry (LSIMS)

FAB and LSIMS ionise samples from a liquid matrix. They are applicable to involatile, high-molecular-weight and ionic species and therefore to most species of surfactant. For this reason a large part of this chapter is devoted to these techniques. Ionisation by FAB and LSIMS is soft. The spectra are dominated by molecular ions and quasimolecular ions. The high level of such ions allows identification of species even from complex mixtures, for example whole-product formulations or environmental matrices. In the simplest case, the sample is mixed with a liquid matrix and is applied to the tip of a direct insertion probe. This probe is then introduced into the source of the spectrometer. Ionisation is the result of a sputtering process at the matrix surface caused by bombardment. In the case of FAB the sample is bombarded with a beam of neutral inert gas (usually xenon) atoms. LSIMS uses a beam of caesium ions. It is claimed that the ability to focus the caesium ion beam results in higher sensitivity for LSIMS.

The matrix acts as a medium to promote ionisation (e.g. by cation attachment) and prolong the life of the sample. As material is lost from the matrix surface, diffusion of sample through the matrix allows a fresh supply of sample to be presented to the ionising beam. Commonly used matrices include glycerol, *m*-nitrobenzyl alcohol and thioglycerol. Anionic and cationic surfactants give intense spectra due to the presence of preformed ions in the matrix. Nonionic surfactants ionise by cation attachment.

These sputtering techniques suffer from two main drawbacks, discrimination and irreproducibility. Highly surface-active species may dominate the spectra as a result of their suppression of other components of the sample at the surface of the matrix. This discrimination may result in poor sensitivity for certain species and makes quantitation difficult (see section 12.6.2). Ionisation of the matrix gives rise to chemical noise, and the spectra may be further complicated by the formation of sample–matrix adduct ions. These effects prevent the use of libraries in the interpretation of FAB and LSIMS spectra.

Both FAB and LSIMS can be used in conjunction with continuous-flow inlet systems to optimise sensitivity, to improve reproducibility and to allow the use of in-line liquid chromatography for LC–MS. In the flow technique an LC pump delivers a continuous flow of solvent (containing a low concentration of the FAB/LSIMS matrix) to the point where the atom or ion beam is applied. The sample is introduced into this solvent flow by an LC injection valve and is delivered to the point of ionisation in a sharp concentrated slug. A good account of the method of continuous flow FAB is given in reference [5].

12.3.4 Thermospray interface for LC–MS

Thermospray is a soft ionisation technique that relies on the production of a spray by direct heating of a metal capillary carrying the LC eluent. The presence of an electrolyte dissolved in the mobile phase causes this spray to consist of droplets carrying a statistical excess charge. Desolvation of these droplets results in the expulsion of ions, which are then extracted into the mass analyser. Thermospray is applicable to many polar materials but can give poor sensitivity for nonpolar species. Other disadvantages of the technique include the fact that the temperature of the inlet is critical to its performance; it is limited to reverse-phase solvents and the dissolved electrolyte may interfere with the chromatography.

Probe temperature can be made less critical and sensitivity improved if the thermospray interface is fitted with a plasma-discharge facility. Plasma discharge also provides a chromatographic advantage due to the increased range of mobile phases (aqueous to organic) that is available to this interface. Both thermospray and plasma discharge give CI-type spectra. Fragmentation is low, often resulting in only the quasimolecular ion. Some typical compounds accessible to thermospray with and without plasma discharge are summarised below. Many of these species are of relevance to raw materials, effluents and whole-product analysis.

Without plasma discharge
- Alcohol ethoxylates
- Alkylphenol ethoxylates
- Amphoterics at low pH
- Cationics (as the amine)

With plasma discharge
- Fatty acids
- Fatty alcohols
- Glycerides
- Ammonium salts of some anionics
- Imidazolinium quaternary ammonium salts

12.3.5 Electrospray interface for LC–MS

Electrospray relies on the deposition of charge on to the LC eluent from a capillary held at high electrical potential. Coulombic forces cause the dispersion of the liquid into a fine spray of droplets. Evaporation of solvent from these droplets increases the electric field at their surface and results in the desorption of ions into the gas phase. Ionic species desorb as the intact ions; polar species may protonate to give quasimolecular ions. Electrospray ionisation of large molecules can result in multiply charged ions, which behave in the spectrometer as if their mass has been reduced by a factor z, where z is the number of unit charges on the ion. By determining the mass-to-charge ratio of the ion and its charge state, the mass of the ion can be calculated. Electrospray can be used in this way to increase the effective mass range of the spectrometer, and this technique is of obvious benefit in biochemical applications, for which most electrospray development has been carried out. The main interest in work on surfactants is, however, in singly charged species.

As electrospray is a very soft ionisation technique, the spectra are dominated by molecular ion- or quasimolecular ion-related species. Ionic, nonionic and zwitterionic surfactants are all amenable to the technique. Unlike FAB or LSIMS, electrospray does not suffer from discrimination effects [6] and it is therefore applicable to the analysis of mixtures. Sensitivity is high; 10 pg of cationic surfactants can be detected with a good signal-to-noise ratio. The spectrum contains peaks from the intact cation. It has been reported [7] that the limit of detection for positive-ion electrospray is of the order of several femtograms (1 fg = 10^{-15} g).

Anionics typically give negative-ion electrospray spectra corresponding to the intact anion [7]. Such ions are present in the spectra of fatty acids, α-olefin sulphonic acids, and linear alkyl benzenesulphonates [6]. In addition to the intact anion, alkyl sulphates also give ions at *m/e* 97 and 80 as in negative-ions FAB (see section 12.6.3). These correspond to the species SO_4^- and SO_3^-. Alkyl sulphonates give the intact anion and the *m/e* = 80 species. In positive-ion electrospray, anionic surfactants can be detected as clusters with two cations (e.g. two sodium ions).

Nonionics and zwitterionics can give positive-ion electrospray spectra by cation attachment [7].

12.3.6 Particle beam interface for LC–MS

Particle beam interfaces use jet separators and differential pumping to separate solvent from the solute before the ionisation step. Once separated from the solvent the sample is ionised using conventional EI/CI sources. The resulting spectra are therefore compatible with library spectra obtained with conventional EI systems.

Particle beam (PB) EI MS is less sensitive than capillary GC–MS. This can be partly compensated for by the fact that much larger volumes of sample can be introduced in the LC–MS method [8]. The method can therefore be applied to the trace analysis of nonionic surfactants. Certain species, which due to their low volatility, high polarity or thermal instability cannot be detected by GC–MS, can be analysed by LC–PB–MS. Similarly, some highly volatile species are difficult to analyse by PB–MS, and the two techniques are therefore complementary.

12.4 Mass analysis

Ions generated in the source of the spectrometer have to be discriminated from each other on the basis of their mass-to-charge ratio. Conventional mass spectrometers usually use one of the following three methods to achieve this.

12.4.1 Magnetic sector mass analyser

Figure 12.1 shows a highly schematic diagram of one form of double focusing mass spectrometer. On leaving the source the ion is accelerated through a potential V, which imparts a kinetic energy to the ion given by:

$$0.5\,mv^2 = zeV \qquad (12.1)$$

where m is the mass of the ion, v its velocity, z the number of unit charges on the ion (usually $z = 1$) and e the charge of an electron. As an ion passes through the magnetic sector the magnetic field causes it to execute a curved trajectory of radius r. The value of r is dependent on the strength B of the magnetic field, the potential V and the mass-to-charge ratio of the ion:

$$r = [(m/ze)(2V/B^2)]^{1/2} \qquad (12.2)$$

For ions to be detected they must pass through the magnetic sector along a fixed trajectory which is set by the geometry of the instrument. Only one value of r allows ions to reach the detector (this is the case for a point detector—a range of r values is allowed for an array detector), therefore only ions of $m/ze = (B^2 r^2)/2V$ are detected. By scanning B or V, ions of differing m/ze are sequentially detected. The resulting plot of the detected signal versus m/ze constitutes the raw data from which the mass spectrum is constructed.

So far only the magnetic sector has been considered; the purpose of the electrostatic sector is to achieve high resolution. Many molecules share the same molecular weight. A spectrum in which all of the masses are accurate to one atomic mass unit (nominal mass) can therefore be very ambiguous, but more specificity can be obtained if the resolution of the instrument can be increased, so that species with the same nominal mass may be resolved on the basis of their exact mass. In addition, accurate mass measurement can be used in conjunction with accurate values of isotopic masses to assign molecular formulae. For example, in a FAB spectrum acquired from a matrix containing caesium iodide (used as an internal standard for mass calibration) a nominal mass peak at $m/e = 404$ could correspond to either of the following structures:

$$[C_{12}H_{25}N(CH_3)_2CH_2COOCs]^+$$
(1)

$$[C_{11}H_{23}CONH(CH_3)CH_2COOCs]^+$$
(2)

The accurate mass corresponding to formula (**1**) is 404.1565, the value for formula (**2**) is 404.1201. Using accurate mass acquisition and a suitable mass calibration, a peak may be assigned to one or other of these formulae. To resolve both species in a single spectrum would require a resolution in excess of 11 000 (an instrument of resolution 1000 can resolve a mass of 1000 from a mass of 1001).

The resolution of a spectrometer depends on the line widths that it can produce; the narrower the peak the higher the resolution. In the case of a mass spectrometer good resolution requires a narrow peak on the m/ze axis. The fact that MS produces finite line widths from individual species is a result of the spread of energies of ions entering the mass analyser. This spread is caused by the ions having thermal energy as well as the kinetic energy imparted to them by the accelerating potential.

To increase the resolution of the spectrometer it is necessary to narrow the spread of energies of the ions entering the magnetic sector. This energy discrimination is achieved in the electrostatic sector of the instrument, which focuses together ions of the same kinetic energy. Commercial double focusing mass spectrometers can achieve resolutions of ~90 000.

12.4.2 Quadrupole mass analyser

This consists of four rod-like electrodes arranged parallel to each other about a central axis. Potentials with a radio-frequency component are applied to these rods in order to create an electrostatic mass filter. Only masses within a specified range can pass down the axis of the filter and be detected. Ions outside this mass range exhibit unstable oscillations and eventually hit one of the electrodes. By varying the potentials applied to the electrodes mass scanning can be achieved.

12.4.3 Ion trap mass analyser

In its most commonly encountered form, the ion trap mass spectrometer is a low-resolution instrument used in dedicated GC—MS systems. Ions introduced into an electrostatic ion trap mass spectrometer perform stable oscillations within a cavity bounded by two hemispherical electrodes and one ring electrode. Ramping the potentials applied to the electrodes causes the oscillations to become unstable and the ions are flung out of the trap in mass (m/e) order.

12.4.4 More advanced mass spectrometric techniques

In some applications, determination of the mass of the ions leaving the source is not sufficient to provide the required confidence in the assignment of the spectrum. Extra information or specificity can be obtained by making use of ions that decompose after leaving the spectrometer source. In a double focusing instrument of the type described in section 12.4.1, ions may decompose between the source and the electrostatic sector (the first field free region) either because they are relatively unstable (metastable ions) or because a collision gas has been introduced into this region to promote decomposition (collisionally induced decomposition). By simultaneously varying the electrostatic and magnetic fields, the mass of both the parent ion (i.e. the ion that leaves the source) and its daughter (the decomposition product) can be determined. Such decompositions can be highly characteristic of a particular compound or class of compounds. They can also be used in conjunction with soft ionisation techniques to provide structural information.

A mass spectrometer containing a single quadrupole mass filter cannot be used to perform this type of experiment. However spectrometers consisting of three quadrupoles in series (triple quadrupole mass spectrometers) are available for this purpose. The first quadrupole is used to select the parent ion, fragmentation takes place in the second quadrupole and daughter ions are selected by the third quadrupole.

12.5 Detection, sensitivity and data processing

Commonly, detection is carried out by an electron multiplier. The resulting electronic signal is then digitised and processed by a microcomputer to reduce noise, detect peaks and display on a mass scale. MS is a very sensitive technique, picogram–nanogram quantities being detected routinely. Smaller amounts can be detected in certain applications. For applications where sensitivity is critical, a selected ion monitoring (SIM) experiment may be performed. By collecting ions for only a few values of m/e rather than sweeping the whole mass range, this technique allows more time to be spent collecting the ions of interest. Similarly, high sensitivity can be obtained over a limited mass range by use of an array detector, which detects all of the ions in its range simultaneously (i.e. a range of r values is allowed—see section 12.4.1) thereby avoiding the sensitivity loss associated with scanning acquisition.

12.6 Applications

12.6.1 Introduction

Table 12.1 provided an indication of the methods used in the analysis of each type of surfactant. Sections 12.6.3–12.6.9 describe the types of ion produced by typical members of each surfactant type in the appropriate ionisation mode. These sections will deal only with EI, CI and FAB/LSIMS spectra. The ions expected in electrospray spectra were dealt with in section 12.3.5. Little fragmentation is seen and there are no discrimination effects. Electrospray may give more molecular ion related responses than FAB when applied to the same sample. There is not as yet extensive literature on the application of electrospray to the analysis of surfactants. However, this situation is expected to change in the future as more laboratories become equipped with this type of interface. Spectra of nonionic surfactants acquired using thermospray with ammonium acetate as the electrolyte give ammonium adduct ions (as in CI using ammonia as the reagent gas). Plasma discharge results in protonated species.

Analyses are often required on surfactant mixtures or on complex matrices containing surfactants. In the latter case extraction procedures may have to be carried out, and for trace analysis, concentration steps will also have to be introduced. In more favourable cases, analysis of surfactant mixtures can be carried out without prior separation. FAB can be applied directly to mixtures of surfactants of all classes.

TLC may be used to separate mixtures of anionics, cationics and nonionics, prior to FAB analysis, which is carried out on silica removed from appropriate regions of the plate. Mixtures of nonionic surfactants may be separated and analysed on-line by SFC–MS or LC–MS, or GC–MS for the low-EO species.

The majority of conventional applications of MS to the analysis of surfactants have been qualitative; the presence of a particular species in the sample of interest is confirmed by the presence of the corresponding ions in the spectrum. In the absence of fragmentation, the spectrum may contain only a single ion per analyte; the analysis of such a spectrum cannot be carried out with a high level of confidence. Accurate mass data will allow discrimination between proposed empirical formulae but cannot be used to assign the elemental composition of a complete unknown. Where the fragmentation pattern of the analyte is known, observation of these ions in the spectrum provides extra evidence for the presence of the analyte in the sample. Similarly, assigning an unknown to a particular chemical class may be aided by identifying characteristic fragments in its mass spectrum.

Analysis of EI spectra may be aided by reference to spectral libraries. However, many ionisation techniques used in the analysis of surfactants do not provide library-searchable spectra, either because the technique is too irreproducible or because the spectra contain too little structure. In any case, libraries should never be used as the only means of spectral interpretation; experience, common sense and knowledge of the chemistry of the sample must also be used. The interpretation of mass spectra is rarely unequivocal, and whenever possible other methods of analysis should be used in conjunction with MS.

12.6.2 Quantitative applications

Quantitation in GC–MS is carried out in a similar way to quantitation in conventional GC. A known quantity of the sample is introduced to the GC–MS system and the intensity of an ion (or series of ions) characteristic of the analyte is monitored. The quantity of analyte corresponding to this response is found from a calibration curve generated by introducing known quantities of the analyte into the GC–MS system. In principle, LC–MS interfaces can be used in the same way. Known quantities of sample are introduced to the mass spectrometer by liquid injection and the characteristic response is monitored. This approach has been used in the determination of straight-chain alkyl benzenesulphonates (see section 12.6.3). A word of warning is needed: quantitative studies using LC–MS with a particle beam interface can be severely affected if species co-elute. Even if the co-eluting species does not contribute directly to the MS response of the analyte, its presence can affect the efficiency of transfer of the analyte through the particle beam interface. In favourable cases, quantitation can be carried out by particle beam MS, as demonstrated in its application to the determination of alkylphenol ethoxylates in drinking water [9].

FAB or LSIMS using a probe inlet does not readily lend itself to quantitative work. Firstly, it is not possible to know how much of the sample has been consumed in the analysis. Secondly, discrimination effects (see section 12.3.3) prevent the comparison of intensities between species of differing surface activity. Semiquantitative results may be readily obtained if discrimination effects are assumed to be constant for the species of interest, for example the determination of homologue distributions in a mixture. For accurate quantitation an internal standard of an isotopically enriched analogue of the analyte should be used. For example, in the determination of cationic surfactants in environmental samples [10], quantitation was achieved by using an internal standard of a trideuterated form of the analyte. In this way the standard will be subject to the same level of discrimination as the analyte. Discrimination effects between different cationic species may also be reduced by adding to the sample an excess of a highly surface-active anionic surfactant. The anionic species will dominate the matrix surface and attract cations into the surface monolayer [10].

12.6.3 FAB/LSIMS of anionic surfactants

Anionic surfactants ionise well by these techniques due to their high polarity and the presence of preformed ions in the matrix. Negative-ion spectra are dominated by the intact anions (denoted L^-) and by adduct ions, for example those of the form $L_nNa^-_{n-1}$, where n is an integer.

Figure 12.2 shows the positive-ion FAB spectrum of sodium dodecylbenzene sulphonate acquired using a glycerol matrix. Species present include $(LNa_2)^+$, $(L_2Na_3)^+$ and $(LNa_2—C_9H_{20})^+$ at $m/e = 371$, 719 and 243. Clusters of the form $L_nNa^+_{n+1}$ in positive-ion spectra can extend up to $n = 7$. The corresponding negative-ion spectrum is presented in Figure 12.3. The peaks at $m/e = 325$ and 673 correspond to the species L^- and L_2Na^-. The ion with $m/e = 197$ is the fragment $CH_2C(CH_3)PhSO_3^-$, characteristic of a branched-chain alkylbenzene sulphonate (ABS). Such ions are used to provide structural information in CID experiments. The corresponding ion for a straight chain ABS, $CH_2{=}CHPhSO_3^-$ ($m/e = 183$), has been used in studies by Hites and co-workers [11] to quantify linear alkylbenzene sulphonates in sewage effluents at the μg/l level. In this application, negative-ion continuous flow FAB was used in conjunction with a triple quadrupole mass spectrometer to identify all species that fragment to the $m/e = 183$ ion.

The negative-ion LSIMS spectrum of sodium dodecyl sulphate (not shown) contains intense responses for $m/e = 265$ (L^-) and 553 (L_2Na^-), and a cluster at $m/e = 841$ due to $L_3Na_2^-$. Ions at $m/e = 80$ (SO_3^-), 96 (SO_4^-) and 97 (HSO_4^-) are typical of the spectra of alkyl sulphates. Alkyl sulphonates can also give the $m/e = 80$ ion in negative-ion spectra. In positive-ion mode a peak at $m/e = 126$ ($Na_2SO_3^+$) is indicative of the sulphonate function.

Similar groups of ions to those described above (L^-, $L_nNa^-_{n-1}L_nNa^+_{n+1}$) are seen in the spectra of α-olefin sulphonates, alkyl ether sulphates, sarcosinates and sulphosuccinates. Fatty acid salts do not give rise to strong positive ions but do give good anion signals in negative-ion spectra.

12.6.4 FAB/LSIMS of nonionic surfactants

Positive-ion FAB spectra of nonionic surfactants (denoted below as M for the molecular species) are dominated by the quasimolecular ions $[M+Na]^+$ or $[M+H]^+$. The identification of polyethoxylated species in complex matrices is aided by the regular series of ions separated by 44 mass units corresponding to the distribution of the degree of ethoxylation.

Polyethoxylated alcohols. This class of surfactant can be ionised from glycerol or thioglycerol. In the presence of sodium ions the dominant species are $(C_nH_{2n+1}O(C_2H_4O)_yH)Na^+$. Sodium salts may be added to the FAB matrix to promote the formation of this type of ion. In flow FAB using methanol–water as mobile phase, $(C_nH_{2n+1}O(C_2H_4O)_yH)H^+$ ions are formed. The typical ion series for this class of surfactants is

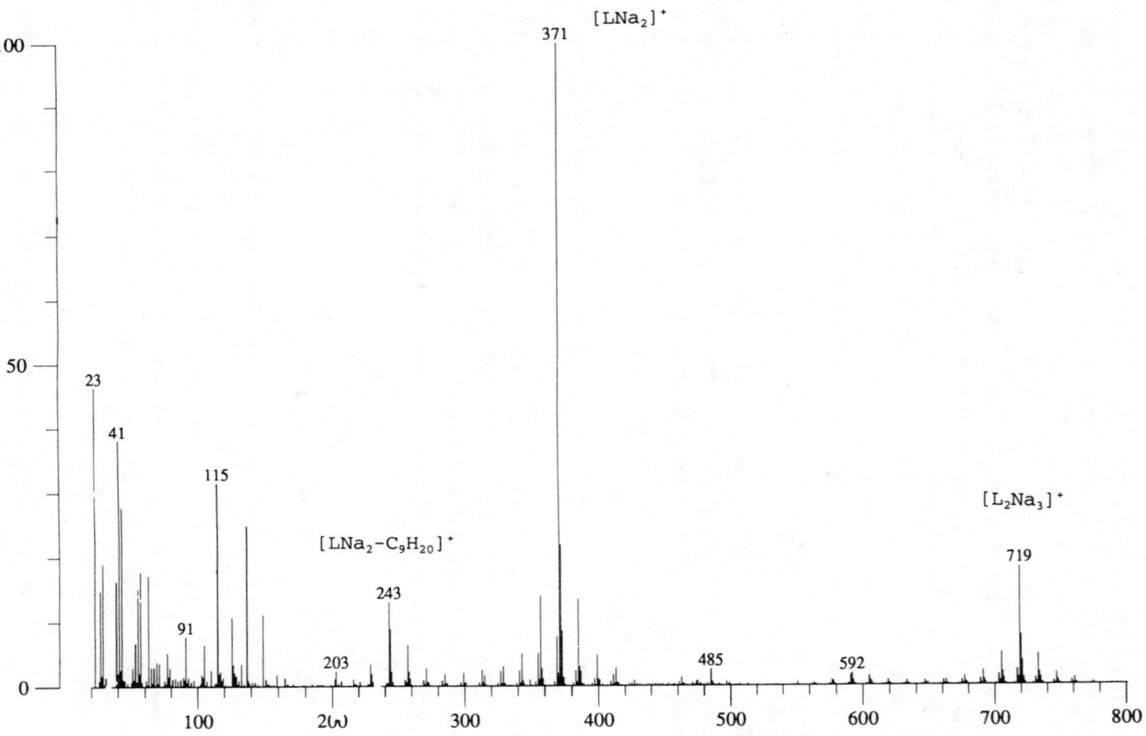

Figure 12.2 Positive-ion FAB mass spectrum of a branched-chain sodium alkylbenzene sulphonate in a glycerol matrix.

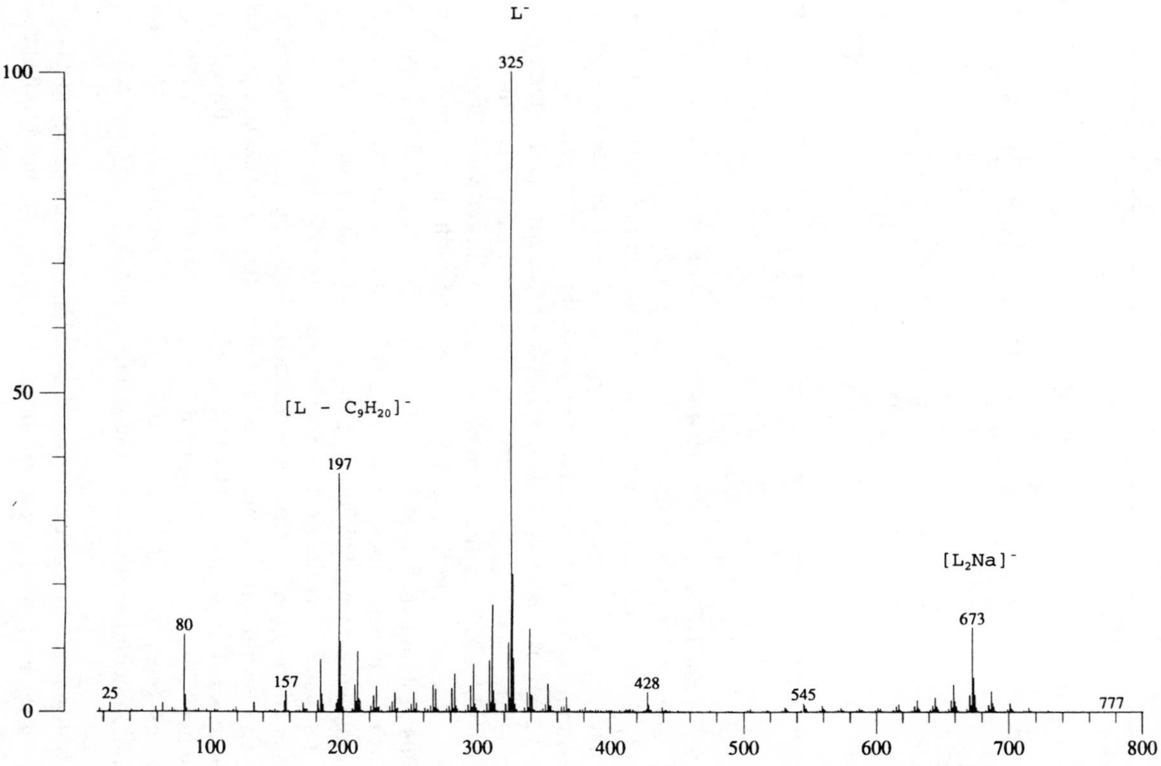

Figure 12.3 Negative-ion FAB mass spectrum of a branched-chain sodium alkylbenzene sulphonate in a glycerol matrix.

Table 12.2 Mass-to-charge ratios for ions formed by cation attachment to polyethoxylated alcohols, $M = (C_nH_{2n+1}O(C_2H_4O)_yH)$

n	y	$(MH)^+$	$(MNa)^+$
12	5	407	429
12	6	451	473
12	7	495	517
12	8	539	561
12	9	583	605
12	10	627	649
14	5	435	457
14	6	479	501
14	7	523	545
14	8	567	589
14	9	611	633
14	10	655	677
n	y	$14n + 44y + 19$	$14n + 44y + 41$

presented in the Table 12.2. This characteristic series of ions separated by 44 mass units may continue to very high masses [12].

A typical surfactant mix may contain a distribution of both alkyl chain length and degree of ethoxylation, giving rise to a complicated spectrum. Analysis may be aided by separation using liquid chromatography; online LC–flow FAB of this type of sample has been reported [13] using a C_{18} column and a mobile phase containing 92% methanol, 7% water, 1% glycerol and 0.005% triethanolamine. Crude separation was achieved on the basis of alkyl chain length. For each chain length the distribution of EO units resulted in a broad LC peak. To identify individual species within these peaks, mass chromatograms may be used. The mass chromatogram is formed by plotting the time dependence of the intensity of an ion characteristic of the species of interest. In this way the selectivity of the mass spectrometer compensates for poor LC resolution.

The only significant fragment ions in the spectra of these compounds are those of the series $(CH_2CH_2O)_nH^+$, diagnostic of ethoxylated compounds. The lack of any fragment ions characteristic of the hydrophobic group (even in CID, see Table 12.1) prevents any structural information being obtained on these groups other than the chain lengths.

Other polyethoxylated species. FAB ionisation of many polyethoxylated species give ions analogous to those described for polyethoxylated alcohols.

A second series of ions in the spectra of polyethoxylated alkyl phenols at m/e = 135, 179, 223, 267 are the result of benzylic cleavage of the alkyl chain. In environmental samples these species can become halogenated as a result of water treatment, substitution occurring on the aromatic ring.

The spectra of these halogenated species are analogous to the non-halogenated compounds but with extra structure arising from the appropriate isotope ratio pattern.

Dominant ions in the spectra of polyethoxylated fatty acids are due to the ring-stabilised 1EO fragment [14]. Fragment ions in the spectra of polyethoxylated fatty amines have structures $[CH_2{=}N(CH_2CH_2O)_mH(CH_2CH_2O)_nH]^+$ and $[C_nH_{2n+1}N(CH_2)(CH_2CH_2O)_nH]^+$.

12.6.5 FAB/LSIMS of cationic surfactants

The positive-ion FAB/LSIMS spectra of cationic surfactants are characterised by the intact cation (L^+) and smaller cluster ions, for example L_2Cl^+. Negative-ion spectra give species of the form LCl_2^-, from which the anion can be identified.

Quaternary ammonium salts and ethoxylated quaternary amines [15]. These compounds ionise well by FAB due to their ionic nature and high surface activity. The spectra are characterised by the intact cation ($R_1R_2R_3N^+CH_2R_4$) and fragment ions of the form $R_1R_2N^+{=}CHR_4$. The ease of ionisation of quaternary ammonium species by FAB has allowed the quantitation of the species $[C_{12}H_{25}N(CH_3)_3]^+$ and $[(C_nH_{2n+1})_2N(CH_3)_2]^+$ at the 10–500 μg/l level from aqueous environmental matrices, following extraction and clean-up [10]. In this application any discrimination effects were eliminated by reference to an internal standard of an isotopically labelled analogue.

Figure 12.4 is the positive-ion FAB spectrum of a commercial mix of polyethoxylated quaternary ammonium salts. The peaks at $m/e = 45$ and 58 correspond to the ions $[C_2H_5O]^+$ and $[CH_2N(CH_3)_2]^+$. The peak at $m/e = 296$ is due to the ion $C_{18}H_{37}N^+CH_2CH_3$. Ions in the range 400 amu to 1400 amu are due to polyethoxylated species. Ions K are due to the intact cation $C_nH_{2n+1}N^+(CH_3)(CH_2CH_2O)_yH(CH_2CH_2O)_xH$. Ions I are due to the fragments $C_nH_{2n+1}N^+(CH_2)(CH_2CH_2O)_xH$. The m/e for these species are presented in Table 12.3 for $n = 18$.

Amines and amine salts. Primary, secondary and tertiary amines protonate to give positive ions corresponding to the amine salt $[R_1R_2R_3NH]^+$. This cation can also lose two hydrogen atoms to give an ion that could be confused with that from an unsaturated analogue. The presence of such ions should be taken into account before the determination of unsaturated species is attempted.

Amine oxides. Positive-ion FAB from acidic media give ions corresponding to the protonated amine oxide ion and its dehydration product.

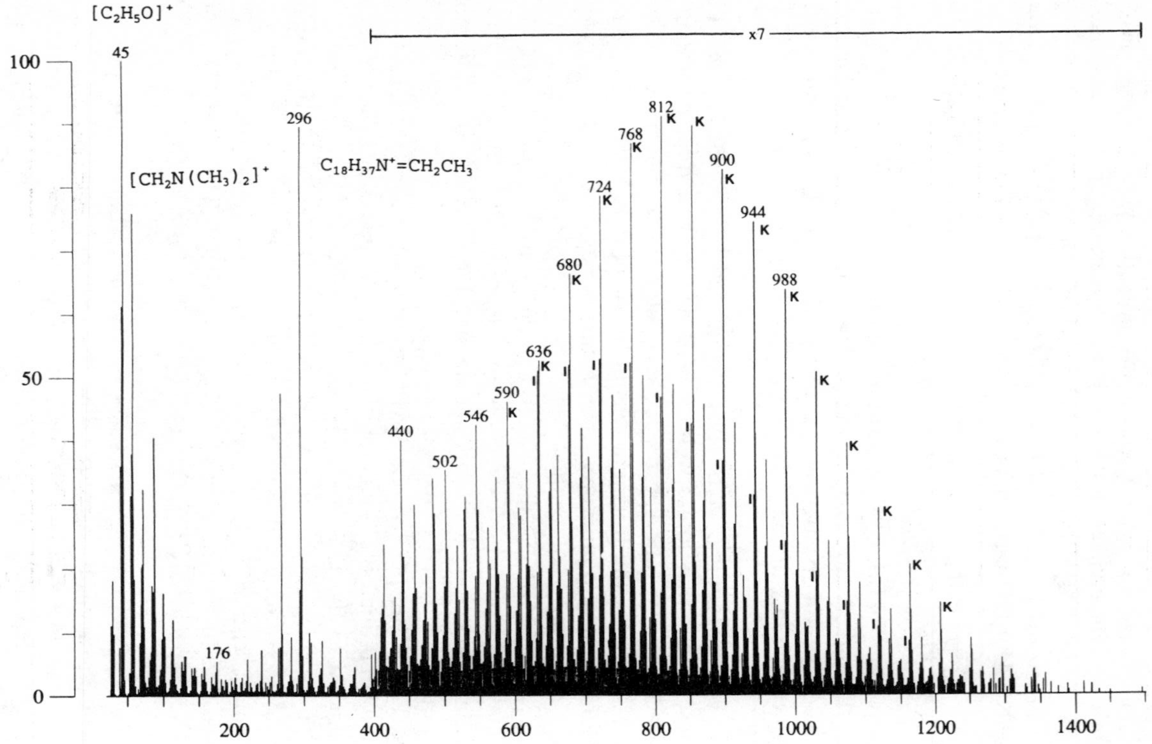

Figure 12.4 Positive-ion FAB mass spectrum of a commercial mix of polyethoxylated quaternary ammonium salts in a glycerol matrix.

Table 12.3 *m/e* Values for ion series I and K in Figure 12.4

x or $x+y$	8	9	10	11	12	13	14	15	16	17	18	19	20
I	634	678	722	766	810	854	898	942	986	1030	1074	1118	1162
K	636	680	724	768	812	856	900	944	988	1032	1076	1120	1164

Table 12.4 Mass-to-charge ratio for amine oxides and their dehydration products

n	*m/e* $C_nH_{2n+1}N^+(CH_3)_2OH$	*m/e* $C_nH_{2n+1}N^+(CH_3)_2OH - H_2O$
12	230	212
14	258	240
16	286	268

Species of the form L_2H^+ may also be present, where L is the amine oxide $C_nH_{n+1}N^+(CH_3)_2O^-$. Table 12.4 gives the *m/e* values for LH^+ and $LH^+ - H_2O$ for various values of *n*.

12.6.6 FAB/LSIMS of amphoteric surfactants

In positive-ion FAB, cocoamidopropyl betaine gives a series of ions corresponding to the $[MNa]^+$ and $[MH]^+$ adducts, and the loss of dimethylglycine ($[M - glyme2]^+$ in Table 12.5) from the intact molecule. These ions are summarised in Table 12.5. Higher-mass ions are also present representing species of the form $[2M + H]^+$, $[2M + Na]^+$.

Impurities in the cocoamidopropyl betaine may be identified by both positive- and negative-ion FAB. Species lacking the CH_2COO^- function protonate to give ions at *m/e* 283, 285, 313 in positive-ion mode. The negative-ion FAB spectrum of cocoamidopropyl betaine is presented in Figure 12.5. The ion at *m/e* = 377 corresponds to $C_{11}H_{23}CONH(CH_2)_3N^+(CH_3)_2CH_2COO^-Cl^-$. Unreacted fatty acids can be detected in this spectrum at *m/e* = 199, 227 and 283.

Table 12.5 Ions in the positive-ion FAB spectra of cocoamidopropyl betaine $M = C_nH_{2n+1}CONH(CH_2)_3N^+ (CH_3)_2CH_2COO^-$

n	11*	11	13	15	17*	17
$[MNa]^+$	363	365	393	421	447	449
$[MH]^+$	341	343	371	399	425	427
$[M\text{-glyme2}]^+$	238	240	268	296	322	324

* = Unsaturated

12.6.7 EI MS of anionic surfactants

Anionic surfactants require modification before they can be ionised in the gas phase. Linear alkylbenzene sulphonates (LAS) may be desulphonated or converted to their sulphonyl chloride, methyl sulphonate or trifluoroethanol derivatives [16]. Alkyl benzenes resulting from desulphonation can be analysed by GC–MS using EI. The main ion is $C_{n+6}H^{+}_{2n+6}$, for an alkyl group containing *n* carbons. Characteristic ions include those from alpha-cleavage at *m*/*e* = 77, 78, 79 ($C_6H_5^{+}$, $C_6H_6^{+}$, $C_6H_7^{+}$) and the more abundant ions from beta-cleavage at *m*/*e* = 91 and 92 ($C_7H_7^{+}$, $C_7H_8^{+}$).

The EI spectra of sulphonyl chloride derivatives give molecular ions corresponding to both of the Cl isotopes. Ions arising from the loss of both the sulphonyl chloride and the longest alkyl chain characterise the point of attachment of the phenyl group on the alkyl chain (see Table 12.6).

Table 12.6 Mass-to-charge ratio for fragment ions in the EI spectra of alkylbenzene sulphonyl chlorides. These ions provide information on the position of attachment of the phenyl group on the alkyl chain

m/*e*	104	118	132	146	160	174
Position of attachment	2	3	4	5	6	7

In CI the main ions are due to $(M+H)^{+}$ and $(M+H-HCl)^{+}$. LAS and dialkyltetralin sulphonates (DATS) may be quantified at the μg/l level in environmental water samples by first converting them to their trifluoroethanol derivatives [16]. The derivatives can then be analysed by negative-ion CI.

12.6.8 EI MS of nonionic surfactants

Nonionics are the only species of surfactant for which EI and CI are major techniques. The sample may be introduced to the EI/CI source by direct insertion probe, GC or LC via a particle beam interface.

The EI spectra of ethoxylated fatty alcohols contain ions, $CH_2{=}O^{+}(CH_2CH_2O)_xH$, $H_2O^{+}(CH_2CH_2O)_xH$ and $^{+}CH_2CH_2O(CH_2CH_2O)_xH$, which are characteristic of ethoxylated species. Information on the alkyl chain length of saturated fatty alcohol species is provided by ions of the form $[C_nH_{2n+1}(OCH_2CH_2)_x]^{+}$ and weak $[M+1]^{+}$ ions of the form $[C_nH_{2n+1}(OCH_2CH_2)_xOH(CH_2CH_2O)_yH]^{+}$ produced by ion–molecule reactions. If the sample is introduced on an insertion probe, some measure of the degree of condensation of these species may be obtained

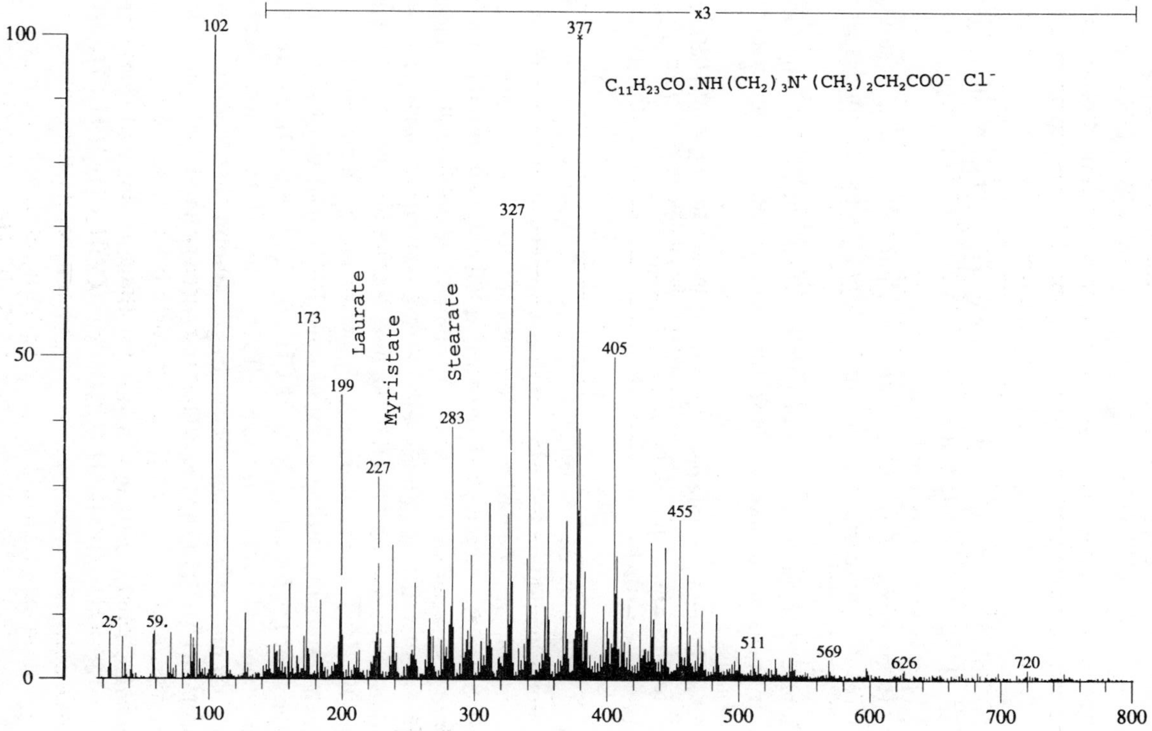

Figure 12.5 Negative-ion FAB mass spectrum of a cocoamidopropyl betaine in an *m*-nitrobenzyl alcohol matrix.

Table 12.7 Mass-to-charge ratio of the molecular ions in EI spectra of ethoxylated alkylphenols. These ions have the form $[C_nH_{2n+1}—Ph—O—(CH_2CH_2O)_xH]^+$

n	8	8	8	8	9	9	9	9
x	0	1	2	3	0	1	2	3
m/e	206	250	294	338	220	264	308	352

by using distillation, spectra being recorded as a function of probe temperature [16]. Ethoxylated unsaturated fatty alcohols give more intense M^+ peaks in place of the $[M+1]^+$ species.

EI spectra of mixtures of ethoxylated alkylphenols contain a series separated by 44 amu corresponding to molecular ions. Examples are presented in Table 12.7.

These ions can fragment by loss of an alkyl fragment followed by loss of water, or by loss of an alkene. Thus, the spectra contain three series of ions at low mass.

For octyl and nonyl phenol species, benzylic cleavage of the alkyl chain gives rise to the ion series $(CH_3)_2C^+PhO(CH_2CH_2O)_xH]$, m/e = 135, 179, 223 ... for x = 0, 1, 2 ...

Ethoxylated fatty amines. Molecules which contain an odd number of nitrogen atoms in addition to C, H and O give molecular ions with an odd value for m/e but fragmentation by loss of a radical gives ions with an even m/e. Since the latter predominates in ethoxylated amines, their spectra are characterised by a large number of even m/e species. Dominant ions at the high-mass end of the spectrum correspond to $M-31$, due to the loss of CH_2OH. Weak ions at $m/e = M-17$ are due to loss of OH. Loss of one of the hydrophilic chains by alpha-cleavage to the nitrogen results in a strong series of ions, $RN^+(CH_2)(CH_2CH_2O)_xH$. Similarly ions of the form $CH_2N^+(CH_2CH_2O)_nH(CH_2CH_2O)_mH$ are the result of alpha-cleavage of the alkyl chain. Other ion series resulting from more complex fragmentation sequences are $HN(CH_2CH_2^+)(CH_2CH_2O)_nH$ and $HN^+(CH_2)(CH_2CH_2O)_nH$.

Alkanolamides. In EI diethanolamides give the following ions characteristic of the R groups, $[M+1]^+$, RCO^+, $RCON^+(CH_2)CH_2CH_2OH$, $[M-18]^+$.

12.6.9 CI MS of nonionic surfactants

Chemical ionisation has been used together with supercritical fluid chromatography–mass spectrometry in the study of ethoxylated species. Ethoxylated alcohols, ethoxylated alkylphenols, ethoxylated alkylamines and ethoxylated fatty acids give spectra dominated by $[MH]^+$ ions. These ions allow confirmation of the alkyl chain length and the degree of ethoxylation. SFC–CI–MS on a triple quadrupole mass spectrometer has been used in structural studies [17].

This chapter represents the view of the author and is not necessarily the view of Unilever Research.

References

1. Stephanou, E., Reinhard, M. and Ball, H. A. *Biomed. Environ. Mass Spectrom.* **15** (1988), 275–282.
2. McEvoy, J. and Giger, W. *Environ. Sci. Technol.*, **20** (1986), 376–383.
3. Lattimer, R. P. and Schulten, H.-R. *Anal. Chem.*, **61**(21) (1989), 1201A–1215A.
4. Zhu, P. and Su, K. *Org. Mass Spectrom.*, **25** (1990), 260–264.
5. Caprioli, R. M. (ed.) *Continuous-flow Fast Atom Bombardment Mass Spectrometry.* John Wiley and Sons, 1990.
6. Hiraoka, K. and Kudaka, I. *Rapid Comm. Mass Spectrom.*, **6** (1992), 265–268.
7. Harvey, G. J. and Dunphy, J. C. Characterisation of cationic, anionic, and nonionic surfactants by positive and negative ion electrospray ionization mass spectrometry. *Proceedings of the 40th ASMS Conference on Mass Spectrometry and Allied Topics*, Washington, DC, May 31–June 5, 1992.
8. Clark, L. B., Rosen, R. T., Hartman, T. G., Louis, J. B. and Rosen, J. D. *Intern. J. Environ. Anal. Chem.*, **45** (1991), 169–178.
9. Clark, L. B., Rosen, R. T., Hartman, T. G., Louis, J. B., Suffet, I. H., Lippincott, R. L. and Rosen, J. D. *Intern. J. Environ. Anal. Chem.*, **47** (1992), 167–180.
10. Simms, J. R., Keough, T., Ward, S. R., Moore, B. L. and Bandurraga, M. M. *Anal. Chem.*, **60** (1988), 2613–2620.
11. Borgerding, A. J. and Hites, R. A. *Anal. Chem.*, **64** (1992), 1449–1454.
12. Facino, R. M., Carini, M., Minghetti, P., Moneti, G., Arlandini, E. and Melis, S. *Biomed. Environ. Mass Spectrom.*, **18** (1989), 673–689.
13. Rockwood, A. L. and Higuchi, T. *Tenside Surf. Deterg.*, **29**(1) (1992), 6–12.
14. Ventura, F., Caixach, J., Figueras, A., Espalder, I., Fraisse, D. and Rivera, J. *Wat. Res.*, **23**(9) (1989), 1191–1203.
15. DeStefano, A. J. and Keough, T. *Anal. Chem.*, **56** (1984), 1846–1849.
16. Trehy, M. L., Gledhill, W. E. and Orth, R. G. *Anal. Chem.*, **62** (1990), 2581-2586.
17. Julia-Danes, E. and Casanovas, A. M. *Tenside Deterg.*, **16**(6) (1979), 317–323.
18. Kalinoski, H. T. and Hargiss, L. O. *J. Am. Soc. Mass Spectrom.*, **3** (1992), 150–158.

Appendix 1 List of BSI and ISO Standards relating to surfactant analysis

British Standards BS 3762 and 6829 deal respectively with the analysis of formulated detergent products and raw materials. Some parts are identical with ISO Standards. Conversely some of the ISO Standards relating to surfactant analysis correspond with parts of these British Standards. The following list shows which Standards from either organisation correspond with those of the other. The words given here describe the subject matter but are only occasionally the official title of either Standard.

BS 3762: Analysis of formulated products		*ISO*
Part 1	Methods of sample division	607
Part 2, Tests 3.2 and 3.3	Qualitative tests for type of surface active matter	—
Part 2, Test 3.4	Qualitative test for nonionic surfactants—cobaltothiocyanate	—
Part 3.1	Anionic-active matter by two-phase titration	2271
Part 3.2	Alkali-hydrolysable and non-hydrolysable active matter	2869
Part 3.3	Acid-hydrolysable and non-hydrolysable active matter	2870
Part 3.4	Cationic-active matter by two-phase titration	2871–2
Part 3.5	Total organic matter by ethanol extraction	—
Part 3.6	Non-detergent organic matter (neutral oil, petroleum ether solubles)	—
Part 3.7	Total nonionic matter by ion exchange	—
Part 3.8	Alkanolamides—Kjeldahl nitrogen	—
Part 3.9	Soap content	—
Part 3.10	Short-chain alkylbenzene sulphonates	—

BS 6829: Analysis of raw materials		*ISO*
Part 1.1	Inorganic sulphate—volumetric	6844
Part 2.1	Total alkane sulphonates	6122
Part 2.2	Mean molecular weight and alkane monosulphonate content by extraction	6845
Part 2.3	Alkane monosulphonate content by two-phase titration	6121

Part 3.1	Mean molecular weight of alkylbenzene sulphonates by gas chromatography	6841
Part 4.1	Cloud point of nonionics	1065
Part 4.2	Total nonionic active and polyethylene glycols—Weibull method	2268
Part 4.3	Total ethylene oxide groups—iodometric method	2270
Part 4.4	Cloud point—volumetric method	4320
Part 4.6	Hydroxyl value—acetic anhydride	4326
Part 5.1	Alkylether and alkylphenylether sulphates—total active matter	6842
Part 5.2	Alkylether and alkylphenylether sulphates—mean molecular weight	8799
Part 5.3	Alkylether and alkylphenylether sulphates—unsulphated matter	6843

Other ISO Standards relating to surfactant analysis having no corresponding British Standard

		ISO
—	Technical alkane sulphonates—Methods of analysis	893
—	Technical primary alkyl sulphates—analysis	894
—	Technical secondary alkyl sulphates—analysis	895
—	Technical alkylarylsulphonates excluding benzene derivatives—analysis	1104
—	Anion-active stable to acid hydrolysis—trace amounts	2868
—	Hydroxyl value—phthalic anhydride	4327
—	Technical ethoxylated fatty amines—methods of analysis	6384

Other Standards dealing with the analysis of soaps and soap products and the determination of non-surfactant components of detergent products are omitted.

Appendix 2 List of relevant ASTM Methods

All ASTM Methods have titles of the form 'Standard Test Method(s) for . . .'. These words are omitted in the following list, which includes all methods for the analysis of surfactants, excluding soaps and soap products and the determination of non-surfactant constituents. All these Methods are in the Annual Book of ASTM Methods, Section 15, Volume 15.04, apart from Methods D 2071 to D 2080 inclusive, which are in the same Book, Section 6, Volume 6.03.

Method No.	*Title*
D 500-89	Chemical Analysis of Sulfonated and Sulfated oils
D 1568-63	Sampling and Analysis of Alkylbenzene Sulfonates
D 1570-89	Sampling and Chemical Analysis of Fatty Alcohol Sulfates
D 1681-83	Synthetic Anionic Active Ingredient in Detergents by Cationic Titration Procedure
D 1768-89	Sodium Alkylbenzene Sulfonates in Synthetic Detergents by Ultraviolet Absorption
D 2023-89	Analysis for Sodium Toluene Sulfonate in Detergents
D 2024-65	Cloud Point of Nonionic Surfactants
D 2071-87	Fatty Nitrogen Products
D 2073-66	(Reapproved 1987) Total, Primary, Secondary and Tertiary Amine Values of Fatty Amines, Amidoamines and Diamines by Referee Potentiometric Method
D 2074-66	(Reapproved 1987) Total, Primary, Secondary and Tertiary Amines by Alternative Indicator Method
D 2075-89	Iodine Value of Fatty Amines, Amidoamines and Diamines
D 2076-64	(Reapproved 1987) Acid Value and Amine Value of Fatty Quaternary Ammonium Chlorides
D 2078-86	Iodine Value of Fatty Quaternary Ammonium Chlorides
D 2080-64	Average Molecular Weight of Fatty Quaternary Ammonium Chlorides
D 2357-74	Qualitative Classification of Surfactants by Infra-red Spectroscopy
D 2358-89	Separation of Active Ingredient from Surfactants and Syndet Compositions

D 2959-89	Ethylene Oxide Content of Polyethoxylated Nonionic Surfactants
D 3049-89	Synthetic Anionic Ingredient by Cationic Titration
D 3673-89	Chemical Analysis of Alpha Olefin Sulfonates
D 4224-89	Anionic Detergent by *p*-Toluidine Hydrochloride
D 4251-89	Active Matter in Anionic Surfactants by Potentiometric Titration
D 4252-89	Chemical Analysis of Alcohol Ethoxylates and Alkylphenol Ethoxylates
D 4337-89	Analysis of Linear Detergent Alkylates
D 5070-90	Synthetic Quaternary Ammonium Salt in Fabric Softeners by Potentiometric Titration

ASTM publications may be obtained from: American Society for Testing and Materials, 1916 Race Street, Philadelphia, PA 19103, USA.

Index